# Springer Proceedings in Physics 16

# Springer Proceedings in Physics

Managing Editor: H. K. V. Lotsch



*Springer Proceedings in Physics* is a new series dedicated to the publication of conference proceedings. Each volume is produced on the basis of camera-ready manuscripts prepared by conference contributors. In this way, publication can be achieved very soon after the conference and costs are kept low; the quality of visual presentation is, nevertheless, very high. We believe that such a series is preferable to the method of publishing conference proceedings in journals, where the typesetting requires time and considerable expense, and results in a longer publication period. Springer Proceedings in Physics can be considered as a journal in every other way: it should be cited in publications of research papers as *Springer Proc. Phys.*, followed by the respective volume number, page number and year.

# Photons and Continuum States of Atoms and Molecules

Proceedings of a Workshop
Cortona, Italy, June 16–20, 1986

Editors: N. K. Rahman, C. Guidotti, M. Allegrini

With 144 Figures

Springer-Verlag Berlin Heidelberg New York
London Paris Tokyo

Professor Naseem K. Rahman
Professor Carla Guidotti
Dipartimento di Chimica e Chimica Industriale
Università di Pisa, Via Risorgimento 35, I-56100 Pisa, Italy

Dr. Maria Allegrini
Istituto di Fisica Atomica e Moleculare del C.N.R.
Via del Guardino 7, I-56100 Pisa, Italy

ISBN-13:978-3-642-71780-2 e-ISBN-13:978-3-642-71778-9
DOI: 10.1007/978-3-642-71778-9


Softcover reprint of the hardcover 1st edition 1987



2153/3150-543210

# Preface

Since 1981 there has been an attempt in Europe to organize a series of small meetings/workshops/tavole rotonde with the aim of bringing together physicists and chemists interested in problems concerning atoms or molecules interacting with external photons where the continua are investigated.

The number of problems that fall into this category turns out to be vast. However, it is not possible to make a strict separation into problems concerning atomic and molecular collisions and those related to the usual spectroscopy. This admixture of two disciplines, discussions on the role of photons and on the interaction of external electromagnetic fields with the continuum provided a central motivation for these workshops.

The fourth of this series of meetings was held at Cortona between June 16 and 20, 1986. It was attended by about 100 researchers in the field and there were 43 presentations, all having equal time. These talks form the subject matter of this volume. The idea of publishing the proceedings of these meetings is not new. It allows one to have small meetings in which the subject matter can be discussed at length in a lively atmosphere. However, after the meeting is over, the speakers can collect their thoughts and produce articles in which the results of their interaction with the other participants can be incorporated.

The last three meetings have resulted in the following volumes:

(a) Photon-Assisted Collisions and Related Topics, ed. by N.K. Rahman and C. Guidotti (Harwood Academic Publishers, London 1982)
(b) Collisions and Half-Collisions with Lasers, ed. by N.K. Rahman and C. Guidotti (Harwood Academic Publishers, London 1984)
(c) Atomic and Molecular Collisions in a Laser Field, ed. by J.L. Picqué, G. Spiess and F.J. Wuilleumier (Les Editions de Physique, Les Ulis 1985)

It is superfluous to make lengthy introductory comments on the contents of this book. The variety of sub-topics that the authors have treated while keeping themselves anchored to the theme of the meeting (as well as the series) is quite remarkable. There are papers regarding structure in the continuum per se either revealed or manipulated by external photons. The collision processes (atom-atom, atom-molecule) are the subject of a wide range of experimental and theoretical studies. Above-threshold ionization continues to flourish, both experimentally with new results and theoretically with even newer interpretations. Ionization by a low frequency field has again come into vogue after a few years in the doldrums and it can be related to both classical mechanics and the fashionable area of chaos. Electron scattering modified with photons continues to occupy a large body of theorists, and here, certainly, there is a crying need for experiments. Absorption of large numbers of photons by polyatomic molecules, with eventual dissociation, continues to survive as an important field of research. The readers of these proceedings will find these topics and more in this volume.

From the reaction and the response of the authors of these articles, and also of the participants of the Workshop at Cortona, it was clear that the subject of these proceedings is in a phase of continuous evolution with many new ideas and results to come in the future. In fact, it was decided that in two years there will be another Workshop to see what has been achieved in the meantime. In conclusion, one may say that the editors will feel their efforts in producing this volume to have been repaid if this volume is remembered as a modest step forward in this area of research.

Pisa, November 1986

*Naseem K. Rahman*
*Carla Guidotti*
*Maria Allegrini*

# Acknowledgements

The Workshop at Cortona was organized under the auspices of the Istituto di Chimica Quantistica ed Energetica Molecolare del Consiglio Nazionale delle Ricerche, Pisa, the Istituto di Fisica Atomica e Molecolare del Consiglio Nazionale delle Ricerche, Pisa and the Facoltà di Scienze Matematiche, Fisiche e Naturali dell'Università di Pisa. The co-operation of several individuals as well as organizations was necessary for its success. Our most sincere thanks go to the other three members of the organizing committee, Prof. E. Arimondo (University of Pisa), Dr. A. Lami (I.C.Q.E.M., C.N.R., Pisa) and Dr. L. Moi (I.F.A.M., C.N.R., Pisa), who constantly cooperated in the organization of the meeting during the entire period of planning and during the meeting itself. The Director of I.C.Q.E.M., Prof. O. Salvetti, and the Director of I.F.A.M.. Dr. M. Vaselli, have given the full support of their respective Institutes in every possible manner. The Director and the staff of Scuola Normale Superiore of Pisa, and in particular Prof. F. Bassani, deserve our gratitude for providing such a lovely venue, the Palazzone, in which to hold the meeting. The financial help from I.C.Q.E.M. under Progetto Strategico Dinamica Molecolare e Catalisi, Comitato Nazionale Scienze Fisiche del Cosiglio Nazionale delle Ricerche, Università di Pisa, Settore di Fisica Atomica e Molecolare del Gruppo Nazionale di Struttura della Materia, Coherent Radiation, dB Electronics Instruments, Banca Popolare di Cortona and Cassa di Risparmio di Pisa is gratefully acknowledged. Among colleagues who have gone out of their way to help us either during the meeting or in helping us with this volume we must mention Prof. V. Aquilanti, Dr. P. Agostini, Prof. J.J. Leventhal, Dr. S. Gozzini and Dr. A. Ferretti. The secretarial help provided by Mr. O. Cosci, Mrs. G. Roventini, Mr. A. Biagi, Mrs. K. Nicotra, Mrs. L. Girolami and Mrs. A.M. Arrighi was instrumental in the smooth running of the Workshop. Finally, the Azienda Autonoma di Soggiorno e Turismo of Cortona is to be thanked for providing the participants with some material and for local help.

# Contents

## Part V **Electronic Collisions in Laser Fields**

# Part I

# Photons and Atomic Continua

# Microwave Excitation and Ionization

*T.F. Gallagher*

Department of Physics, University of Virginia,
Charlottesville, VA 22901, USA

## I. Introduction

Atoms may be ionized by static fields in excess of a threshold field $E \approx 1/16n^4$ where n is the principal quantum number [1]. On the other hand photoionization has a frequency threshold, requiring photons with energy in excess of the binding energy W. The rate of photoionization, though, is proportional to the intensity of the radiation, or $E^2$. Midway between these two processes is multiphoton ionization. Ionization by m photons requires that the photon energy be in excess of W/m, and in lowest order perturbation theory the ionization rate is proportional to $I^m$ or $E^{2m}$. This scaling as $I^m$ has been observed for m = 21. For higher values of m, or even for m = 21, it is difficult to measure an $I^m$ dependence due to dynamic range considerations. Furthermore, it is not clear that the region in which the lowest order solution is valid is physically very interesting, for appreciable ionization will only occur where it is not valid. In these cases it is more useful to describe the multiphoton processes as having field or intensity thresholds. Thus multiphoton ionization qualitatively resembles field ionization.

To generate a more solid understanding of the connection between photoionization and field ionization quantitative studies of the highly nonlinear multiphoton ionization process are required. One approach is to use tightly focused high-powered pulsed lasers, and this has been done in the pioneering work at Saclay. However the spatial and temporal intensity variations inherent in the experiment make the study of highly nonlinear processes a challenge to say the least. Another approach is to use Rydberg atoms and microwave fields. As the microwave fields may be measured with a high degree of accuracy this approach allows very quantitative measurements to be carried out. This realization prompted Bayfield and Koch to begin the study of the microwave ionization of the Rydberg states of H [3].

Here we describe measurements of microwave excitation and ionization in alkali atoms. Alkali atoms, or any atom other than H, exhibit dynamical phenomena that can not be seen in H, under normal experimental conditions. For example in He, Na, K, and Ba the ionization by microwave fields requires a threshold field of $E = 1/3n^5$ [4-7].

## II. Approach

The experimental approach [8] we use can be understood by inspecting Fig. 1. A thermal atomic beam passes into a microwave cavity through a hole in its sidewall, and the atoms in the cavity are excited by two pulsed dye laser beams entering through a hole in the opposite sidewall. For Na the lasers drive the transitions 3s → 3p and 3p → ns or nd. In the top of the cavity there is a small hole through which ions produced

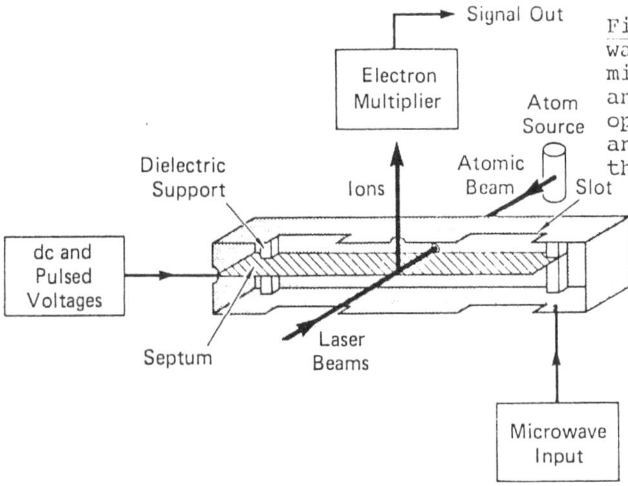


Fig. 1 View of the microwave cavity cut down the middle. Note that the laser and atomic beams enter on opposite sides of the cavity and the ions are extracted through the top

can be extracted. The hole is located so that we only extract ions from an antinode of the microwave field ensuring the spatial homogeneity of the microwave field. As shown by Fig. 1 there is a septum in the cavity which allows the application of a static field to tune the atomic levels and extract ions produced by microwave ionization. A high voltage pulse may also be applied to the septum to selectively field ionize the excited atoms. We have used several microwave cavities at frequencies from 2-15 GHz of the general design shown in Fig. 1. These cavities have Q's in the range of $10^3$ and allow us to produce fields of ~ $10^3$ V/cm with 20W of input power.

With the apparatus shown in Fig. 1 it is possible to conduct a variety of investigations. Microwave ionization measurements are the most obvious, but the ability to do selective field ionization allows us to study as well microwave multiphoton transitions between levels.

## III. Observations

The striking observation of an unexpected $1/3n^5$ threshold field for microwave ionization of Na by 15 GHz fields and similar results for He with 9 GHz fields showed that these atoms behaved very differently from H. The origin of this $1/3n^5$ dependence becomes clear upon examining Fig. 2, where we show how an initially excited Na 20d atom is ionized. The microwave pulse is applied to the cavity ~ 100 ns after the laser excitation, and in the filling time of the cavity, 100 ns, the atoms are redistributed over all the $n = 20$ $\ell \geq 2$ Stark states which are degenerate with 20d at zero field. Some, ~ 1/20 th, of the atoms are in the highest lying Stark state, and they move up and down in energy as the microwave field oscillates. If the field amplitude is sufficient to reach the point where the highest lying $n = 20$ Stark state intersects the lowest lying $n = 21$ Stark state, the atom may undergo a Landau-Zener transition from one to the other at the avoided crossing of the levels shown in the inset of Fig. 2. The field at which this crossing occurs (for $m = 0$ states) is $1/3n^5$. Since the $n \rightarrow n+1$ crossing field becomes lower for higher n, once the atom has gone from $n = 20$ to $n = 21$ it can easily continue making such transitions and finally ionize.

Although it is possible for the Landau-Zener transition to occur, for it to be most likely the magnitude of the avoided crossing, $\omega_o$, must be

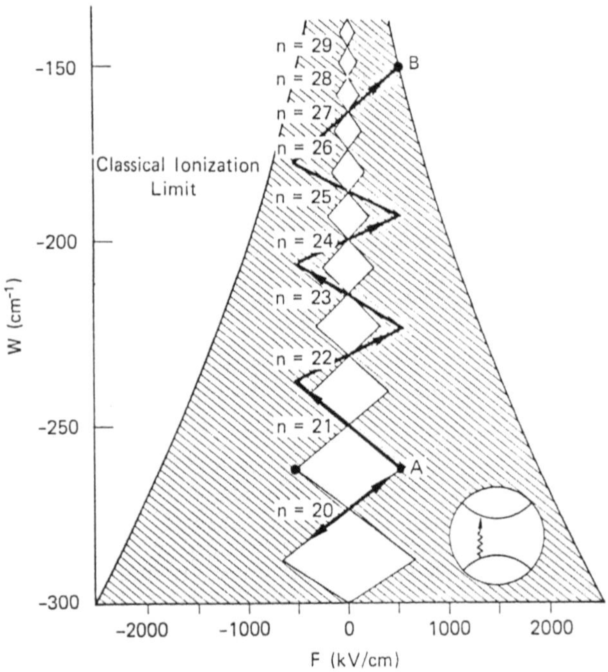


Fig. 2 Schematic diagram of the ionization of Na n=20 atoms by a 600 V/cm microwave field. Of the initially excited n=20 atoms some are in the highest energy Stark state and are brought to point A by the oscillating field as shown by the bold arrow. At point A the atoms can make a Landau-Zener transition to the lowest n=21 Stark state as shown by the inset. Further transitions to successively higher n states ensure culminating in ionization when it becomes classically allowed at point B

comparable to the microwave frequency $\omega$ [4,5,8]. Due to the large number of microwave cycles in any real experiment this requirement is rather loose. Nonetheless we note that with a microwave frequency of 15 GHz this form of ionization occurs for the Na $|m| = 0$ and 1 states, for which $\omega_0 \sim 1\ cm^{-1}$ at $n = 20$. For the $|m| = 2$ states, which have $\omega_0 \sim 0.01 cm^{-1}$, it does not occur. In fact for Na the Na $|m| = 0$ and 1 states, frequencies of 2-15 GHz are equally effective in producing the $1/3n^5$ ionization.

The $1/3n^5$ field is of course the field for the rate-limiting $n \rightarrow n+1$ transition. The transitions from Na s and p states of non-zero quantum defect to nearby Stark manifolds also exhibit threshold fields in accord with this model. These are all transitions, presumably multiphoton, to real states, and one must ask if they can be observed in such a way as to clearly exhibit this character. Preliminary surveys in Na did not reveal this; however a more recent experiment with K has.

In Fig. 3 we show a K energy level diagram in an electric field. We have observed the transition from the 18s state to the n = 16 manifold. Specifically we have concentrated on the transitions to the lowest states of the manifold as shown by Fig. 3. In the experiment we populate the 18s state and observe the population transferred to the 16 manifold state using selective field ionization. The cavity fixes the microwave frequency, and we tune the atomic levels by sweeping the static field. As shown by Fig. 4 we see the series of 18s - lowest Stark state resonances, separated by 28 V/cm, corresponding to 1,2,3... photon transitions. For any given scan the congestion at higher static field is due to transitions to the numerous higher Stark states. For low microwave powers these are readily identified, but for higher powers the overlap is too severe. An analysis of the microwave field dependence shows that the one and two photon transitions occur for low microwave power, but beyond that point the number of photons absorbed

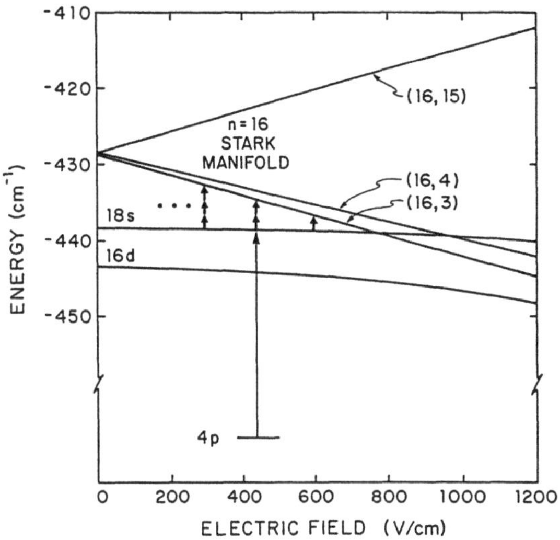


Fig. 3 K level diagram in an electric field. The 16s state is populated by the laser, and the entire sequence of microwave multiphoton transitions to the (16,3) Stark state is observed by selective field ionization of the (16,3 state. Transitions to higher Stark states such as (16,4) are also observed

varies as $E_{mw}$, the microwave field amplitude. This is well out of the perturbative regime. However it is completely consistent with expectations based on the creation of sidebands of the Stark state by the microwave field [9-11].

To show the relation to the Landau-Zener model we have measured the threshold for the ns→(n-2) manifold transition as a function of the microwave field amplitude (with zero static field). The non-resonant microwave field required is ~ 800 V/cm, about 10% higher than the crossing field of 753 V/cm, in good agreement with the Landau Zener model. Note though that the resonance near zero static field only requires a microwave field of 460 V/cm, 60% of $E_c$. However if we examine the $E_{mw}$ = 410 V/cm scan we see that the onset of serious overlap

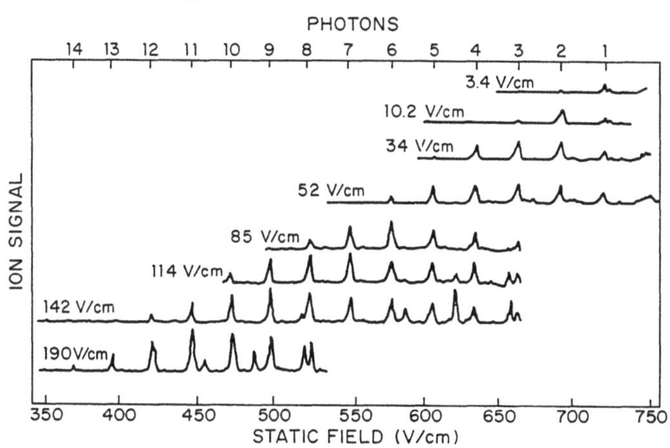


Fig. 4a Multiphoton 18s-(16,3) transitions observed by scanning the static field with a microwave frequency of 10.353 GHz and the microwave field amplitudes given by each trace. The resonances corresponding to the absorption of each additional photon are 28 V/cm apart. Note that for microwave fields of 114 V/cm, 142 V/cm and 190 V/cm that extra lines appear which are part of the analogous 18s-(16,4) series of resonances

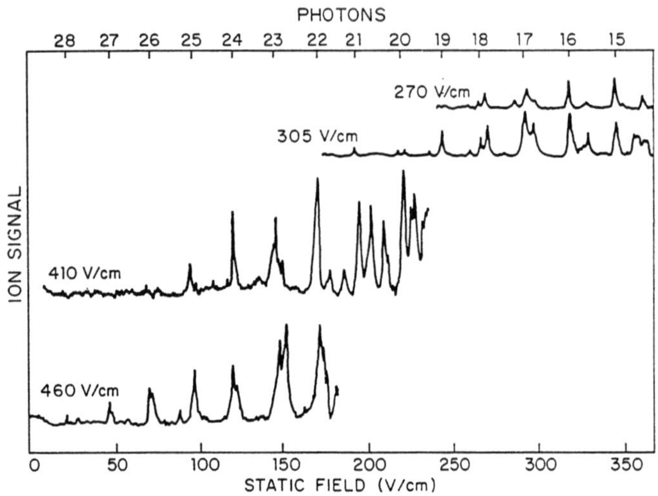


Fig. 4b Multiphoton 18s-(16,3) resonances at higher microwave fields than Fig. 4a, allowing zero static field to be reached. Note the spectral congestion in the trace for a microwave field of 410 V/cm. This is due to overlapping transitions to higher states of the n=16 Stark manifold

in the resonances occurs at $E_s$ = 200 V/cm. This occurs at a lower static field for $E_{mw}$ = 460 V/cm and finally occurs at zero static field for $E_{mw}$ = 800 V/cm.

## IV. Conclusion

At this point it is clear that the Landau-Zener picture is intimately related to a multiphoton resonance description. This is not really surprising as the resonance picture is simply a many cycle, coherent Landau-Zener picture. An aspect that is definitely unclear is the effect of other competing transitions on the observed resonances. For example the n = 16 manifold states are easily mixed by microwave fields of a few V/cm, yet the 18s - 16 manifold resonances are still very evident in Fig. 4. This and other questions are currently under study.

## Acknowledgements

This represents the efforts of several people, among them P. Pillet, H.B. van Linden van den Heuvell, J.L. Dexter, L.A. Bloomfield and R.C. Stoneman. The research has been supported by the Air Force office of Scientific Research under grant AFOSR-85-0016.

## References

1. F.B. Dunning, R.F. Stebbings: In Rydberg States of Atoms and Molecules; ed. by R.F. Stebbings, F.B. Dunning (Cambridge University Press, Cambridge, 1983) p. 315
2. A. L'Huillier, L.A. Lompre, G. Mainfray, C. Manus: J. Phys. B 16, 1363 (1983)
3. J.E. Bayfield, P.M. Koch: Phys. Rev. Lett.
4. P. Pillet, W.W. Smith, R. Kachru, N.H. Tran, T.F. Gallagher: Phys. Rev. Lett. 50, 1042 (1983)
5. D.R. Mariani, W. van de Water, P.M. Koch: Phys. Rev. Lett. 50, 1261 (1983)

6. L.A. Bloomfield, R.C. Stoneman, T.F. Gallagher: (unpublished)
7. U. Eichmann, J.L. Dexter, E.Y. Xu, T.F. Gallagher: (unpublished)
8. P. Pillet, H.B. van Linden van den Heuvell, W.W. Smith, R. Kachru, N.H. Tran, T.F. Gallagher: Phys. Rev. A 30, 280 (1984)
9. S.H. Autler, C.H. Townes: Phys., Rev. 100, 703 (1955)
10. J.H. Shirley: Phys. Rev. 138, B979 (1965)
11. J.E. Bayfield, L.D. Gardner, Y.Z. Gulkok, S.D. Sharma: Phys. Rev. A 24, 138 (1981)
J.R. Rubbmark, M.M. Kash, M.G. Littman, D. Kleppner: Phys. Rev. A 23, 3107 (1981)

# Nonlinear Electron Dynamics Within the Stretched Hydrogen Atom in Intense Microwaves

*J.E. Bayfield*

Department of Physics and Astronomy, University of Pittsburgh, Pittsburgh, PA 15260, USA

The classical nonlinear dynamics of a charged particle in a Coulomb electric field and in an intense time-oscillating electric field exhibits interesting nonperturbative features, including a period-doubling bifurcation sequence for periodic trajectories [1] and a transition to largely stochastic trajectories at high values of classical action [2]. Because quantum numbers can be large, a highly excited hydrogen atom in an intense microwave field can be expected to exhibit quasiclassical behavior, with the time-development of an appropriate ensemble of classical electron trajectories perhaps not being dominated by the quantal modifications of electron wave effects and quantum tunneling. Although semiclassical methods for connecting quantum and classical physics are not yet extended to the case of classically stochastic systems, yet we can make comparisons of nonperturbative quantum calculations with predictions of classical calculations and with experiments. This approach to the search for quantum chaos [3] has been feasible only for the nearly one-dimensional case where the atom is electrically polarized ("stretched") along the direction of an applied static electric field; otherwise, accurate numerical quantum calculations of the electron's time-development have not been possible.

The one-dimensional SSE classical model [2] for our problem describes the motion of the electron in a classical electromagnetic field as that of an explicitly time-dependent Hamiltonian system with one degree of freedom. By extending the usual two-dimensional phase space $(x,p_x)$ to also include energy and time, the problem is believed mathematically equivalent to that of near-integrable autonomous Hamiltonian systems with two degrees of freedom, a class of problems studied in considerable detail using the methods of modern classical nonlinear dynamics [4]. The problem is transformed to action-angle space $(I,\theta)$, a large sample of electron trajectories is calculated numerically, and their Poincare sections obtained by periodic sampling and plotting in that space. A characteristic of these near-integrable systems is intermingled regions of regular (periodic and quasiperiodic) motion and chaotic (stochastic) motion in the Poincare section pictures. In addition, there is a transition, here with increasing action, from one region dominated by regular trajectories to another region dominated by chaotic ones; this transition seems associated with a breakdown in the time-averaged energy of the electron. If the system is in the chaotic region, the atom on time-average absorbs more and more energy from the external oscillating field, ultimately leading to ionization. As the Poincare section displays a time-sequence of random points for a chaotic trajectory, the energy absorption classically is described by a diffusion in action space obeying the Fokker-Planck equation [4].

A question of great interest concerns the behavior of such a near-integrable system when it is physically microscopic and displays the characteristics of a quantum system. For highly excited hydrogen in microwaves of low field strength, the behavior of the system is described by perturbative quantum mechanics and appears very nonclassical. Yet at high field strengths, the ionization probability is described well by classical theory [5]. The quantum picture of this is presently incomplete, but at high fields it contains the idea of many free-atom bound and continuum states being strongly coupled together to produce electron wavefunction delocalization in a basically nonresonant way. There is a quantum delocalization initial action threshold (or initial quantum number threshold) for this, which for many possible values of the system's parameters nearly coincides with the classical initial action threshold for chaos [6].

Most quantum calculations for this problem show that in the absence of an external static electric field the classically chaotic microwave absorption is transient, with the diffusion stopping after a time [7,8,9]. As static fields produce rapid quantum tunneling ionization of atoms sufficiently highly excited, a diffusion due to combined microwave and static fields might not be transient if high enough quantum numbers are reached.

The quantum suppression of classical behavior has also been predicted to separate the delocalization and chaos thresholds for certain ranges of microwave field strength, microwave frequency, and atom initial energy [6].

As the microwave absorption seems to be transient, the way the microwave field is switched on is important since transients remember initial conditions. If the switching is sudden as is essentially assumed in most of the quantum calculations, then the initial quantum state of the atom is a single physical state with definite quantum number, which is a linear combination of quasienergy states of the atom-field system. If the switching is adiabatic [10], then the system is in a single (field-strength dependent) quasienergy state which at the full field value must contain a portion of the physical continuum in order to produce ionization.

What is observed in our experiments are the final results of an exposure of an ensemble of highly excited atoms to a microwave pulse 3000 oscillations long, with 100 oscillation switching times both field on and field off. The final action distribution is just the final atom quantum number probability distribution for (one-dimensional atom) bound states, plus a distribution of continuum states that is experimentally summed over to produce a measured overall ionization probability. Final bound-state quantum-number probability distributions are measured by state-selective electric field ionization, for various microwave frequencies and field strengths [11,12].

The experiments utilize state-selective laser excitation of excited hydrogen atoms in a fast atomic beam formed by charge exchange collisions [13]. A monodirectional static electric field is present during the entire two microsecond time history of each atom. Atoms initially formed by chance in the electrically polarized n=7 state (parabolic quantum numbers n, $n_1$, $n_2$, m = 7, 0, 6, 0) are selectively excited to the 10, 0, 9, 0 state in a region where the static field is strong, and further excited to the 63, 0, 62, 0 state in a weak field region. This two-stage stretching out of the atom via laser transitions between stretched atom states results in a Stark-localized electron wave packet peaked about 4000 Angstroms to one side of the atomic nucleus, with a transverse width of about 400 Angstroms. The atom then traverses a microwave region, the microwave electric field being along the stretching direction. Under these conditions, the electric dipole coupling matrix element between stretched atom states (with different n but with $n_1$ and m remaining zero) is a factor of n larger than that for $n_1$ or m changing by minimal amounts [14]. Thus, to a good approximation, the atom's quantum time development within the microwave field region is described by a time-dependent mixing of stretched atom states of different n.

There are three major features of the measured final-state distributions. First there are frequencies where resonant many-photon absorption occurs, with the energy of an integral number k of microwave photons being close to an energy level separation of that atom [10,12]. At high values of n these quantum resonances are almost the same as the classical resonances (islands in the Poincare sections [2,3]) associated with periodic electron trajectories; the trajectory winding number is equal to the rational number $\Delta n/k$, where $\Delta n$ is the quantum number change associated with the relevant energy level separation [15]. At low microwave field strengths, this sharply resonant absorption can be understood using multiphoton perturbation theory [14]; at high fields, this absorption saturates.

A second and more interesting feature is a frequency-independent microwave absorption that occurs at high fields [16], see Fig. 1. We call this absorption "nonresonant," although it is the result of many of the one-photon microwave quantum resonances in the atom being microwave power-broadened beyond recognition. When the microwave field strength is high enough to produce

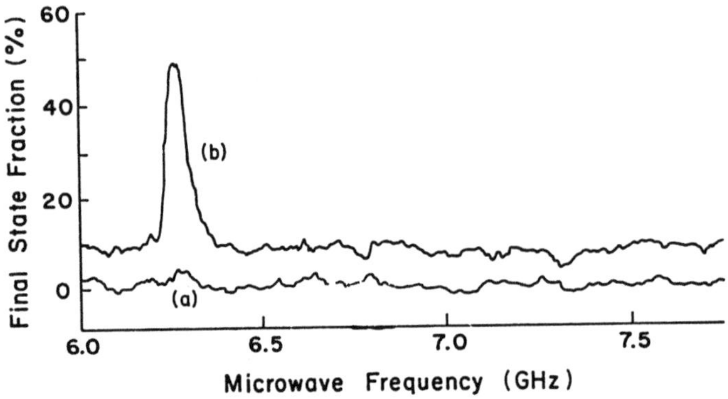


Fig. 1 Experimental data (b) for n = 63 → n = 62 resonant and "nonresonant" microwave n-changing transitions, with the noise and zero measured by curve (a), see reference 16

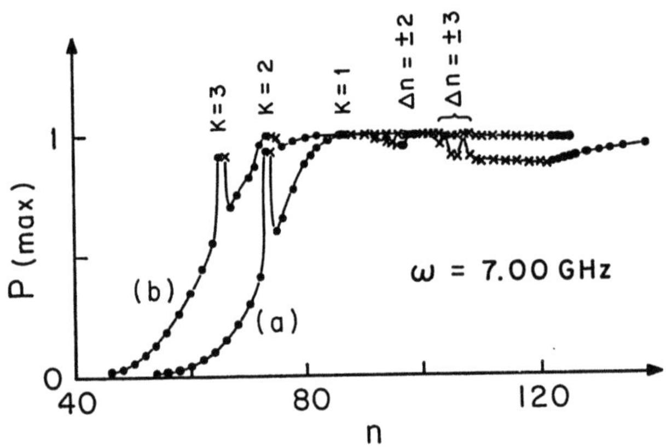


Fig. 2 Maximum two-state Rabi-flopping population-changing amplitudes for electrically polarized highly excited hydrogen atoms in different states n (a) at a microwave field strength of 3 V/cm, weakly ionizing at $n_0$ = 63, and (b) 10 V/cm, strongly ionizing. The dots (crosses) are up (down) n-changing. Resonance perturbations of the one-photon, Δn = ±1 curves are indicated

at least 10% ionization probabilities, the one-photon Rabi-flopping population-changing probability amplitudes are 50% or higher for pairs of adjacent stretched atom energy levels at or above the level of the initial n-63 state [15], see Fig. 2. The one-photon Rabi-flopping frequencies are high, about that of both the microwave itself and also the free-atom orbital frequency. This many-state "nonresonant" many photon absorption seems to correspond to the non-island quasiperiodic classical electron trajectories so prevalent in the corresponding Poincare projections of the trajectories in action-angle phase space. At high enough fields, these trajectories become unstable, leading to a classically stochastic energy absorption from the microwave field that terminates with ionization of the atom, the electron being far from the nucleus and having positive quiver-compensated total energy [1,2].

The experimental time-cutoff of the microwave absorption after 3000 oscillations catches some of the atoms part way along their path to ionization, leaving them in a more highly excited state associated with the absorption of a fraction of the 100 or so microwave photons needed for ionization, see Fig. 3a. Although the final state quantum number probability distribution for these atoms is often perturbed by resonance effects, it still tends to be relatively smooth and characterized

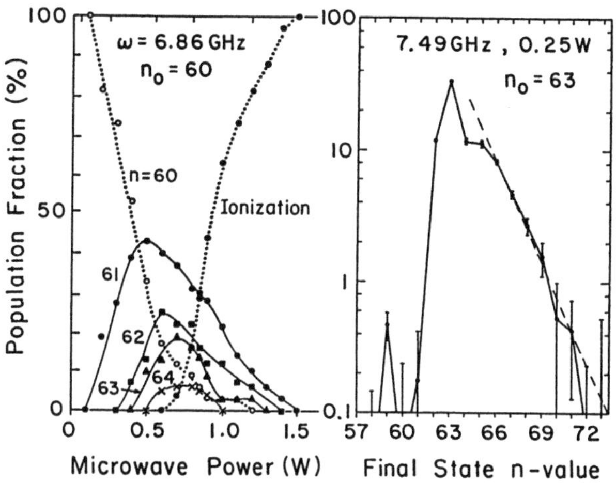


Fig. 3 (a) On the left is the variation with microwave power of an experimental final-state quantum number distribution, see reference 11. (b) On the right is one such distribution at one fixed value of power. There is a central peak in the distribution and an exponential tail at high n, see reference 16

by an exponential dependence with increasing n, see Fig. 3b. This third and very interesting feature seen in the experimental data appears associated with the classical phenomenon of diffusion in action space, which can produce exponential action distributions that also appear as exponential quantum number distributions in the quasienergy states [17]. It may be that the exponents of the quantum number distributions are proportional to the Lyapunov exponent which, when positive, characterizes the exponential divergence of adjacent classical trajectories for classically stochastic systems [4].

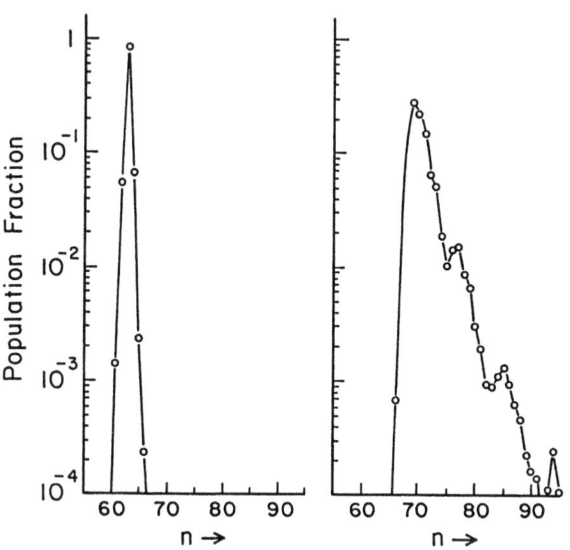


Fig. 4 Quantum-theoretical calculations of final-state quantum-number distributions, left, for initial action below the classical threshold for stochastic behavior, right, for above it. The calculations close-coupled 45 bound electrically polarized hydrogen atom states, see reference 8

All three of the experimental features mentioned above, the resonances, the "nonresonant" microwave absorption, and the exponential final-state distributions, have been confirmed by purely quantum mechanical many-state numerical close-coupling calculations [8,16]. In addition, the quantum calculations show an excellent correspondence with the threshold values of initial action classically required for stochastic behavior; above the corresponding initial quantum number value, exponential final-state distributions develop, while for lower values they do not, see Fig. 4. In addition, the quantum quasienergy states of the atom-field system are relatively delocalized in quantum number if and only if they are built up from free-atom states with quantum numbers above the threshold value.

The observed threshold and final-state distribution features of classically stochastic behavior are a result of the breakdown of "conservation" of the time-averaged action or energy. This does not directly show that there is a new type of randomness in quantum mechanics that is associated with stochastic behavior in the classical limit. Thus, in a sense, our search for quantum chaos is inconclusive. What remains to be done are studies of the statistical properties of the time-development of ensembles of atoms while they are within the microwave field.

REFERENCES

1. J. G. Leopold, D. Richards: J. Phys. B 18, 3369 (1985).
2. R. V. Jensen: Phys. Rev. Lett. 54, 2057 (1985) and references therein.
3. J. E. Bayfield: in Fundamental Aspects of Quantum Theory, V. Gorini, A. Frigerio, ed. (Plenum Press, New York, 1986).
4. A. J. Lichtenberg, M. A. Lieberman: Regular and Stochastic Motion, Springer-Verlag, New York, 1983.
5. K. A. H. Van Leeuven, G. V. Oppen, S. Resnick, J. B. Bowlin, P. M. Koch, R. V. Jensen, O. Rath, D. Richards, J. G. Leopold: Phys. Rev. Lett. 55, 2231 (1985).
6. G. Casati, B. V. Chirikov, D. L. Shepelyansky, I Guarneri: Phys. Rev. Lett. 57, 823 (1986).
7. G. Casati, B. V. Chirikov, D. L. Shepelyansky: Phys. Rev. Lett. 53, 2525 (1984).
8. B. Sundaram: Ph.D. Dissertation, University of Pittsburgh, 1986, and papers to be published.
9. R. V. Jensen: private communication.
10. G. L. Peterson, C. D. Cantrell: Phys. Rev. A 31, 807 (1985).
11. J. E. Bayfield, L. A. Pinnaduwage: Phys. Rev. Lett. 54, 313 (1985).
12. L. A. Pinnaduwage: Ph.D. Dissertation, University of Pittsburgh, 1986, and papers to be published.
13. J. E. Bayfield, L. A. Pinnaduwage: J. Phys. B 18, L49 (1985).
14. J. N. Bardsley, B. Sundaram: Phys. Rev. A 32, 689 (1985).
15. J. E. Bayfield, L. A. Pinnaduwage: paper presented at the "First International Conference on the Physics of Phase Space," College Park, MD, 20-23 May 1986, Springer-Verlag Lecture Notes in Physics, to be published.
16. J. N. Bardsley, B. Sundaram, L. A. Pinnaduwage, J. E. Bayfield: Phys. Rev. Lett. 56, 1007 (1986).
17. D. L. Shepelyansky: Phys. Rev. Lett. 56, 677 (1986).
18. A. Wolf, J. B. Swift, H. L. Swinney, J. A. Vastano: Physica 16D, 285 (1985).

# One-Dimensional Atomic Physics

*R.V. Jensen and S.M. Susskind*

Mason Laboratory, Yale University, New Haven, CT 06520, USA

Recent experiments on the microwave ionization of highly excited hydrogen atoms are well described by a one-dimensional classical theory in which the ionization is associated with the onset of ***chaos*** in the nonlinear electron dynamics. Motivated by the success of the one-dimensional classical theory and the challenge of uncovering the quantum mechanical mechanism responsible for the stochastic ionization, we have studied the numerical solution of the Schrödinger equation for a ***one-dimensional hydrogen atom*** in an oscillating electric field. Our numerical algorithm describes the interaction of a large number of bound states with a discrete representation of the continuum in terms of Coulomb wave functions. Preliminary results of these quantal calculations are compared with the classical predictions and with the results of recent experimental measurements of the microwave ionization of highly excited hydrogen atoms in order to address some of the issues raised by the problem of "quantum chaos". Moreover, since this numerical approach provides an explicit calculation of the energy distribution of the liberated electrons, this work may also provide valuable insight into the closely related problem of ***Above-Threshold Ionization.***

## 1. Introduction

In recent years, the interaction of intense electromagnetic radiation with atoms and molecules has been an important subject of experimental and theoretical research. The development of powerful sources of radiation for experimental studies and of high-speed computers for theoretical work has greatly stimulated and aided the investigation of this new frontier of quantum theory where the interactions are too strong for conventional time-dependent perturbation theory to be useful. The dominant experimental emphasis has been on the interaction of many-electon atoms with intense lasers; however, one of the simplest physical systems for studying the behavior of bound electrons in intense fields is provided by the experiments on the interaction of highly excited hydrogen atoms with strong microwave fields.[1-4]

In their pioneering work, Bayfield and Koch [1] found that the microwave ionization of highly excited hydrogen atoms exhibited a very strong dependence on the microwave field strength and a relatively weak dependence on the microwave frequency. More recently, van Leeuwen and Koch [3-4] have completed a detailed study of the ionization of hydrogen atoms with principal quantum numbers ranging from $n = 32$ to 90 which were exposed to $\sim 300$ periods of a $\Omega/2\pi = 9.92$ GHz field as they passed through a microwave cavity. Their preliminary measurements of threshold microwave fields for the onset of significant (10%) ionization are plotted in scaled field units, $n^4F$, versus scaled frequency, $n^3\Omega$, in Fig. 1 for each $n$. These results exhibit a gradual decrease in the threshold fields with increasing $n$ and scaled frequency, interupted by distinct peaks at rational values of $n^3\Omega = 1, 2/3, 1/2, 2/5, 1/3, 1/4, 1/5$.

The quantum theory for these experiments has proven to be very difficult since one hundred or more microwave photons are required for ionization and the field strengths are very large, $\sim 10\%$ of the Coulomb binding field. However, these remarkable experimental results are well described by a nonperturbative classical theory of the nonlinear electron dynamics in which the ionization is associated with the onset of chaotic, classical electron trajectories.[5-6] As a consequence, these remarkable experimental studies have been the focus of considerable interest since they provide a unique laboratory for investigating the quantum behavior of a dynamical system which exhibits the irregular, unpredictable motion known as ***chaos*** in the classical limit.[7-8]

A hydrogen atom in an oscillating electric field corresponds, classically to a nonlinear oscillator subject to a periodic perturbation. When the perturbation exceeds a critical threshold, this classical oscillator exhibits a transition from regular to chaotic behavior. Since the regular electron orbits remain confined to the nucleus, while the chaotic trajectories can escape, the onset of classical chaos results in the ionization of the atom. Classical calculations [3-6] of the chaotic ionization have been performed by integrating the nonlinear equations of motion for a microcanonical ensemble of initial conditions with energies appropriate

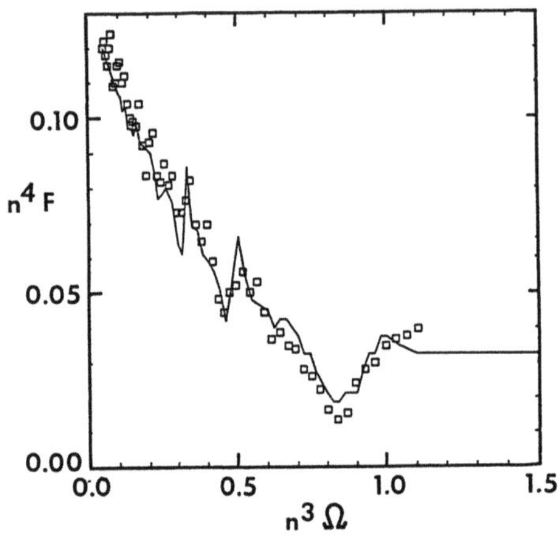


Fig. 1: A comparison of the theoretical predictions for the threshold field strengths, $n^4F$, for the onset of chaos (solid line) with experimental values for 10 % ionization (squares) is shown as a function of scaled frequency, $n^3\Omega$. (The error bars $\sim 5-10\%$ associated with this preliminary theoretical and experimental data have been suppressed to simplify the comparison.)

for the given quantum state. By counting the number of classical trajectories which became chaotic and wandered away from the nucleus, the ionization probability can then be determined as a function of the scaled microwave field and frequency.

Although classical calculations were performed in one [3,4,6], two, and three spatial dimensions.[3-4,9] the *one-dimensional* model, defined by the Hamiltonian

$$H(x,p,t) = p^2/2 - \begin{cases} -1/x, & x > 0; \\ \infty, & x \leq 0 \end{cases} + xF(t)cos\Omega t \quad (a.u.), \tag{1}$$

where $F(t)$ models the slow time variation of the microwave field amplitude as the atom enters and leaves the cavity, appeared to be adequate for predicting the threshold fields for the onset of ionization. Preliminary theoretical predictions for the critical fields for the onset of chaotic ionization in this one-dimensional model are also displayed in Fig. 1 showing the striking agreement between the threshold fields for significant ionization and the onset of classical chaos. In particular, the classical theory exhibits the distinct peaks in the threshold fields near resonant values of the scaled frequency.[6] The threshold fields given by two- and three- dimensional classical calculation show similar agreement with the experimental measurments.[3-4,9] This agreement of one-dimensional theory with the two- and three- dimensional theory and with the experiment can be attributed to the fact that the nearly one-dimensional orbits, elongated along the field direction, are generally the easiest to ionize.[10]

The challenge now remains to understand the quantum mechanisms which *simulate* the classical chaos and which account for the observed ionization. Fortunately, the success of the one-dimensional classical theory suggests that a quantum treatment of a one-dimensional hydrogen atom in an oscillating electric field may also succeed in capturing the essential physics of the chaotic ionization. In this case, the quantum dynamics are determined by the solutions of the Schrödinger equation corresponding to the one-dimensional Hamiltonian, Eq. (1),

$$i\frac{\partial \psi}{\partial t} = -\frac{\partial^2 \psi}{\partial x^2} - \frac{1}{x}\psi + xFcos\Omega t\psi, \quad x \geq 0. \tag{2}$$

This one-dimensional theory also has direct application to other physical systems which approximate a one-dimensional hydrogen atom such as surface-state-electrons bound to the surface of liquid helium by their image charge [5] or electrically polarized hydrogen atoms prepared in extremal Stark states [2](see also the article by J.E. Bayfield in this volume). A three-dimensional quantum theory is required for a complete description of the ionization experiments of van Leeuwen and Koch [3-4]; however, since a factor of $\sim n^2$ more states are involved, the full theory is much more difficult.

At present, there are a number of research groups investigating the quantum theory for the one-dimensional hydrogen atom in an oscillating electric field using a variety of different techniques ranging from the summation of high-order perturbation theory [11] to direct numerical solutions of the time-dependent Schrödinger equation. [12-16] As a result our understanding of the quantum dynamics for this strongly perturbed quantum system has been progressing rapidly. These one-dimensional calculations are beginning to show quantitive agreement between the threshold fields for significant excitation to continuum with the critical fields for the onset of classical chaos and with the measured fields for 10% ionization. However, Casati et al. [12] predict that for scaled frequencies somewhat larger than those studied in the present experiments, $n^3\Omega > 1$, and for lower quantum levels, the quantum threshold for significant ionization should be higher than the threshold for the onset of classical chaos due to a remarkable quantum localization effect which is related to Anderson localization in disordered solids. [17] (For a review some of these numerical efforts see Ref. [18] and the article by G. Mantica contained in this volume.)

The primary purpose of this paper is to present a progress report on our preliminary investigations of the quantum behavior of one-dimensional atoms in intense fields. One particular numerical algorithm for the approximate solution of Eq. (2), which includes a large number of bound states coupled to a discretized continuum, is outlined in Section 2.1 This approach is essentially equivalent to placing the one-dimensional hydrogen atom in a large box. Using the exact Coulomb wavefunctions for both the bound states and the continuum states we can explore the role of bound-bound, bound-continuum, as well as continuum-continuum coupling on the excitation of an initially bound state to the continuum. One advantage of this numerical procedure is that by varying the size of the box and the numbers of bound and discretized continuum states it is easy to assess and control the accuracy of the calculation. In addition, the quantum evolution is wholly described in terms of the familiar concepts of the unperturbed bound and continuum states.

In Section 2.2, we present preliminary results of calculations of the microwave ionization of one-dimensional hydrogen atoms using this numerical algorithm which provide a clear picture of a two-step excitation-ionization mechanism which appears to be responsible for the chaotic ionization observed in the numerical simulations as well as the experiments.. In addition, a simple analytical treatment of the latter stages of this ionization process is described which provides important insight into the coupling of the excited bound states to the continuum. The qualitative picture which emerges from this analysis is in agreement with the mechanism originally proposed by Blümel and Smilansky [15] and is consistent with the extensive numerical studies of Casati et al. [12] and Bardsley et al. [13-14].

Finally, motivated by the success of the one-dimensional classical and quantum theories for this strongly perturbed quantum system, Section 3 concludes with some speculative comments on the possible applications of such studies of *one-dimensional atomic physics* to other problems involving the interaction of intense radiation with atoms which have defied conventional perturbative analysis. One important example is the problem of *Above-Threshold Ionization* (ATI) in which atoms exposed to intense laser radiation preferentially absorb more photons than the minimum number needed for ionization.[19-20] A variety of elabotate theories have been devised in attempts to account for this behavior.[20] However, since our numerical algorithm for the solution of the one-dimensional atom in an oscillating electric field resolves the spectrum of liberated electrons, it is easy to use this simple model to study the changes in the free-electron energy spectrum as the strength of the perturbation is increased. In particular, a simple calculation of the coupling of a single bound state to 200 discretized continuum levels clearly exhibits ATI-like behavior.

## 2. Quantum Theory for the Intense Field Ionization of 1-D Atoms

### 2.1 Numerical Solution of the 1-D Schrödinger Equation

The most direct approach to the numerical solution of the time-dependent Schrödinger equation, Eq. (2), for a one-dimensional hydrogen atom in an oscillating electric field is to expand the wavefunction, $\psi(x,t)$, in terms of the bound and continuum eigenfunctions of the unperturbed hydrogen atom

$$\psi(x,t) = \sum_{n=1}^{\infty} a_n(t)\phi_n(x) + \int_0^{\infty} dk\ a_k(t)\phi_k(x). \tag{3}$$

Then in the interaction representation with $b_n(t) = a_n(t)e^{-iE_nt}$ and $b_k(t) = a_k(t)e^{-iE_kt}$, where $E_{n,k}$ are the corresponding hydrogenic energy levels, the Schrödinger equation can be formally expressed as

$$i\frac{db_n}{dt} = \sum_m W_{nm}b_m + \int_0^{\infty} dkW_{nk}b_kdk, \tag{4}$$

$$i\frac{db_k}{dt} = \sum_m W_{km}b_m + \int_0^{\infty} dk'W_{kk'}b_{k'}dk', \tag{5}$$

where $W_{ab} = -e^{i\omega_{ab}t}F(t)\cos\Omega t\ x_{ab}$, depend on the values of the Bohr frequencies $\omega_{ab} = E_a - E_b$ and the dipole matrix elements $x_{ab} =< a|x|b >$ for the bound and continuum wavefunctions with $a, b = n, k$. However, to solve this formally infinite set of ordinary differential equations, we must first truncate the number of bound states to a large but finite number and then introduce some approximations to represent the interaction with the continuum.

A variety of different numerical approaches to dealing with the continuum are currently being pursued ranging from neglecting it altogether[12-13] to the use of projection operators [15] or Sturmian basis functions. [12,14-15] However, these latter methods have the disadvantage that the details of the bound-continuum coupling are obscured. In contrast we have chosen to introduce a discrete representation of the continuum in terms of the natural Coulomb wavefunctions of the unperturbed hydrogen atom.

For the purposes of this paper we will neglect the continuum-continuum coupling in Eq. (5). Although these effects can be included in numerical calculations (as described in Ref. 16), our preliminary calculations suggest that they do not modify the qualitative features of the excitation of bound states to the continuum. If we then divide the $k$-axis between $[0, k_{max}]$ into $N_c$ segments with width $\Delta_j$, each centered 0 at $k_j$, $j = 1, N_c$, the integral over the continuum states in Eq. (4) can be approximated by

$$\int_0^{\infty} dk W_{nk} b_k \simeq \sum_{j=0}^{N_c} W_{nk_j} b_{kj} \Delta_j \qquad (6)$$

as long as $W_{nk}b_k$ varies little on the scale of $\Delta_j$.

With this scheme, we can treat the discretized continuum states in the same way as the bound states. If all of the $\Delta_j$ are chosen to be equal, then this prescription is essentially equivalent to putting the one-dimensional hydrogen atom in a large box of length $\sim 1/\Delta_j$. More generally we have the flexibility of varying $\Delta_j$ as well as $k_{max}$ to optimize the coverage of the range of $k$ where most of the coupling from the bound states to the continuum occurs. In addition, we find in practice that resonant multi-photon processes from the bound states give rise to prominent peaks in the distribution over the continuum states which grow in time. Consequently, the assumptions used to discretize the continuum fail after a period of time which can be increased by decreasing $\Delta_j$ and increasing the number of states. In fact any numerical treatment of the continuum, such as the use of a discrete basis of Sturmian functions, will suffer this defect for large times. However, by carefully monitoring the evolution of the probability distribution over continuum states we can easily detect and remedy any failures in the approximations.

### 2.2 Numerical Results and Analysis

Using this numerical algorithm we have begun to study the quantum evolution of electrons initially prepared in an excited hydrogenic state with $a_{n_0}(0) = 1$ which are perturbed by an oscillating electric field. For example, Fig. 2 shows some typical results for the probabilities, $|a_n(t)|^2$ and $|a_k(t)|^2$, for exciting the $n_0 = 20$ state to various bound and continuum states after 40 perturbation periods. In order to exhibit the resonance structures associated with multi-photon processes, these probabilities are plotted as a function of the energies, $-1/2n^2$, for the bound states and, $k^2/2$, for the continuum states. Note, in particular, that the distribution over the continuum states exhibits distinct peaks separated by the "photon" energy $\Omega$.

A close study of the distributions over excited states, displayed in Fig. 2 for two different perturbation strengths $n_0^4F = .06$ and $n_0^4F = .1$ at a frequency of $n^3\Omega = 1.$, suggests a simple qualitative picture of the physical mechanisms responsible for ionization in strong oscillating fields which was first proposed by Blumel and Smilansky.[15] In both cases the perturbation excites a broad range of bound states. However, by summing over the continuum state probabilities we see that increasing the field strength from $n_0^4F = .06$ to $n^4F = .1$ results in a $\sim$ 100 times more ionization.

Since a minimum of 10 $n^3\Omega = 1$ "photons" must be absorbed for a direct excitation of the $n_0 = 20$ state to the continuum, lowest order time-dependent perturbation theory predicts that the ionization rate should increase very rapidly with increasing field like $F^{20}$. However, even at these large perturbation strengths, which are comparable to 10% and 20% of the Coulomb binding field, the magnitude of this rate is too small. This is easily verified by performing a numerical calculation which couples a single $n_0 = 20$ bound state to the continuum. Therefore, as indicated by Fig. 2 and the calculations of other research groups, the ionization appears to be the result of a two-step process where the initial state is first excited to more highly excited states which are ionize more rapidly.

The first excitation step cannot be described by perturbation theory, since many quantum levels are strongly coupled. However, numerical studies [12-13] of this problem, which integrated the equations of

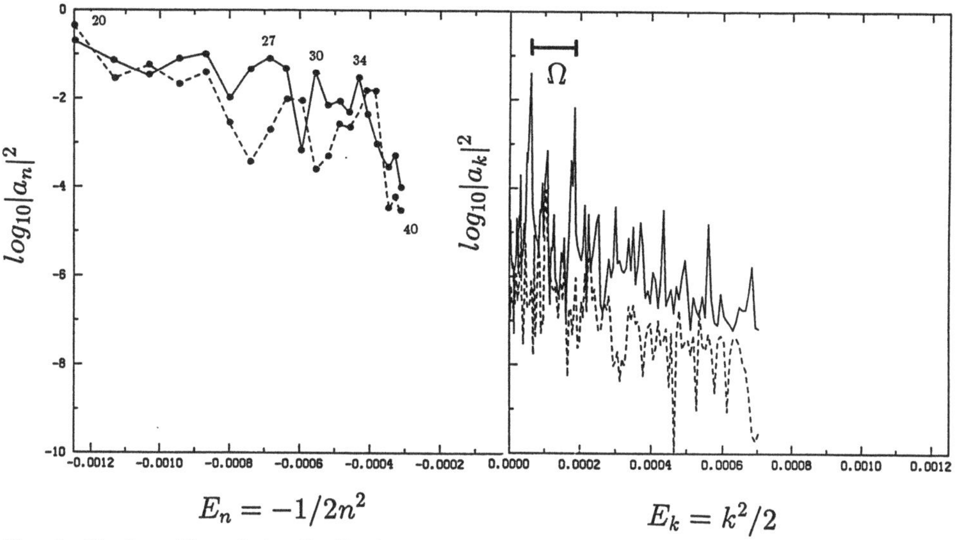

Fig. 2: The logarithm of the distributions over bound and continuum states is displayed as a function of the energy of the states after 40 periods of a $n_0^3\Omega = 1$ perturbation with two different field strengths $n_0^4F = .1$ (solid line) and $n_0^4F = .06$ (dashed line). The initial bound state was $n_0 = 20$. The solid dots on the left half of the graph represent specific bound states with $n = 20 - 40$

motion for a large number bound states and initially neglected the continuum states, have shown that the first excitation step results from the delocalization of the distribution of the evolving wavepacket over bound states as the perturbation is increased. For weak fields the wavepacket remains highly localized about the initial state, coupling to only a few of its neighboring states; however, as the perturbation is increased above a critical threshold the distribution spreads rapidly to higher and higher states. For the range of quantum states and scaled frequencies studied in the experiments by van Leeuwen and Koch [3-4],this delocalization threshold appears to be in good agreement with the onset of classical chaos as well as the measured threshold for significant ionization,(although the calculations of Casati et al. [12] indicate that this agreement will not necessarily persist for higher scaled frequencies, $n^3\Omega > 1$.)

If the wavepacket is sufficiently delocalized to populate bound states which are strongly coupled to the continuum via multi-photon processes, then significant ionization will result. The numerical results for this second step in the ionization process are suprisingly well described by a simple extension of the conventional time-dependent perturbation theory which provides quantitative estimates for the probabiltity of ionization from these excited states and of range of states which must be heavily populated to give significant ionization in a $\sim$ 100 periods of the perturbation.

Bardsley and Sundaram [13] first noted that the permanent dipole moments, $x_{nn}$, of the states of the one-dimensional hydrogen atom play a dominant role in multi-photon processes between bound states. In particular, it is easy to show that, within the rotating wave approximation, a $s$-photon, Rabi-flopping frequency between two states with permanent dipole moments is given by

$$W_{nm}^{s} = Fx_{nm}(2sJ_s(\beta)/\beta), \tag{7}$$

where $\beta = F(x_{mm} - x_{nn})/\Omega$ is proportional to the difference in the permanent dipole moments.[21] In the small field limit, the Bessel function $J_s$ can be expanded for small argument to reduce Eq. 7 to the formula for the multiphoton Rabi frequency given in Eq. 1 of Ref. 13 which in turn reduces to the usual single photon rate, $W_{nm}^1 = Fx_{nm}$, for $s = 1$. Therefore, for weak fields the $s$-photon excitation rate, $\propto W_{nm}^2$, exhibits the usual $F^{2s}$ dependence on the oscillating field. However, Eq. (7) indicates that for large fields the rates saturate at a maximum value, for $\beta \sim s$, and then exhibit a damped oscillatory behavior for larger fields.

Unfortunately, it is not yet clear how the higher order terms in the expansion of the Bessel function are modified by corrections to the rotating wave approximation. However, we nevertheless find that Eq. (7) gives a very good description of the numerical results for the excitation rate between bound states and the discretized continuum states. In particular, the introduction of the effective multiphoton coupling given by Eq. (7), into the Fermi Golden rule for the multiphoton ionization rate per period of the perturbation of the $n^{\text{th}}$ bound state gives

$$\Gamma_n^s = \frac{2\pi}{k\Omega}\left(\frac{F}{2}\right)^2 x_{kn}^2 (4s^2 J_s^2(\beta)/\beta)^2, \tag{8}$$

where the bound-continuum matrix elements are given to very good approximation by the formula [16]

$$< n|x|k > \simeq -\frac{2^{4/3}\,3^{1/6}}{\pi}\Gamma(2/3)\frac{n^{11/6}\,k^{1/2}}{(1+n^2k^2)^{5/3}}. \tag{9}$$

Since we have neglected the continuum-continuum terms in Eq. (5) including the contributions of the permanent dipole moments, $x_{kk}$, of the continuum states, the argument of the Bessel function,

$$\beta = \frac{3}{2}n(n^4F/n^3\Omega), \tag{10}$$

is simply proportional to the permanent dipole moment, $x_{nn} = \frac{3}{2}n^2$, of the bound state. Although the permanent dipole moments of the continuum states, $x_{kk} = 1/k^2$ can be very large, the dominant contributions to the ionization rate come from values of $k$ near the extremal value of $x_{kn}$, which is easily determined from Eq. (9) to be $k_0 \simeq 1/2.38n^2$. Therefore, since $x_{k_0k_0} \simeq 5.67n^2$, the value of $\beta$ would be larger but still of the same order of magnitude. As a consequence, the inclusion of the continuum-continuum effects will make significant quantitative corrections but should preserve the overall qualitative picture of the ionization process. However, these issues clearly require more extensive numerical and analytical study.

Nevertheless, within the present numerical model, which neglects the continuum-continuum contributions, Eq. (8) does a remarkably good job of accounting for the rates of ionization from individual bound states. In particular, it provides an estimate of the range of excited states with the highest ionization rates for a given perturbation strength. For small fields the expansion of the Bessel function for small argument provides the perturbative result for the multi-photon ionization rate which increases very rapidly as the $2s$ power of the field. However, for large fields with $\beta \sim s$ the ionization rate saturates at a maximum value. This condition can be used to determine the value of $n_c$ at which the ionization rate first achieves its maximum value. Since the minimum number of photons to ionize the state $n$ by an oscillating field with frequency $\Omega$ is $s = n/2(n^3\Omega)$, we can equate this expression with Eq. (10) to find

$$n_c = n_0\left(\frac{1}{3n_0^4F}\right)^{1/4}, \tag{11}$$

which is the same as Blumel and Smilansky's estimate for the lower edge of the "window" of states which are strongly coupled to the continuum.[15]

For the two calculations displayed in Fig. 2, Eq. (11) gives $n_c = 27$ for the large field and $n_c = 30$ for the small field. In fact a close examination of Fig. 2 shows that for the larger field the states $n$ =27, 30, and 34, which are approximately 6,5, and 4 photons from the resonance peaks in the continuum, all contribute to the ionization. In contrast for the lower field only the n=35,36 states, which are more weakly populated, appear to contribute to the resonance peaks in the continuum through a 4 photon excitation. As a consequence, we see dramatically more ionization for the higher field.

Although this two-step ionization process appears to be a very promising candidate for the physical mechanism responsible for the chaotic ionization observed in the experiments with real hydrogen atoms, much work remains to be done to assess the quantitative effects of the continuum-continuum coupling and to extend the full numerical calculations to the high quantum states, $n \sim 30-90$, have been studied in the experiments. [3,4] Despite the simplicity of this one-dimensional problem these investigations require many hours of super-computer time. However, since a number of different research groups are attacking this problem from several different directions, a more complete quantum picture of the intense field ionization of highly excited atoms and its relationship to the classical theory of chaos should soon be forthcoming.

### 3. Above-Threshold Ionization

Since this one-dimensional quantum theory appears to have captured the essential physics of the mechanisms responsible for the ionization of highly excited hydrogen atoms, this model may also prove useful in developing a physical intuition for understanding other interesting phenomena involving the strong interaction of radiation with matter. In particular, since our numerical algorithm for the solution of the one-dimensional Schrödinger equation resolves the free electron energy spectrum, this approach can also be used to explore the quantum mechanisms which underly the process of Above-Threshold Ionization (ATI) which is currently of great interest to researchers studying the interaction of intense laser radiation with atoms.[19-20]

ATI refers to the experimental observation that for very intense fields the energy spectrum of ionized electrons exhibits multiple peaks associated with the absorption of more photons than the minimum number to reach the continuum. In particular, as the laser field is increased the higher order photon absorption becomes more probable. In fact, the peak associated with the minimum number of photons required to reach the continuum actually disappears. A number of elaborate and controversial theories [20] have been proposed to account for this remarkable phenomena which, like the microwave ionization of highly excited hydrogen atoms, cannot be explained by simple perturbation theory. However, such multiple peaks are clearly evident in the one-dimensional calculations of the distribution over continuum states displayed in Fig. 2, which suggests that the study of the one-dimensional atomic physics may provide a useful laboratory for understanding the physical origin of ATI in real atoms.

The results of a series of numerical calculations using a very simple model of a single $n_0 = 20$ state coupled to 200 discretized continuum states are displayed in Fig. 3. These results clearly show the rise of the resonance peaks associated with higher order photon absorption as the perturbing field strength is increased from $n_0^4F$=.36 to 1.0 at a frequency of $n_0^3\Omega = 3.2$. At the lowest field, which is already in excess

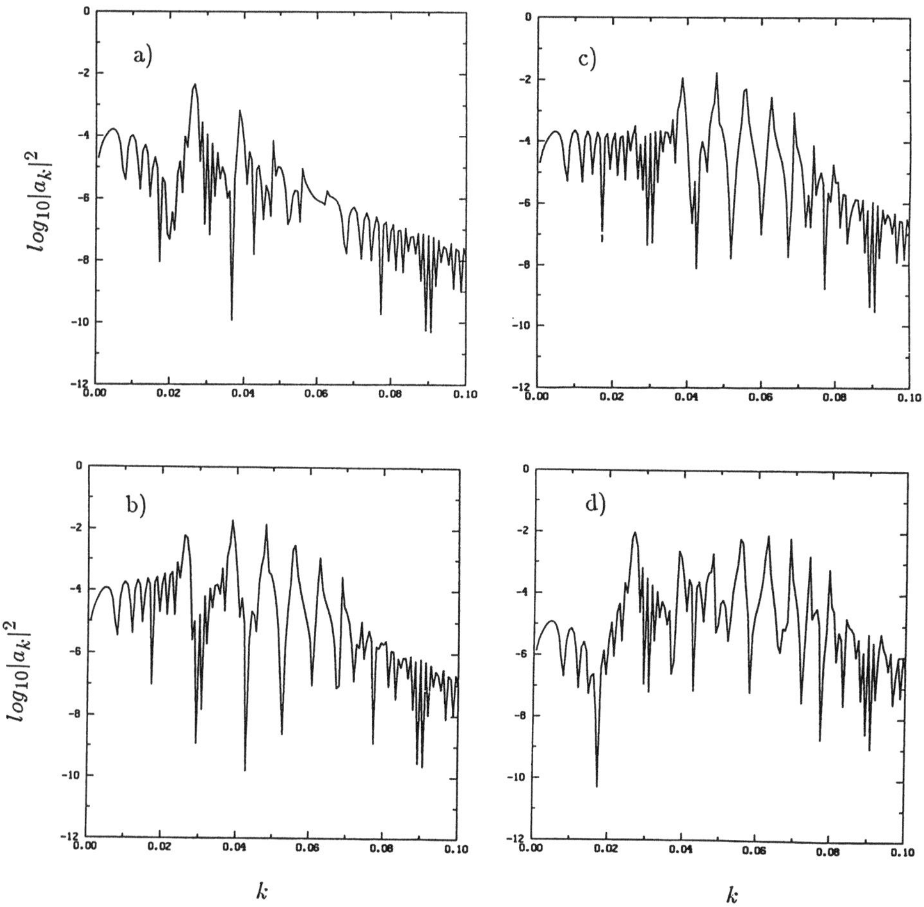


Fig. 3: The logarithms of the probabilities of exciting an $n = 20$ state to the continuum states after 10 periods of a $n^3\Omega = 3.2$ perturbation are plotted as functions of the momentum $k$ for increasing field strengths a) $n^4F = .36$, b) $n^4F = .72$, c) $n^4F = .8$, and d) $n^4F = 1.$. The first peaks in a), b) and d)correspond to resonant 4-photon transitions from the $n = 20$ state.

of the Coulomb binding field, for $n_0 = 20$, the dominant peak in the electron momentum spectrum shown in Fig. 3a corresponds to the absorption of 4 "photons". The 5 and 6 photon peaks are clearly visible but smaller in size. At $n_0^4 F = .72$, Fig. 3b, additional peaks corresponding to the absorption of 7,8, and 9 photons emerge and the absorption of 5 and 6 photons becomes more probable than the 4 photon process. For a still larger field of $n_0^4 F = .08$, Fig. 3c, the 4-photon peak vanishes. Thus far the numerical results of this very simple model show a remarkable similarity to the experimental observations. However, for a still higher field of $n_0^4 F = 1.0$, Fig. 3d, the first peak reappears in our model calculations,which apparently has not been observed in any experiments.

We note that these effects,which closely mimic the experimental measurments,can be easily explained using the simple theory for multi-photon ionization described in Section 2.2. Equation (8) shows that the $s$-photon ionization rate is proportional to the square of a Bessel function $J_s(\beta)$. For low fields, such that $\beta$ is small, the Bessel function can be expanded in a power series in $\beta$ with leading term $J_s(\beta) \sim (\beta/2)^s$. In this case the absorption of additional photons is reduced by a factor of $\beta^2$. However, as the perturbing field is increased the value of the Bessel function reaches a maximum and then decreases. In this case the rates for the lower order processes will decrease while the higher order processes will continue to rise. In fact, the $s$-photon ionization rate can actually vanish when $J_s(\beta)$ passes through a zero. This accounts for the disappearance of the 4-photon peak in Fig. 3c since $\beta = 7.5 \sim$ the first zero of $J_4$. However, in this model problem the peaks return when $\beta$ moves away from the zeros of the Bessel functions.

Clearly, this model is too simplistic to be compared directly with experiment. Much more extensive numerical studies are required for a reasonable test of the relevance of the one-dimensional atomic physics to the real problem. However, these provocative results suggest that further investigations including additional bound states as well as the continuum-continuum coupling may prove fruitful.

The authors are very grateful to K.A.H van Leeuwen and P.M. Koch for providing the preliminary experimental data, shown in Fig. 1, prior to publication, and to R. Blümel, G. Casati, and B. Sundaram for many stimulating discussions of these problems. This work was supported by the National Science Foundation and the Alfred P. Sloan Foundation.


1. J.E. Bayfield and P.M. Koch, Phys. Rev. Lett. *33*, 258 (1974).
2. J.E. Bayfield and L.A. Pinnaduwage, Phys. Rev. Lett. *54*, 313 (1985).
3. K.A.H. van Leeuwen, G.v. Oppen, S. Renwick, J.B. Bowlin, P.M. Koch, R.V. Jensen, O. Rath, D. Richards, and J.G. Leopold, Phys. Rev. Lett. *55*, 2231 1985).
4. P.M. Koch, "Chaotic Ionization of Highly Excited Hydrogen Atoms by a Microwave Electric Field", to appear in *Fundemental Aspects of Quantum Theory*, ed. V. Gorini and A. Frigerio (Plenum, London, 1986).
5. R.V. Jensen, Phys. Rev. Lett. *49*, 1365 (1982); R.V. Jensen, Phys. Rev. *30A*, 386 (1984).
6. R.V. Jensen, to appear in the Proceedings of the Adriatico Research Conference on Quantum Chaos, Trieste, 1986 published by Physica Scripta.
7. A.J. Lichtenberg and M.A. Lieberman, *Regular and Stochastic Motion*, (Springer, New York, 1983). See also R.V. Jensen, "Classical Chaos", (to appear in American Scientist, March 1987) and references therein.
8. See articles in *Chaotic Behavior in Quantum Systems*, edited by G. Casati, (Plenum, New York, 1985).
9. O. Rath and D. Richards, private communication.
10. R.V. Jensen and M.M. Sanders, in preparation.
11. S.Graffi, V. Grecchi, H.J. Silverstone, private communication.
12. G. Casati, B.V. Chirikov, and D.L. Shepelyansky, Phys. Rev. Lett. *53*, 2525 (1984); G. Casati, B.V. Chirikov, D.L. Shepelyansky, and I. Guarneri, Phys. Rev. Letters, *57*, 823 (1986).
13. J.N. Bardsley and B. Sundaram, Phys. Rev. A *32*, 689 (1985).
14. J.N. Bardsley, B. Sundaram, L.A. Pinnaduwage, and J.E. Bayfield, Phys. Rev. Lett. *56*, 1007 (1986); J.N. Bardsley and M.J. Comella, to appear in J. Phys. B.
15. R. Blümel and U. Smilansky, Phys. Rev. A *32*, 1900 (1985), R. Blümel and U. Smilansky, preprints (submitted to Phys. Rev. Lett. January, 1986 and August, 1986).
16. S.M. Susskind and R.V. Jensen, submitted to Phys. Rev. A, October 1986.
17. R.E. Prange, D.R. Grempel, and S. Fishman, in *Chaotic Behavior in Quantum Systems*, edited by G. Casati, (Plenum, New York, 1985) p. 205.

18. R.V. Jensen, "Chaos in Atomic Physics", to appear in the *Proceedings of the Xth International Conference on Atomic Physics(ICAP-X)*, ed. H. Narumi, (North Holland, Amsterdam, 1987).
19. P. Kruit,J. Kimman, H.G. Muller, and M.J. van der Wiel, Phys. Rev. A *28*, 248 (1983) and references therein.
20. A. Szöke, J. Phys. B *18*, L427 (1985) and references therein.
21. M.A. Kmetic and W.J. Meath, Phys. Lett. *108A*, 340 (1985).

# Is Subthreshold Ionization a Chaotic Phenomenon?

*G. Mantica*†

School of Physics, Georgia Institute of Technology,
Atlanta, GA 30332, USA

The dynamics of strongly perturbed quantum systems is discussed in this paper through the study of a simple model. Aiming at a clarification of the role of classical chaos in quantum mechanics, the mechanism of strong field subthreshold ionization is illustrated in different situations.

## 1. INTRODUCTION

One of the leading theoretical problems of today physics is commonly known as "Quantum Chaos", and is related to the manifestation in quantum mechanics of any erratic, unpredictable kind of motion [1,2]. Quantum mechanics has an inherent statistical nature due to the probabilistic interpretation of the wave function, in the line of the School of Copenhagen. Nevertheless, the Schroedinger equation provides a deterministic dynamical evolution, that can be studied with the techniques of modern dynamics.

A basic question arises if we consider the implications of classical chaos to the "Correspondence Principle": let's focus our attention on the quantized version of a classical K-system, a billiard, for example. Loosely speaking, the correspondence principle states that as the Planck constant goes to zero, the predictions obtained by quantum mechanics should converge to their classical analogue. Now, the latter is determined by the chaotic characteristics of the motion. This opens the question whether in quantum mechanics any chaotic process could possibly exist. The first systems to be analyzed in this light were the quantized chaotic billiards. Being these systems bounded and autonomous, their energy spectrum is forced to be pure point. Since chaotic motion is necessarily related to a continuous spectrum of the evolution operator, stricto sensu their quantum dynamics cannot be chaotic. Nonetheless, some interesting spectral properties have been detected for these systems, and related to the theory of Random Matrix Ensembles [3].

Much richer spectral properties are possible in a different context: the class of externally perturbed quantum systems, where dynamical characteristics resembling chaos might possibly manifest [4]. Among these, an interesting example has attracted much attention, since the early work of Bayfield and Koch [5]: a Rydberg hydrogen atom in the field of a microwave cavity [6-14]. For high intensities of the microwave field, a strong subthreshold ionization is experimentally detected. To explain this fact, a purely classical description has been invoked, with satisfactory results [11,13]. The major semiclassical phenomenon seems however to reside in the transition from "quantum localization" to "delocalization" [8,10].

In this paper I introduce a simple model that presents a clear picture of the mechanism at work in strongly perturbed quantum systems. By letting the Planck constant be a variable parameter, the dynamics goes from that of a "deep" quantum region to that of a semiclassical regime. I present mainly numerical results showing the corresponding transition from direct multiphoton ionization to ionization after quantum delocalization.

† Permanent Address: Dipartimento di Fisica, Via Celoria 16, 20133 Milano, Italia

## 2. A SIMPLE MODEL

The model I shall consider is defined by the following one-dimensional Hamiltonian:

$$H = H_0 + \varepsilon W(x,t) = - \hbar^2/2\cdot\partial^2/\partial x^2 + V(x) + \varepsilon W(x,t) \ , \tag{1}$$

where:

$$V(x) = \begin{cases} -V_0 , & x \in (-\pi,\pi) \\ 0 , & x \notin (-\pi,\pi) \end{cases} , \qquad W(x,t) = \cos(x)\cdot\cos(\omega t) \ . \tag{2}$$

In words, H describes a particle in a square well potential, acted upon by an external oscillating field. When $\varepsilon = 0$, the spectrum of the unperturbed Hamiltonian has a pure point component related to a finite number of negative energy states, exponentially localized inside the well. A continuous component is present as well, and it consists of the positive real axis. In the numerical simulations, it resulted convenient to introduce confining barriers at a distance L, $L >> \pi$. These change the continuum of positive energy states into a high density distribution of energy levels.

The advantages of this model are several. First, notwithstanding its simplicity, it accounts for the relevant phenomena we are studying. Second, it allows us to vary its internal parameters to reproduce different physical situations, and finally it is suited for numerical treatment, particularly for the study of the effect of the continuum component of the spectrum.

## 3. THE QUANTUM MECHANICAL DESCRIPTION

Let's now consider the quantum motion given by the Hamiltonian (1). Since the external field varies periodically on time, we examine the period evolution ( Floquet ) operator U defined by:

$$(U\psi)(x,t) = \psi(x,t - 2\pi/\omega) \ . \tag{3}$$

The spectrum of U (and of the corresponding quasi-energy operator) is continuous, because of the possibility of multiphoton transitions. The stationary states of the unperturbed system become quasi-stationary states,with a finite lifetime depending on $\varepsilon$. Following the mathematical literature [15], I shall also call these latter "resonances".

Resonances are of primary interest in discussing the "deep" quantum region. This situation is obtained by letting $\hbar = 1$ in such a way that only few negative energy states are present. Then let's choose a value for the external frequency below the one-photon threshold and let's study the quantum evolution of such a system, initially prepared in the ground state. The dynamics, despite the continuum spectrum of U, depends on the few metastable states. A quantitative description of this fact can be obtained as follows: consider the resolvent $R_z(K)$ of the quasi-energy operator of the system (1), $K = -i\ \partial/\partial t + H$ :

$$R_z(K) = ( K - zI )^{-1}, \quad z \in C \ . \tag{4}$$

Then, quasi-energies are defined as the poles in the analytic continuation of $R_z(K)$ in the complex z-plane. Perturbation theory predicts that the imaginary part $\Gamma$ of each quasi-energy (which is inversely proportional to the "lifetime" of the metastable state) should behave like:

$$\Gamma \propto \varepsilon^{2p} \ , \tag{5}$$

where p is the number of photons energetically required for direct ionization of the corresponding stationary state. The quantities $\Gamma$ can be calculated from the quantum evolution as long as the resonances determine the dynamics. The calculation consists of a Pade' summation of an appropriate series. In such a case the motion is also numerically reversible [14]. Fig. 1 shows that resonances are detectable even in the re-

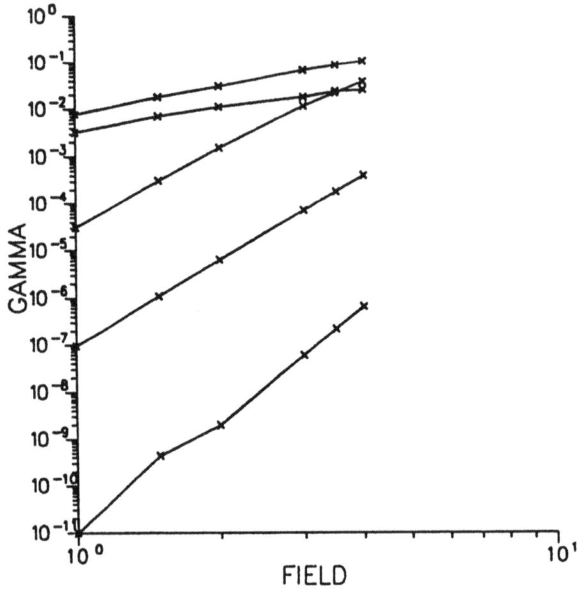


Fig.1. Imaginary part of the quasi-energies of the few quasi-stationary states in the case $\hbar = 1$, $V_0 = 20$, $\omega = 5$, versus the field intensity $\varepsilon$

gion of strong fields. The slopes of the curves in Fig. 1 agree with (5) in an extended region.

## 4. THE SEMICLASSICAL REGIME

Before turning the discussion to the semiclassical region, some results from the classical analysis of the Hamiltonian (1) are needed. Let us consider the evolution of negative energy states. Because of symmetry, parity is conserved, and we can restrict our consideration to odd states. In this case, the classical analogue of the model (1) is given the Hamiltonian:

$$H(p,\theta) = p^2 / 2 + \varepsilon\cos(\theta)\cdot\cos(\omega t) , \qquad \theta = x \bmod(2\pi) , \tag{6}$$

where action-angle variables have been introduced.

The presence of the external perturbation ($\varepsilon>0$) renders the Hamiltonian (6) not integrable. However, for small values of $\varepsilon$, K.A.M. Theorem assures that at least some of the tori - regular orbits - of the unperturbed motion are still present, even if in a somewhat deformed fashion. As the perturbation gets stronger and stronger, most tori are destroyed, and the system becomes more and more chaotic. A way to predict the onset of chaos is given by the "Chirikov Resonance Overlapping Criterion". In our case, two resonances appear explicitly in (6). A straightforward application of the above criterion predicts that as

$$\sqrt{(2\varepsilon)} / \omega > 1 , \tag{7}$$

a chaotic domain is created in phase space. In this case, classical motion is truly chaotic, so that a statistical description is most appropriate. This may be obtained by considering the distribution function $f(p,n)$ of the action p after n periods of the external field, the evolution of f being provided by the classical trajectories. In presence of chaotic dynamics, f obeys a diffusion equation [6,12]. In particular, a distribution peaked initially at $p_0$ will spread in a time n over a width $\Delta p$ in action space given by:

$$\Delta p(n) \simeq (D(p_0)\cdot n)^{1/2}, \tag{8}$$

where D(p) is the appropriate diffusion coefficient. It must be remarked, however, that diffusive behavior is strongly influenced by the presence of islands of regular motion, which are particularly important for values of $\varepsilon$ not too high.

Classical estimates are important in developing the theory of quantum localization in action space. This was described originally as a stochastic phenomenon related to the Anderson localization in disordered materials [16]. Since the basic assumptions of randomness on which such description rests turned out to be too strong, and inadequate to describe properly a variety of situations [17], I refer here to a different picture of the localization mechanism [18]. A primitive explanation of it may be sketched as follows: let us consider the Hamiltonian (1), when $\hbar^{-1} >> 1$. Let's suppose that the condition (7) for chaos in the classical analogue is verified. In this regime the evolution of a quantum wave function initially peaked at the value p of the action will follow the classical evolution of f(p,n), at least as long as the quantum nature of the spectrum becomes evident. This occurs at a "break-time" $\tau_b$ , as described by the theory of "Transient Stochasticity" [19]. $\tau_b$ is related to the number N of quasi-energy states significantly excited in the evolution:

$$\tau_b \simeq N / \hbar . \tag{9}$$

It is easy to give an intuitive picture of the estimate (9): $2\pi/N$ is the mean quasi-energy spacing, which can be related to the time "necessary" to the system to "realize" its quantum nature via the uncertainty relation between energy and time. In a less anthropomorphic view, $\tau_\hbar$ is the time required for quantum interference effects to set up and stop the "classical" diffusion. If the diffusion coefficient is constant, it follows from (8),(9) that the quantum evolution is localized over a finite number of energy states, whose amplitude $\Delta p$ in action is:

$$\Delta p \simeq D / \hbar \equiv \ell . \tag{10}$$

The above relation shows the essence of the localization phenomenon: quantum interference bounds the classical diffusion on a finite number of states. The quantity $\ell$ is then called "localization length". In general, and in our model as well, the diffusion coefficient is not constant, and a different relation from (10) holds. Possibly, the localization length may be infinite, meaning a "delocalization" of the wave function on the unperturbed basis set.

The model (1) is well suited to test the above picture. To this aim I simulated numerically the quantum evolution of such system, when the initial wave function corresponds to the ground state of the unperturbed Hamiltonian. In Fig. 2 the superposition of a classical surface of section and of a quantum distribution function is plotted. On the x-axis is reported the action, in common units for the two plots. The intensity of the field is just above the classical threshold (7), and a regular region of classical motion is still present. The spreading of the quantum wave function is mainly limited to eigenfunctions in the region of higher classical instability. An increase in the intensity of the external field has the effect to reduce the amplitude of the classical regular region. After doing this, we can vary $\hbar$ from "deep" quantum to semiclassical values. Fig. 3 shows the corresponding localization length and ionization current. We see that direct multiphoton transitions are quenched by the diminishing value of the Planck constant; while an ionization appears in the semiclassical limit, when the localization length is comparable with the whole span in action of the bound (negative energy) states.

The transition from localization to delocalization appears in this model to be "quantitative" more than "qualitative": more resonances participate in the dynamics, as can be seen in Fig. 4. Any small square corresponds to a value of the quasi-energy z whose phase is defined as $\mathrm{Re}(z)\cdot 2\pi/\omega\cdot\hbar \mod(2\pi)$. Due to the presence of resonances, the quantum motion is much stabler than the classical, even for small values of $\hbar$. In a way more clear than correct, we can say that resonances make the system behave to some extent as if its spectrum were pure point - a correct characterization should

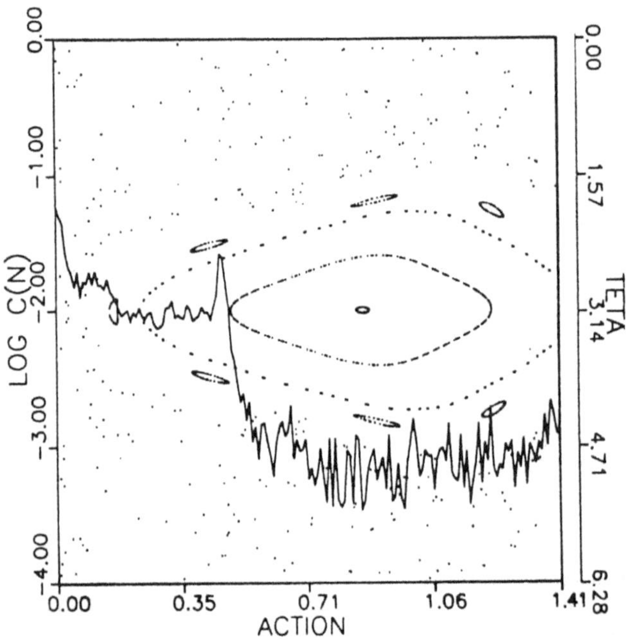


Fig.2. Classical surface of section (dots) and square coefficients of the projection of the wave function over negative energy states (continuous line), plotted as a function of the action. Parameters are: $\hbar = 1/120$, $\varepsilon = .5$, $\omega = 1$, $V_0 = 1$. The quantum distribution is averaged over 80 cycles of the external field, being peaked at time zero on the ground state.

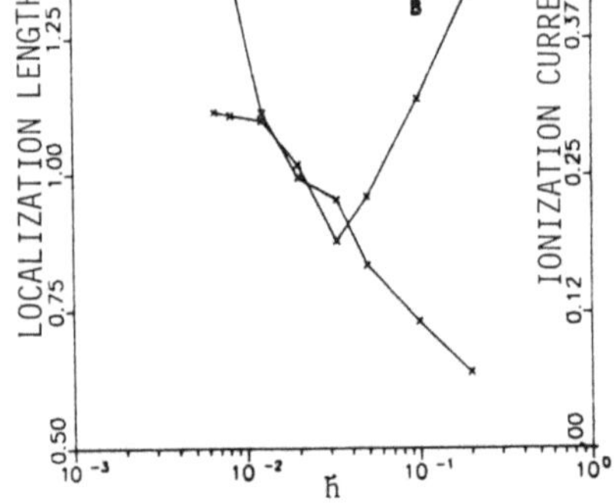


Fig.3. Localization length (A) and ionization current (B) - defined as projection of the wave function over positive energy states - as a function of the Planck constant. Parameters are: $\varepsilon = .7$, $\omega = 1$, $V_0 = 1$, cycles of the external field = 80

be stated in terms of "spectral concentration", however. Indeed, simple tests like numerical time reversal [14] indicate that the "qualitative" transition needed for a properly defined chaotic dynamics has not taken place. On the contrary, the above "delocalization" ionization may as well be one of many "milder" instabilities of physical significance characterizing quantum mechanics.

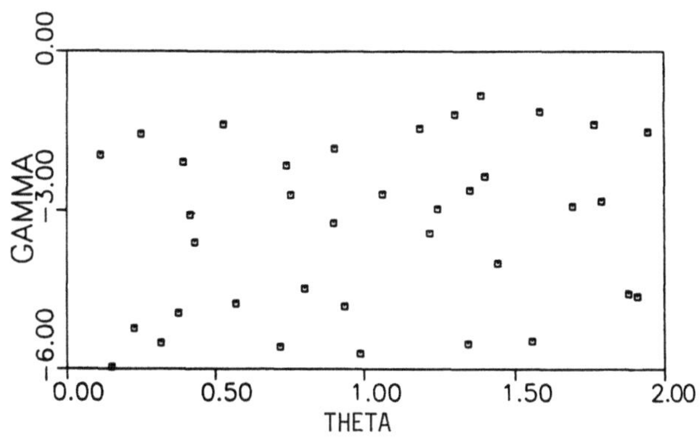


Fig.4. Resonances for the case $\hbar = 1/120$, $\varepsilon = .7$, $\omega = 1$, $V_0 = 1$. The x-axis reports the phase in units of $\pi$, while on the y-axis the imaginary part is reported in logarithmic scale

## 5. CONCLUSIONS

This research aims to provide a clear illustration of the dynamics of strongly perturbed quantum systems. Its major questions are related to the actual degree of randomness of quantum dynamics. However, it could provide results interesting from other viewpoints, as a by-product. As an example, detecting regions of localized quantum evolution might be a key feature in the problem of selective excitation in molecules, where good control can be gained by dealing with a finite number of quasi-energy eigenfunctions.

## REFERENCES


1. Proc. Int. Conf. on Quantum Chaos, Como. G. Casati Ed. (Plenum, 1985).
2. J. Ford, "Chaos, Solving The Unsolvable, Predicting The Unpredictable", in Chaotic Dynamics and Fractals, M. Barnsley and S. Demko Eds. (Academic Press, New York 1986).
3. G. Casati, I. Guarneri and G. Mantica, in Proc. Int. Conf. on Quantum Chaos, Como 1983; also Physica 21 D (1986) 105.
4. J. Bellissard, "Stability and Instability in Quantum Mechanics", in Trends in The Eighties, Lect. Notes in Math., P. Blanchard Ed. (1985).
5. J.E. Bayfield and P.M. Koch, Phys. Rev. Lett. 33 (1974) 258.
6. B. Delone, V.P. Krainov, D.L. Shepelyansky, Sov. Phys. Uspeky 26 (1983) 551.
7. D.L. Shepelyansky, in Ref. 1.
8. G. Casati, B.V. Chirikov, D.L. Shepelyansky Phys. Rev. Lett. 53 (1984) 2525.
9. J.E. Bayfield, L. Pinnaduwage, Phys Rev. Lett. 54 (1985)
10. G. Casati, B.V. Chirikov, I. Guarneri, D.L. Shepelyansky, Phys. Rev. Letters 57 (1986) 823.
11. J.G. Leópold, I.C. Percival, Phys. Rev. Lett. 41 (1978) 944; also J. Phys. B 12 (1979) 709.
12. R.V. Jensen, Phys. Rev. A 30 (1984) 386.
13. K.A.H. Van Leeuven et al., Phys. Rev. Lett. 55 (1985) 2231.
14. G. Casati, B.V. Chirikov, I. Guarneri, D.L. Shepelyansky, Phys. Rev. Letters 56 (1986) 2437.
15. K.J. Yajima, Comm. Math. Phys. 87 (1982) 331; S. Graffi, V. Grecchi, H.J. Silverstone, Ann. Inst. Poincare 42 (1985) 215.
16. S. Fishman, D.R. Grempel, R.E. Prange, Phys. Rev. Lett. 49 (1982) 509; also Phys. Rev. A 29 (1984) 1639.
17. G. Casati, I. Guarneri, Comm. Math. Phys. 95 (1984) 121; also G. Casati, J. Ford, F. Vivaldi, I. Guarneri, Phys. Rev. A 34 (1986) 1413.
18. B.V. Chirikov, D.L. Shepelyansky, Preprint 85-29 INP Novosibirsk 1985.
19. B.V. Chirikov, D.L. Shepelyansky, F.M. Izrailev, Sov. Scient. Rev. 2C (1981) 209.

# Stochasticity in Hydrogen in Presence of dc and ac Fields

*M.H. Nayfeh and D. Humm*

Department of Physics, 1110 West Green Street,
University of Illinois at Urbana-Champaign, Urbana, IL 61801, USA

We have analyzed Stochasticity in highly excited atomic hydrogen in the presence of a microwave field and a dc electric field using a classical one-dimensional model similar to that of an electron system over a helium surface. We have determined in detail the effect of the dc field on the threshold of global stochasticity and on the number of states trapped in the nonlinear resonces. Our results indicate that the number of trapped states may not always increase, and may decrease with increasing dc field contrary to the case of a system of surface electrons.

## 1. Introduction

There has been a lot of interest in the question of existence of chaotic behavior in quantum mechanical systems whose classical analogs are known to be non-integrable and exhibit chaotic behavior[1]. The interest stems from the fact that the allowance for quantum effects in the systems gives rise to substantial anomalies in the manifestation of stochasticity and even when the initial population of the system is quasiclassical. The periodically kicked classical and quantum rotors have been analyzed extensively by many researchers[2-4]. In some studies it was found that the quantum correlation functions attenuate much more slowly than in the classical limit, reducing in the end the rate of quantum diffusion. In another study[4] it was noted that one cannot tell for sure whether a quantum state of the rotor describes a chaotic or a regular state beyond the quantum resolution $\sqrt{\hbar}$. Other studies of quantum systems, however, have shown that under certain conditions and for evolution over finite times, quantum dynamics can be brought closer to the classical stochastic approach[1].

The feasibility of chaotic behavior in the response of atoms and molecules to radiation fields are underway in a number of laboratories including ours[5-8]. The collisionless dissociation of polyatomic molecules that interact with coherent IR laser radiation is one system. Another system that is under study is the ionization of highly excited atomic hydrogen by a microwave field[6-8].

Because of the large degrees of freedom involved in the hydrogen system (the electron moves around in a six-dimensional phase space), detailed numerical and analytical investigations of the process is time consuming and remain incomplete respectively[9].

A system that is amenable to theoretical analysis is the one-dimensional hydrogen atoms (involving two-dimensions in phase space) that we have been experimenting with recently[10]. These atoms are prepared by laser excitation of hydrogen in the presence of strong dc electric fields. They offer not only the nearest approximation to the ideal one-dimensional problem, but in addition, systems of low quantum numbers as well as high ones, so that a wide range of quantum numbers can be studied.

The classical non linear oscillator model that need to be solved is

$$H(\vec{r},\vec{p},t) = \frac{p^2}{2} - \frac{Z}{r} + F_0 x + xF_1 \cos \Omega t \tag{1}$$

where $F_0$ is the strength of the dc field and $F_1$ is the amplitude of the microwave field. Because of the cylindrical symmetry about the direction of the fields, the system can be reduced from six to four dimensions in the action-angle phase space.

In the case of strong $F_0$ one can approximate the Hamiltonian of the nonlinear oscillator to

$$H(x,p_x,t) = \frac{p_x^2}{2} - \frac{Z}{x} + F_0 x + F_1 x \cos \Omega t \qquad (2)$$

which reduces the system to two dimensions in the action-angle phase space. How good this approximation is remains to be investigated, since there is always a residual motion in the other directions. The effective potential of the combined dc and coulomb potentials is shown in Figure 1.

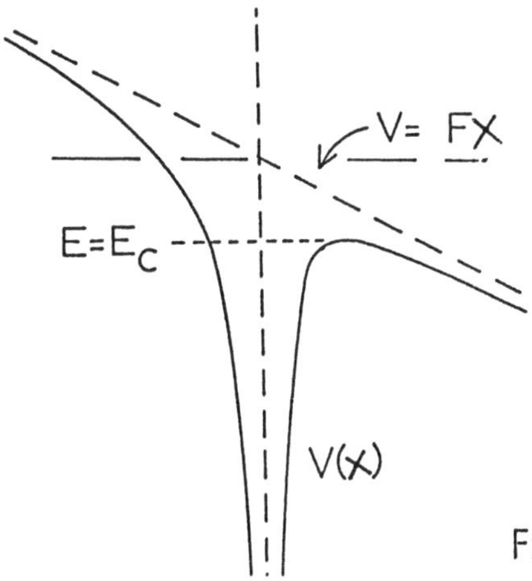


FIGURE 1

The system, in the absence of the external dc field ($F_0 = 0$) was extensively analyzed[9]. It describes a system of an electron located over a liquid-helium surface and interacting with an oscillatory field polarized perpendicular to the surface. Recently some aspects of the effect of a positive clamping dc field on the electron-surface system have been studied. In specific the studies were oriented towards the determination of the number of quantum levels that the nonlinear resonances trap, and their dependence on the clamping field. Asymptotic results were found giving a quadratic dependence in the limit where the clamping field dominates the coulomb field[11].

## 2. Surface Electrons

Here we analyze the one-dimensional model of surface electrons in the presence of the dc field classically. We study it for both positive and negative "clamping" fields and for a wide range of magnitudes including the regime where neither the coulomb nor the clamping field dominates. In our analysis we use a numerical technique that is based on the method that Jensen used in his analytical study of the unclamped case[9].

We use a canonical transformation to transform the Hamiltonian given in Eq, 2 to

$$K = \alpha(I) + F_1 x(\theta,I) \cos \Omega t \qquad (3)$$

where I and $\theta$ are the action and angle variables. The constant $\alpha$ is

$$\alpha = \frac{p_x^2}{2} - \frac{1}{x} + F_0 x \qquad (4)$$

The variables $\theta$ and I are chosen such that $x(\theta,I)$ is periodic in $\theta$ with a period of $2\pi$. Thus we expand $x(\theta,I)$ in the Fourier series[12]:

$$F_1 x = \sum_{m=-\infty}^{\infty} V_m(I)\, e^{im\theta} \text{ with } V_m = = V_{-m} \tag{5}$$

which when substituted in Eq. (3) and rearranging the terms while realizing that x is symmetrical about $\theta = 0$, that is $V_m = V_{-m}$ gives

$$K = \alpha(I) + \sum_{m=0}^{\infty} V_m(I) \cos(m\theta - \Omega t) \tag{6}.$$

The problem is first solved for weak time-dependent fields. The effect of higher intensities will be determined as a distortion to the weak field solution. If $V_m$ is weak and since $\cos(m\theta - \Omega t)$ in general oscillates rapidly, then the microwave field can only affect the system significantly when the phase $m\theta - \Omega t = 0$. This condition is called the $m^{th}$ resonance. Thus when the system is excited to near the $m^{th}$ resonance, it can only interact strongly with this resonance in the weak field limit, and consequently the expansion in Eq. 6 can be separated as follows:

$$K = \alpha(I_m) + V_m(I_m)\cos(m\theta - \Omega t) + f(t) \tag{7}$$

where

$$f(t) = \sum_{n \neq m}^{\infty} V_n(I_n)\cos(n\theta - \Omega t) \tag{8}$$

contributes very little to the interaction in the weak field limit, and hence can be neglected. The system of Eq. 7 with f(t) neglected can now be solved by making another transformation to get rid of the explicit time dependence. The new variables $\Delta$ and $\psi$ are defined as follows:

$$I - I_m = m\Delta \quad , \text{ and } \psi = m\theta - \Omega t \tag{9}$$

the new corresponding generating function will be chosen as

$$F(\Delta,\theta,t) = (I_m + m\Delta)\,[\theta - \frac{\Omega t}{m}] \tag{10}$$

with the transformation equations:

$$I = \frac{\partial F}{\partial \theta} \quad , \quad \psi = \frac{\partial F}{\partial \Delta} \tag{11}$$

The transformed Hamiltonian becomes

$$K' = \alpha(I) - \Omega I/m + V_m(I)\cos\psi \tag{12}$$

It is to be noted that the new variable $m\Delta$ is just the small change in the action of the $m^{th}$ resonance caused by the weak microwave field. The new variable $\psi$ measures the phase difference between the phase of the system $m\theta$ and that of the external field $\Omega t$. In order to see the nature of K' further, we can examine it in the limit of very small $V_m$, that is in the limit of small $\Delta$ and $\psi$. Taking

$$V_m(I) \simeq V_m(I_m)$$

$$\alpha(I) \simeq \alpha(I_m) + \left.\frac{d\alpha}{dI}\right|_{I_m} m\Delta + \frac{1}{2}\left.\frac{d^2\alpha}{dI^2}\right|_{I_m} m^2\,\Delta^2$$

$$\frac{d\alpha}{dI} = \frac{d\theta}{dt} = \omega$$

we get:

$$K' = C_1(\Delta^2 + V_m\cos\psi) + C_2 \tag{13}$$

where $C_1 = \frac{1}{2}\, m^2 \frac{d\omega}{dI}$ and $C_2 = \alpha(I_m) - \frac{\Omega}{m} I_m$ are constants. Hamiltonian (13) is a pendulum Hamiltonian, where $\Delta$ and $\psi$ can be associated with the angular momentum and the pendulum angle respectively, and $C_1$ and $V_m$ have the units of $mass^{-1}$ $length^{-1}$ and -(mass) (gravity) respectively.

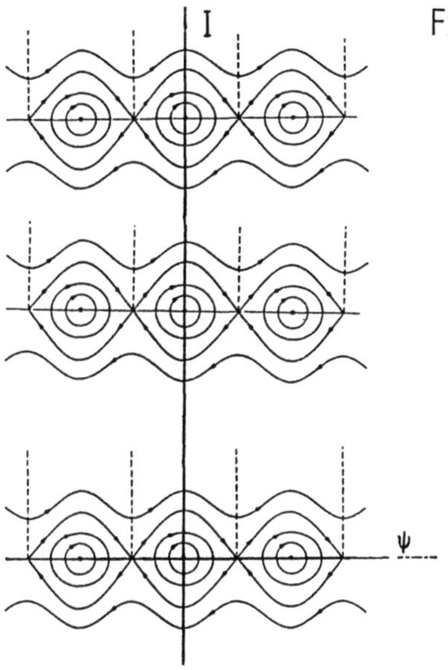

FIGURE 2

From K' we can determine the phase and the angular momentum of the system near a resonance. Fig. 2 shows a typical plot of $\Delta$ versus $\psi$ (pendulum response). Actually I is plotted instead of $\Delta$. Although the plotting of I versus $\psi$ is somewhat misleading since the $\psi$ variable is different for each resonance, it is qualitatively correct. The separatrix divides two kinds of motion, where on one of its sides the system goes one way, while on the other side it goes the other way. Moreover each separatrix is symmetrix and its width is given in terms of the maximum $\Delta$. Since $\Delta$ maximizes for $\psi$ = 0 and $\pi$, one can show that near the $m^{th}$ resonance:

$$(I - I_m)_{max} = 2\sqrt{V_m/(d\omega/dI)} \qquad (14)$$

We can now discuss the case where the amplitude of the microwave field $F_1$ is not weak, such that the interaction Hamiltonian f(t) (given in Eqs. 7 and 8) which we neglected so far becomes non-negligible. First of all the width of the resonances given by Eq. 14 will rise since $V_m$ is proportional to $F_1$. Secondly the f(t) term acts as an additional time-dependent perturbation embodying the interaction between the otherwise non-interacting resonances.

What happens to this system when we include the time-dependent perturbation f(t)? Trajectories far away from the separatrix are affected very little. They remain within a very small region of where they would have been anyway. Trajectories near the separatrix are strongly affected, because the time-dependent external force can push them to one side or the other of the separatrix, making their motion completely different. For trajectories very close to the separatrix, which side of the separatrix they end up on in the next cycle depends not at all on which side of the separatrix they start on, but rather which way the force happens to be pushing them along the way. Since different trajectories take different amounts of time to complete a cycle, especially near the separatrix, this pushing can be completely different for two very close trajectories. Very close to the separatrix, the trajectories get all twisted up. Two nearly the same trajectories may end up going in opposite directions, and two different ones in the same direction.

If one simulates this motion on a computer, it would look irregular and disordered. Researchers have searched through their computer results for some kind of correlation and order in this motion but found none. All correlations that have been examined die out within a fairly short time. It appears that a random motion has arisen from a deterministic system. This is called "chaos".

The transition to global chaos[12] occurs when the sum of the two adjacent $\Delta I_m$'s is equal to the difference between their $I_m$'s. The pendulum approximation to K' is not always that good. It is particularly bad when the transition to global chaos has occurred for all resonances but one is trying to see if the m=1 separatrix extends down low enough to encompass states of low I (otherwise, those states will still be stable, even if global chaos occurs). For this problem, one can dump the pendulum equation and figure out the actual separatrices of the Hamiltonian K'. We follow Jensen and do this in the one case and use the pendulum equation in all other cases[9].

In our present studies we use this criterion for global chaos[12]. Thus we need to calculate $V_m(I_m)$ and $d\omega/dI$ at $I_m$ for all m. In the absence of the dc field, it is possible to get analytical expressions for these quantities. However, in the presence of the dc field it is not possible to do so except in some asymptotic regions where the dc field dominates the Coulomb field or when the Coulomb field dominates the dc field. Since in our application we are interested in field strengths that are comparable to the Coulomb field strength, we calculated these numerically.

The calculation of $\alpha$ and $I_m$ can be done by taking (from Eq. 4)

$$H_o(x,p_x) = H_o(x,\partial W/\partial x) = \alpha \qquad \text{or}$$

$$\frac{1}{2}(\partial W/\partial x)^2 - \frac{1}{x} + Fx = \alpha \tag{15}$$

where $P_x = \partial W/\partial x$ is evaluated for constant $\alpha$ and

$$I = \frac{1}{2\pi}\oint P_x\,dx = \frac{1}{2\pi}\oint(\partial W/\partial x)\,dx \tag{16}$$

Moreover we have the fact that $P_x = 0$ when x reaches its maximum value $X_{max}$. From this, we have $\partial W/\partial x = 0$ at $X_{max}$, or $\alpha - Fx + 1/x = 0$ which can be solved for $X_{max}$. In addition, at resonance we have $m\theta - \Omega t = 0$ or $m\partial\theta/\partial t = \Omega$ which gives $m\omega = \Omega$. But $\omega = d\alpha/dI$, thus $d\alpha/dI = \Omega/m$. Thus by concentrating on the first half of the cycle where x goes from 0 to $X_{max}$ (corresponding to $0 \leq \theta \leq \pi$), we can get the result for the full cycle because of the symmetry. We first calculate I in terms of $\alpha$ and then we can invert to get $\alpha(I)$.

We now present our results on the effect of the dc field on the dynamics of the system. Fig. 3 gives a plot of the microwave threshold for global stachosticity as a function of I for a number of values of the dc field: -40, -20, -10, 0, 10,20, 40, 60, 70, and 80 V/cm. The frequency of the microwave field is 30.5 GHz. We will make the following observations making use of the curve corresponding to $F_0 \doteq 0$. It is apparent that the threshold drops essentially in steps as the binding energy of the electron decreases (or equivalently as I increases). As I increases from low values, the threshold drops dramatically till I approaches the first resonance $I_1$, where it levels off and a plateau develops and lasts for $I \leq I_1$. For I larger than $I_1$, but smaller that $I_2$, the action of the second resonance the threshold starts dropping till $I_2$ is approached where it levels off,then it starts dropping when I becomes larger than $I_2$ and so on.

Another observation is the fact that for negative dc fields or a zero field, the plateaus get narrower as I increases,whereas for positive fields those plateaus get wider as I increases. Thus if the resonances get closer to each other as I increases, then once overlap takes place between the lowest two responses, the higher resonances will also overlap and global stochasticity takes place. This is such a case for a pure Coulomb field, and a Coulomb field plus a negative clamping field. But for a positive clamping field, the plateaus get wider and eventually the drop in threshold as a given resonance is passed, turns into an increase,as illustrated in Table I for a number of dc fields and for a number of plateaus.

Although local stochasticity over a few resonances can be achieved at certain intensities of microwaves, global stochasticity cannot be reached and the surface electron will never ionize because it is bound in an infinitely high well.

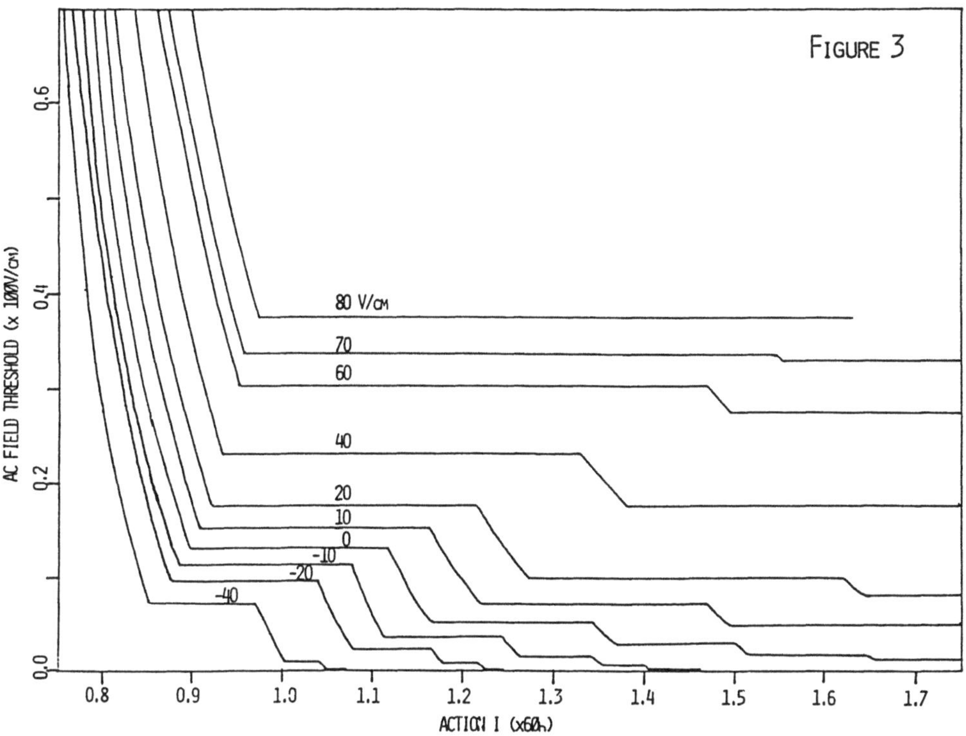


Table I.
AC Field Overlap Threshold (V/cm)

| $F_o$(V/cm) | 12 | 23 | 34 | 45 | 56 | 67 |
|---|---|---|---|---|---|---|
| -40 | 7.02 | 1.10 | .21 | | | |
| -20 | 9.65 | 2.41 | .8 | .29 | | |
| -10 | 11.27 | 3.53 | 1.52 | .74 | .39 | .21 |
| 0 | 13.10 | 5.14 | 2.78 | 1.8 | 1.27 | |
| 10 | 15.19 | 7.19 | 4.87 | | | |
| 20 | 17.52 | 9.83 | 8.0 | | | |
| 40 | 23.00 | 17.23 | 17.28 | 18.68 | 20.09 | 21.24 |
| 60 | 29.58 | 26.97 | | | | |
| 70 | 33.27 | 32.52 | 35.9 | | | |
| 80 | 37.21 | 42.14 | 42.46 | 45.17 | 46.67 | |

It is interesting to examine the width of a given plateau as a fraction of the dc electric field. Fig. 3 shows that the width increases for increasing positive fields, but decreases as the magnitude of a negative field increases. Fig. 4 gives a plot of the width of the first plateau taken from Fig. 3 as a function of the dc field.

We calculated the number of quantum states that are trapped in a given nonlinear resonance. We choose to present those trapped in the first resonance. Fig. 5 gives the number of trapped states as a function of the dc field with a microwave field of

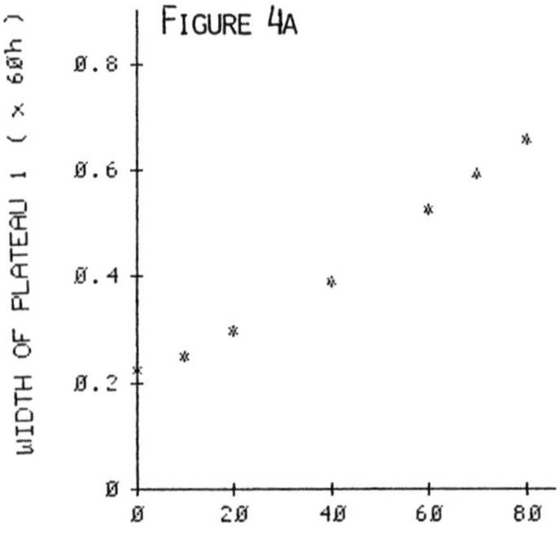


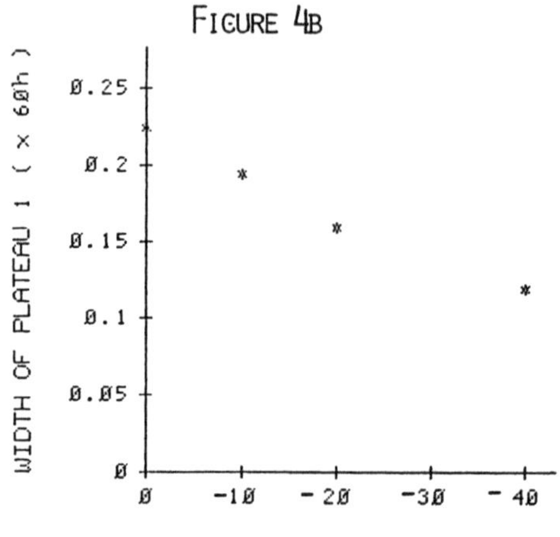


$\Omega$ = 30.5GHz which results in overlap with the next resonance (given in Fig. 3). Note that $F_1 th/F_0$ is kept constant in these cases. We now make the following observations. The number of trapped states increases as the positive field increases, but it decreases as the negative field increases in magnitude. Secondly, the number of trapped states, in the positive field case, increases quadratically in the field amplitude for large fields that dominate the Coulomb field. This power dependence has been analytically calculated by Berman et al[11]. However, for low dc fields where it is comparable to the Coulomb field, the power dependence is closer to being linear. Figure 6 gives similar results for a microwave frequency of $\Omega$ = 7GHz. In Figs. 5 and 6 we also give the prediction of the asymptotic result for large positive fields[11] (shown as solid lines).

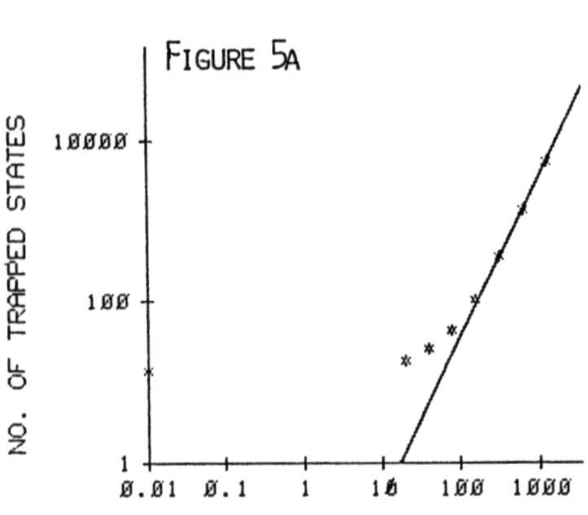


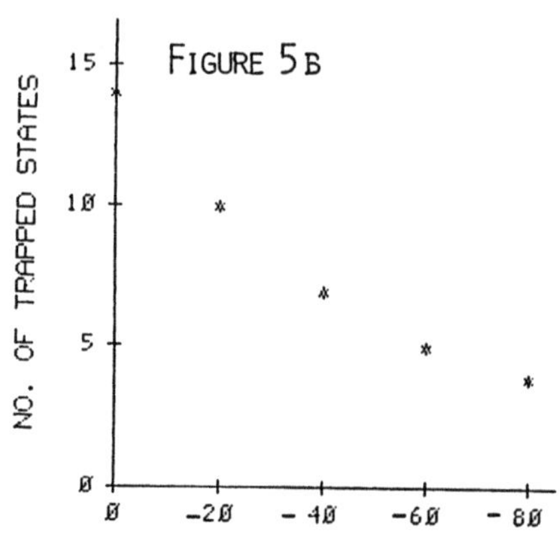


3. One-dimensional Hydrogen Atoms

When trying to study the one-dimensional hydrogen atom in the presence of microwaves, one notes that the presence of a dc field creates states that may be extended upfield, down field or symmetric[10]. The shape of these states and their degree of extension and direction depends on their location in the energy spectrum Our previous results indicate that the energy of these states is given by a fraction of the nuclear charge $Z_1 \leq 1$ rather than by the total charge, whereas their ionization life times are determined by $1-Z_1$, the rest of the fraction of the nuclear charge. The degree of extension, energy and ionization life times can be classified

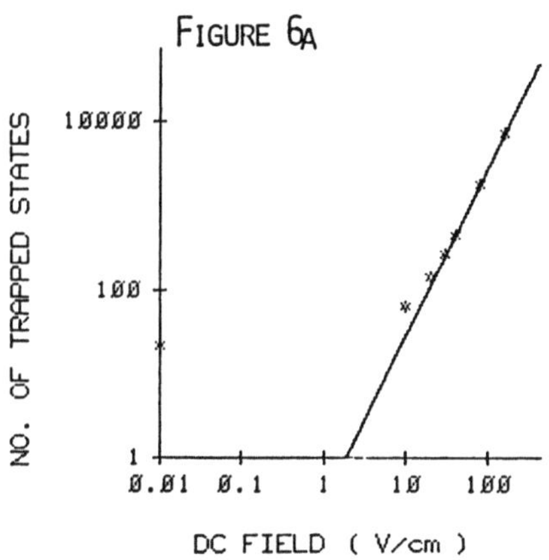


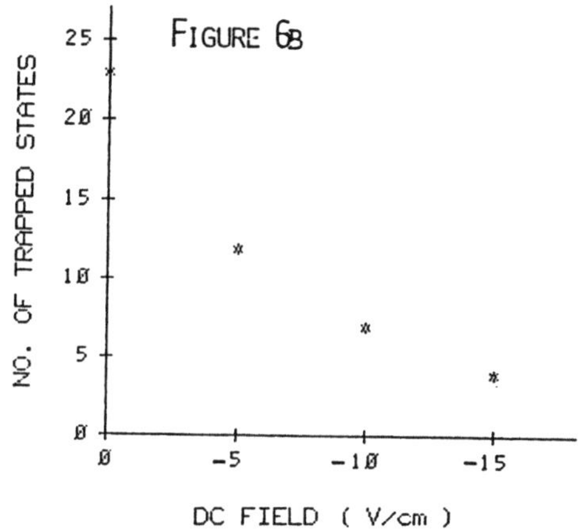


according to their $Z_1$ values. For example, the fractions of the charge Z = 0, 1/4, 1/2, 3/4, and 1 define four quarters. The electronic charge is found to be stretched opposite to the field in the first and third, stretched along it in the other two quarters, and symmetric at the common boundaries.

Thus for states that are highly stretched along the field, a surface electron model with a positive clamping field would be appropriate, whereas for states that are highly stretched opposite to the field, a surface electron model with a negative clamping field would be appropriate. For states which are symmetric, it may not be possible to describe them by such a model.

Therefore if one wants to increase the number of trapped states in a nonlinear resonance, one would want to pick a portion of the spectrum where the states are "positively clamped" and one would have to keep track of this as the magnitude of the dc field is changed.

References

1. G. Casati, B. V. Chirikov, F. M. Israelev, J. Ford: in Stochastic Behavior in Classical and Quantum Systems, ed. by G. Casati and J. Ford, Vol. 93 of Lecture Notes in Physics, (Springer, New York, 1979).
2. D. R. Grempel, S. Fishman, and R. E. Prange: Phys. Rev. Lett. 49, 833 (1982).
3. M. V. Berry: Physica (Amsterdam) 10D, 369 (1984).
4. S. J. Chang and K. J. Shi: Phys. Rev. Lett. 55, 269 (1985).
5. E. V. Shuryak: Zh. Eksp. Teor. Fiz. 71, 2939 (1975) [Sov. Phys. JETP 44, 1070 (1976)]. P. I. Belobrov, G. P. Berman, G. M. Zaslavski, and A. P. Slivinskii: ibid. 76, 1960 (1979) 49, 993 (1979).
6. J. E. Bayfield and L. A. Pinnaduwage: Phys. Rev. Lett. 54, 313 (1985).
7. P. M. Koch: J. Phys. (Paris), Colloq. 43, C2-187 (1982).
8. D. Humm and M. H. Nayfeh: to be published.
9. R. V. Jensen: Phys. Rev. A 30, 386 (1984); in Chaotic Behavior in Quantum Systems, ed. by G. Casati, Plenum Publishing Corporation, London (1985).
10. M. H. Nayfeh, D. Yao, Y. Ying, D. Humm, K. Ng, and T. Sherlock: in Advances in Laser Science-1, AIP Conference Proceedings NO. 146, 370 (1986), M. H. Nayfeh, K. Ng. and D. Yao, in Laser spectroscopy, VII ed. by T. Hänsch and R. Shen, Springer-Verlag (1985).
11. G. P. Berman, M. G. Zaslavskii, and A. R. Kolovskii: Zh. Eksp. Teor. Fiz 88, 1551 (1985) [Sov. Phys. JETP 61, 925 (1985)].
12. G. M. Zaslavskii and B. V. Chirikov: USP Fiz. Nauk. 105, 3 (1971) [Sov. Phys. USP 14, 549 (1971)].

# The Problem of Laser-Induced Tunnel Ionization

*Gy. Farkas*

Central Research Institute for Physics, P.O. Box 49, H-1525 Budapest, Hungary

## 1. Introduction

The general interpretation of the interactions of photons with the electrons of atoms or solids in arbitrary light intensity range is one of the fundamental tasks of QED. In this context there has recently been renewed interest in these processes in the extremely high laser intensity range. In connection with these processes we very often meet in the literature the expression "optical tunnelling". In the following, we shall discuss the problem without any claim to completeness, rather to give a short outline and to expose it for discussion.

After a brief summarizing formulation of the topics we present several characteristic preliminary experiments and describe our recent works carried out to investigate the intense field tunnelling induced by strong $CO_2$ laser pulses for both noble gas atoms and a gold surface.

## 2. Summarizing formulation of the topics

The origin of the concept of optical tunnelling goes back to the early years of quantum electronics when the possibility of using lasers to ionize atoms and solids was first investigated. These first qualitative expectations imagined that ionization took place in the form of one of two possible alternative processes: either the electron absorbs a sufficiently high number of photons, or the electron tunnels through the atomic potential barrier which is narrowed by the strong oscillating laser electric field strength (Fig. 1).

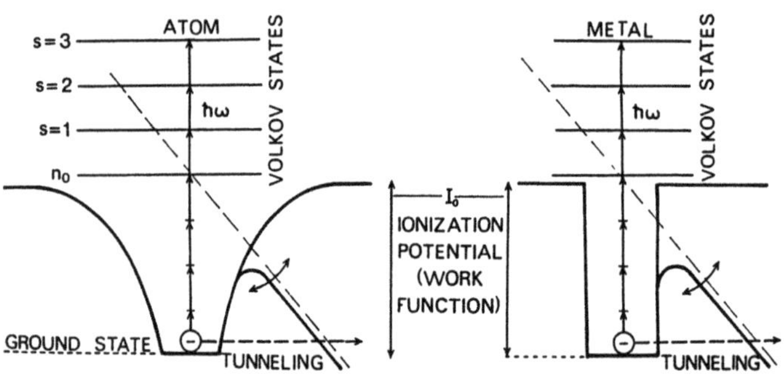


Fig. 1 Illustration of the multiphoton and the laser-induced optical tunnelling processes

Based on these expected processes, the theories proceeded along two directions:

(A) Perturbative theories based on the higher order perturbation approach resulting in multiphoton-type interaction involving a well-defined low number of photons ("quantum limit").

(B) Non-perturbative theories based on semiclassical approaches resulting in classical tunnel-type interactions involving an increasingly high number of photons ("classical limit").

Historically, activities began with (B); the predicted intensities, however, proved to be too high. In view of this all research activity then became concentrated on (A) up till recent years after which lasers have been developed to furnish intensities high enough to observe optical tunnelling (B). Let us now survey the mean predictions of (A) and (B).

## 2.1. Perturbative theories (A)

The programme and basic procedure used here is to calculate the transition probability from a given initial state to a final state. In the preliminary version the final state is a free electron state, while the more developed theory takes into account the coupling with the continuum states, too. The result is the excellent description of the processes in the relatively low laser intensity range, which is summarized in [1] by the most prominent specialists, furnishing nearly all necessary formulae and predicting at the same time the well-known "above threshold ionization" (ATI) phenomenon observed first by AGOSTINI et al. [2]. Meanwhile, however, unexpected and unexplained questions have been raised by these ATI experiments, themselves viz. the multiple ion production, etc., at higher intensities [3,4]. Although the related suggestions and interpretations are exciting, most of them are not settled on the common basis of (A) alone; they are of rather special and sometimes sophisticated character. Therefore we will now consider what the nonperturbative theories (B) have to say concerning these problems.

## 2.2. Nonperturbative theories (B)

These theories treat the interaction processes by calculating the transition probability from the initial atomic state into Volkov final states (i.e. no longer into a free electron state), in which latter states (Fig. 1) the field effects are considered to be predominant but the atomic binding effects are neglected. Therefore, these theories are expected to hold at high intensities. The first formulation of this procedure was published in the early work of KELDYSH [5], using a short-range binding model potential (e.g. the case of $H^-$). Later on, many theories - based on Keldysh's work - were proposed: for atoms [6-13], and for metals [13-15]. The most recent elaboration of Keldysh's idea for atoms was published by REISS [16]. At the same time others [17-19] approached the nonperturbative problem using other kinds of special calculating methods which have led to the same conclusions. Our following considerations will be based on Keldysh's original theory and its modification by Reiss. A very useful account can also be found in the review work of EHLOTZKY [20] and the most comprehensive summary is given in [21] by DELONE and KRAINOV.

The main results of this theory may be summarized essentially in the form of a very general closed formula obtained for the ionization probability W which has turned out to be a closed function of the atomic and laser characteristics plus a useful parameter of both, defined as $\gamma \equiv \omega\sqrt{2mI_0}/eE_L$. Here $\omega$ is the laser frequency, $E_L$ is the laser electric field strength, $I_0$ is the ionization potential of the system, e and m are the electron charge and mass, respectively. It turned out that W has two extreme approaches depending on the value of $\gamma$. Parameter $\gamma$ furnishes the perturbation criterion condition $2\,\Delta/I_0 \sim \gamma^{-2}$, where $\Delta \equiv e^2E_L^2/4\,m\omega^2$ is the average oscillation electron energy in the laser field, or $\gamma$ may also be expressed as $\gamma \sim \tau/T$, where $T \sim \sqrt{2mI_0}/eE_L$ is the tunnelling time and $T = 2\pi/\omega$ is the half-laser period.

Case (A). When $\gamma \gg 1$ we reobtain the formulae of the perturbation theories which are characteristic for the multiphoton or quantum limit, when the ionization rate or the number of created ions or electrons $N_{i,e}$ reads

$$W \propto N_{i,e} \propto |E_L|^{2n_0} \propto I^{n_0}$$

where $I \propto |E_L|^2$ is the laser intensity, $n_0 \equiv [I_0/\hbar\omega+1]_{int}$ the minimum photon number necessary for ionization.

Case (B). When, however, $\gamma \ll 1$, W furnishes automatically the formulae of the non-perturbative theory in the classical limit ("optical tunnelling"), when

$$W \propto N_{i,e} \propto f(E_L)\exp[-E_0/E_L]$$

where $f(E_L)$ depends on light polarization and $E_0$ is the atomic field strength. These two extreme cases presented here in the form of somewhat qualitative statements may also be followed in an exact rigorous way on the basis of Keldysh's theory. Then the total ionization rate is determined in the form of the sum of individual n order elementary multiphoton processes of $w_n$ probability, for which

$$W = \sum_{n=n_0}^{\infty} w_n \delta(I_0+p_n^2/2m+e^2E_L^2/4m\omega^2-n\hbar\omega)$$

The delta function leads to the condition

$$E_n \equiv P_n^2/2m = n\hbar\omega-(I_0+\Delta);\ n = n_0+s;\ s = 0,1,2,...$$

where $E_n$ and $P_n$ are the electron energy and momentum, respectively, after absorbing n photons. In the course of the calculations it turns out that $w_n$ depends on - among other factors - the $J_n(a)$ Bessel function of which argument "a" contains the atomic and laser parameters ("a" is proportional to $E_L$), or $\gamma^{-1}$. The above-mentioned approaches then correspond more or less to the mathematical behaviour of $J_n(a)$ at different "a" (or $\gamma$) values.

Case (A). When $a \ll 1$, that is, $\gamma \gg 1$ we have

$$W \sim \sum_{n=n_0}^{\infty} J_n^2(a) \sim \sum_{n=n_0}^{\infty} (a/2n!)^{2n} \sim \sum_{n=n_0}^{\infty} I^n$$

Case (B). When $a \gg 1$, that is, $\gamma \ll 1$, we have (as a rough illustration only)

$$W \sim \sum_{n=n_0}^{\infty} J_n^2(a) \sim \sum_{n=n_0}^{\infty} (\pi a/2)^{-1/2} \cos(a-n\pi/2-\pi/4)\ .$$

Our task now reduces to the proper summation of Bessel functions.

Case (A). Here we may write the sum as

$$W \propto \sum_{n=n_0}^{\infty} I^n = I^{n_0} + \sum_{s=1}^{\infty} I^s$$

together with the energy condition

$$E_n = n\hbar\omega-(I_0-e^2E_L^2/4m\omega^2) \equiv n\hbar\omega-(I_0-\Delta)$$

where $n = n_0+s$, $s = 0,1,2,...$

If we retain only the $I^{n_0}$ term, we reobtain the lowest order perturbation results. If, however, we do not neglect the terms corresponding to s = 1,2,..., we see that the theory furnishes the ATI results automatically in a quite natural way, including the ATI electron energy spectrum formula for $E_n$. Even the term $\Delta$ appears, which causes the suppression of the first few electron peaks, when $I \sim E_L^2$ is slightly increased (see Fig. 2).

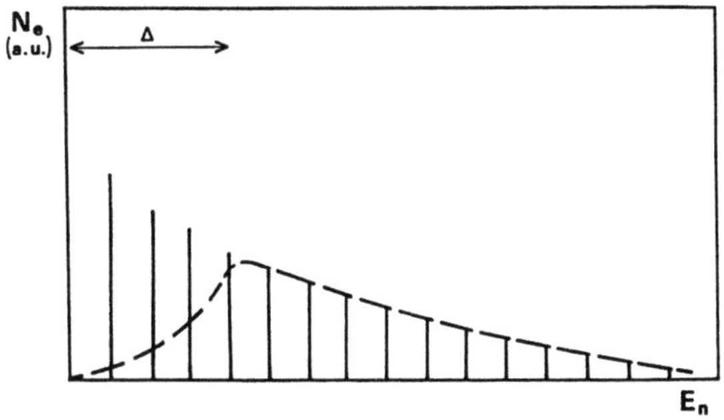


Fig. 2 The ATI electron energy spectrum in the $\gamma \gg 1$ case

Case (B). The expression of W may be approached here in another way, replacing the summation by integration. We then get an analytic final expression for W which recalls the classical tunnel formula, as mentioned above

$$W = f(E_L)\exp[-E_0/E_L]$$

together again with the same expression for the electron energy spectrum

$$E_n = n\hbar\omega-(I_o-e^2E_L^2/4m\omega^2) \equiv n\hbar\omega-(I_o-\Delta) .$$

We should like to underline that through this process it turns out that the so-called "optical tunnelling" formula is nothing other than the mathematical expression of the properly summarized elementary multiphoton interaction processes, to which we may attach the spectacular "tunnelling" picture.

A very interesting feature is that in the electron spectrum only the extremely high $E_n$ values occur (due to the very high Δ values here); the lower energy part is completely suppressed. The resulting spectrum shape, as a whole, is shifted into the high-energy range (Fig. 3).

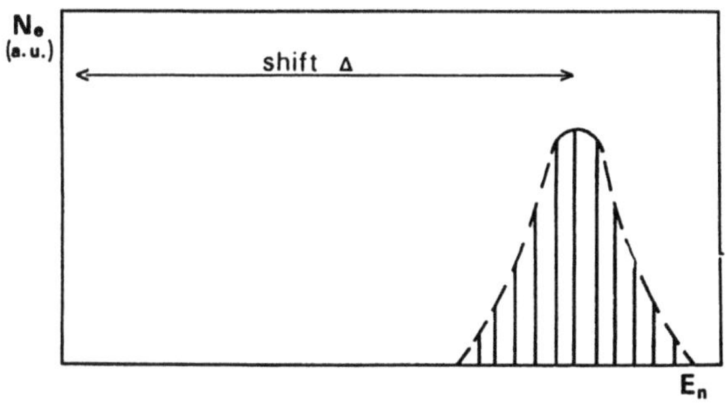


Fig. 3 The ATI electron energy spectrum in the $\gamma \ll 1$ case

This phenomenon reflects the fact that in this case the coupling to the laser field is predominant, in contrast with case (A) where the coupling to the atom is essential. Although none of the cited theories mention the problem, it is instructive to support the attached tunnelling picture by assuming that due to the periodic modulation of the potential barrier during the tunnelling, the occurrence of bunching of the emitted electrons may be expected at ω,2ω,ω3,..., due to the strong nonlinearity. This effect was predicted [22] (and verified) before the advent of lasers by microwaves showing even the shift of the spectrum shape towards the higher energies.

The overall intensity dependence of the total (W) and partial ($w_n$) ionization rates may be summarized in an illustrative way (Fig. 4).

When we use the term "tunnelling", however, we have always to keep in mind whether we are speaking of the total W or of the individual $w_{n_0+s}$ probabilities.

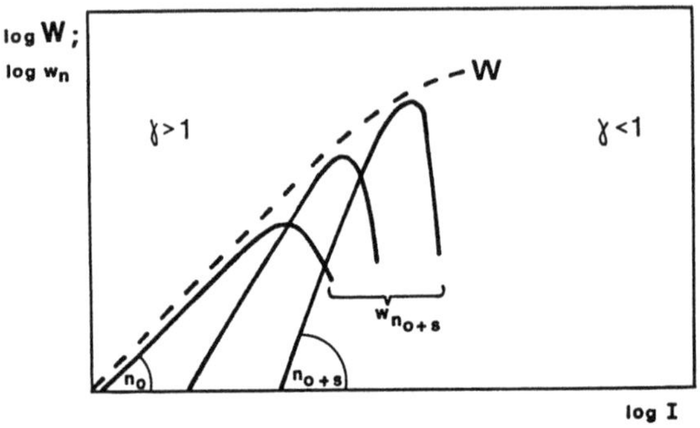


Fig. 4 General qualitative dependence of W and $w_n$ on the laser intensity I

The reasonability of the presented basic ideas of the KELDYSH theory [5,21] has recently been confirmed by other nonperturbative theories [17-19] which arrived at similar conclusions using different calculating procedures.

After dealing with these general characteristics we continue our discussion with the predictions of the nonperturbative theories [5,7,8,16,21] on the role of polarization and light statistics. As for the cases of linear or circular polarization, the theory predicts that $W_{circ}/W_{lin} \sim \frac{1}{2}$, when $\gamma > 1$ (case (A)), whereas when $\gamma < 1$ (case (B)) this ratio strongly increases with increasing I and may reach a value of 100. To illustrate qualitatively this phenomenon we imagine that for linear polarization the oscillating laser field may be effective for short periodic time fractions only whereas for circular polarization the atom constantly "feels" the total absolute value $|E_L|$ of the laser field. In this latter case at high fields (case (B)) the situation coincides with the static field emission. This effect may manifest itself, for example, in suppressing the first peaks in the ATI electron energy spectrum more easily using circular polarization.

In the following we summarize and compare the tunnelling formulae (always in case (B)) for the different polarizations and for a static field. It should, however, be kept in mind that Keldysh's theory is restricted to the case of a short-range model potential, $V(r) \sim \delta(r)$, which is valid, for example, in the case of $H^-$. Therefore we present its recent generalization [23] for real atoms of long-range (Coulomb) potential. In this case, using $n^*$, $l^*$ real effective quantum numbers, and the Z atomic number, we may write the general form of W (in atomic units) for the case when $\ell=0, |m|=1$, with $n^* = Z/(2I_0)^{1/2}$ as

$$W = (3e/\pi)^{3/2} 2eZ^2/n^{*5.5} (4eZ^3/E_L n^{*4})^{2n^*-2.5} \exp(-2Z^2/3n^{*3}E_L) \text{ ; here } e = 2.7182...$$

Using this expression, we summarize the values of W for different cases in Table 1.

TABLE 1

| Potential | Static field | Circ. osc. field | Lin. osc. field |
|---|---|---|---|
| Short range ($\delta(r)$; $H^-$) | $\sim E \exp(-const/E)$ | $\sim E_L \exp(-const/E_L)$ | $\sim E_L^{1/2} \exp(-const/E_L)$ |
| Long range (Coulomb) | $\sim E^{-1} \exp(-const/E)$ | $\sim E_L^{-1} \exp(-const/E_L)$ | $\sim E_L^{-1/2} \exp(-const/E_L)$ |

We see the predicted respective coincidences between W (static; short range) and W (circ; short range); the same between W (static;long r.) and W (circ; long range). Another important consequence also follows from the data: for the different oscillating cases the formulae differ only in their preexponential factors. Therefore the W(I) dependence is determined primarily by the preexponential factors for real atoms (which may be stronger than the exponential dependence) [21,23].

The role of light statistics in the W(I) ionization probability is studied in [24,25] on the basis of the simplified KELDYSH model [24,25] using the well-known formula

$$W = \int W(I)\, P(I) dI$$

where P(I) is the light distribution function. It turns out [24] that with any (multiphotonic, or tunnel) kind of W(I), for chaotic P(I) distribution the integral takes its preponderant value at $I = I_{eff} \sim n_0 \langle I \rangle$. Therefore for low intensities (case (A)) we reobtain again the well-known statistical factor $g_{n_0}$ of the perturbative (multiphoton) theories, when

$$g_{n_0} \equiv W_{chaotic}/W_{single\ m.} = \langle I^{n_0} \rangle / (\langle I \rangle)^{n_0} = n_0!$$

For high intensities (case (B), tunnel limit) we have

$$g_{n_0} \equiv W_{chaotic}/W_{single\ m.} \sim \exp[-2\sqrt{3}\, I_0/(\langle I \rangle)^{1/3}]/\exp[-4/3\sqrt{2}\, I_0^{3/2}/I^{1/2}] = g_{n_0}(I)$$

that is, $g_{n_0}$ depends on I. (Let us recall that $I_0$ is the ionization potential.) It is mentioned that for chaotic light $\gamma$ (i.e. the ranges of the multiphoton and tunnelling approach) is also slightly influenced:

$$\gamma_{chaotic} \sim \omega\sqrt{2m\hbar\omega}/e(\langle I \rangle)^{1/2} \lesssim \gamma_{single\ m.};\ \text{from multiphoton approach}$$

$$\gamma_{chaotic} \sim (m\hbar\omega^2/e^2\langle I \rangle)^{1/3} \lesssim \gamma_{single\ m.};\ \text{from tunnelling approach.}$$

The results are represented in a qualitative form in Fig. 5.

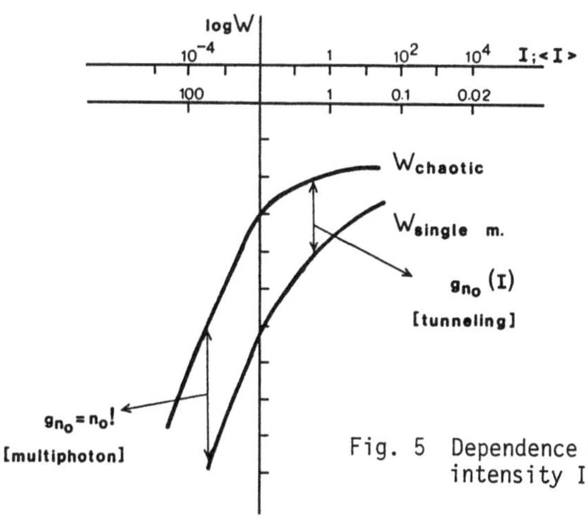


Fig. 5 Dependence of $W_{chaotic}$ and $W_{single\ m.}$ on the laser intensity I

We see that in contrast with the multiphoton case the statistical factor is no longer constant but depends on I, slowly decreasing with increasing I.

To close this section we note that to be able to achieve the tunnelling condition we have either to increase I (that is, $E_L$) or to decrease $\omega$, as follows from the $\gamma \equiv \omega(2mI_0)^{1/2}/eE_L < 1$ condition. With lasers of high $\omega$, however, when I is increasing during the laser pulse, multiphoton ionization may first saturate, before the appearance of tunnelling, as was predicted [26,27]. In view of this, only the use of ultrashort laser pulses of low $\omega$ (e.g. $CO_2$ laser) may ensure the $\gamma < 1$ tunnelling condition during the whole pulse. The other possibility is to reduce artificially $I_0$ bringing first the electron dynamically into a highly excited state by using microwaves, additional laser light [28], by fast heating, or by the inverse Bremsstrahlung process, obtaining in this way $I_0'(t) < I_0$ and $\gamma'(t) < \gamma$ values to realize a certain "dynamic" tunnelling [27].

## 3. Experimental results

### 3.1. Preliminary experiments

Preliminary experiments were carried out both for gas atoms and for metals to demonstrate possible tunnelling effects. In the first experiments for gases [29], when the condition $\gamma < 1$ was realized with $CO_2$ laser, no detectable ionization was found presumably due to the low detection efficiency. In another experiment [30] RF radiation and highly excited atoms were used: owing to the numerous competing multiphoton mechanisms occurring here no definite conclusion could be drawn. In the first quantitative careful experiment by LOMPRE et al. [31] for the total ionization rate W the pure multiphoton type $W \propto I^{n_0}$ relation was found both for $\gamma > 1$ and $\gamma < 1$. As we have seen, however, the $n_0$ slope dependence of the total rate W is less meaningful for the failure of the perturbation approach than the information furnished by the partial $w_{n_0+s}$ dependences and $E_{n_0+s}$ energies, as was later shown by the same authors [32] and others.

As for metals, the $\gamma \ll 1$ condition could not have been achieved with the generally used Nd mode-locked short pulses because the high intensities needed exploded the metal. Therefore only the $\gamma \lesssim$ situation was considered. The experiments [33] performed by us showed that the observed surface photoeffect current j, had the $j \propto W \propto |E_L|^{2n} \propto I^n$ dependence with $n \lesssim n_0$; that is, the breakdown of the perturbation law appeared. This was attributed tentatively to the very weak appearance of the possible tunnelling. As it was discussed earlier, we mention here the interesting work [22] performed long ago - before lasers - which is perhaps not generally known to laser people, which may be called true electromagnetic field-induced tunnelling. RF field of $\omega \sim 10$ GHz with $E_L \sim 10^3$ V/cm was applied to a field emission cathode metal tip on which $E_L \sim 10^8$ V/cm originated, achieving in this way the value $\gamma \sim 10^{-4} \ll 1$. Strong oscillating tunnel emission current was observed and analysed showing the "classically" predicted dependences even for the higher harmonics of $\omega$, with the corresponding electron "bunches". In addition, an electron energy distribution shape shifted to the very high energy range (corresponding to the $E_n$ formula) was also demonstrated. In an interesting way these results reflected in advance the predictions of our analysis.

### 3.2. More recent experiments for gas atoms

The first observation of the ionization of Xe and Kr atoms at high laser intensities with $\gamma \sim 0.1 < 1$ and $n_0 > 100$ was made by us [34] using $CO_2$ laser pulses of low $\omega$ to eliminate the possible multiphoton predominance during the pulse [26,27]. In the course of the subsequent experiments [35] all noble gas atoms have been ionized under similar conditions at $I \sim 10^{14}$ Watt/cm$^2$ with easy multiple ion production, as well. Figure 6 shows the characteristic results.

It can be seen that by increasing I the number of the created ions increases with a slope $n \ll n_0$ which continuously decreases up to the maximum ion yield value at which it becomes negative and the production of double ions starts. The same is true for the double ion creation, etc. The observed ion yield is higher than predicted by the simplified KELDYSH theory [5].

To interpret the results a direct numerical comparison was carried out with both multiphoton and tunnel formulae. No satisfactory agreement was found: the measured W is higher and the measured n is lower than predicted. In another attempt it was taken into account that the ionization increases extremely rapidly in both the multiphoton and the tunnelling case. When I increases either in a given space or time interval the ionization saturates immediately, then the next interval saturates, etc. Integrating over the whole space or time interaction volume another fit was obtained which has not proven to be good either. In these procedures, however, always the simplified Keldysh formula based on formal model potential was used which certainly

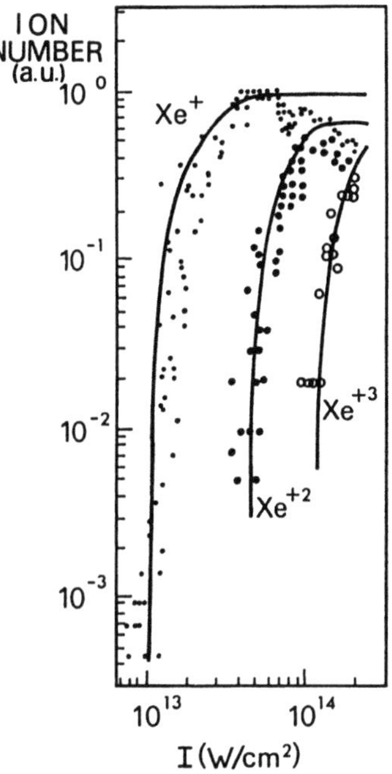


Fig. 6 Typical dependence of the observed number of ions (points) on the laser intensity I. Full line: fit with the simplified Keldysh theory

cannot be correct. Therefore using the tunnel formula improved for a real atom as discussed above, DELONE et al. [23] repeated the fit. They calculated the theoretical $I_{theor}$ values which corresponded to a given W value and to the $10^{-1}$ W value, respectively, and then compared them with the experimental (absolute) $I_{exp}$ values.

Good agreement is found except for the cases where the experimental reading of W was uncertain. These positive results suggest that the nonperturbative theories in their improved forms for real atoms describe correctly the interaction at very high intensities at long (10 μ/ wavelengths.

The main conclusion of these experiment is that at $\gamma < 1$ ionization does exist and the perturbation (multiphoton) character is no longer observed, therefore the "optical tunnelling" may be studied at these long wavelengths.

On the other hand, however, the results of the very careful experiments performed recently [32] at very high intensities for medium wavelengths (Nd laser) also support, even numerically, many predictions of the nonperturbative theories outlined here (e.g. wavelength and polarization-dependent shift of the ATI electron energy spectrum as a whole into the high-energy range, observation of considerable high $E_{n_0+s}$ energy electrons, etc.). At the same time it is more difficult to find such good agreements with the new results obtained with very short wavelength lasers [36]. It will certainly be interesting to find how the nonperturbative theory works in the short (VUV) wavelength region [19].

### 3.3. More recent experiment for metals

In the case of metals we may investigate the problem in a slightly different context. The number of metallic electrons capable of ionization is of the order of $10^{22}/cm^3$ instead of $10^{22}/cm^3$ as in the case of gases, eliminating thus the immediate saturations discussed previously. At the same time an additional competing process may occur: the laser-induced thermionic (Richardson) emission. Bearing in mind these processes in our experiments [37] we demonstrated the characteristics of the surface photoelectron emission (when $\vec{E}_L$ was perpendicular to the surface) from a gold surface induced by strong nsec $CO_2$ laser pulses. In this case: $I_0$ = 4.679 eV; $\hbar\omega$ = 0.117 eV; $n_0$ = 40. The experimental setup is shown in Fig. 7.

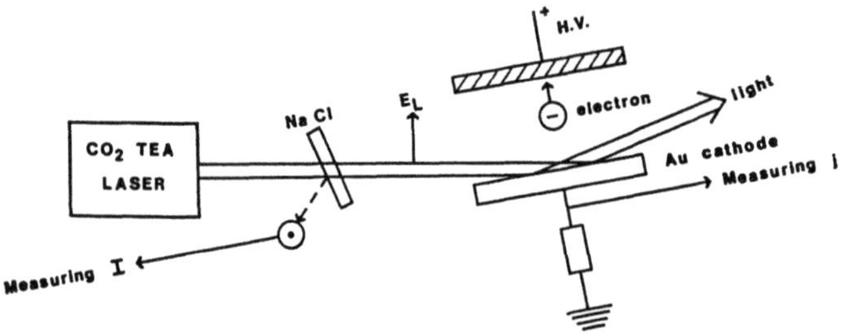


Fig. 7 Experimental setup for observation of laserinduced electron emission process

We measured the current j emitted from the gold surface, for which $j \propto W$, then we determined the $n(I) \equiv d \log W / d \log I$ slope values at different I laser intensity values. In the first experiment [37] single 4 nsec laser pulses were used in the 60 M Watt/cm$^2$ < I < 4 G Watt/cm$^2$ range. Considering the $\gamma = 6.68$ value here, the possible processes might be the $n_0 = 40$ order multiphoton photoeffect, the Richardson emission and the discussed dynamic tunnel emission. The results (Fig. 8) exhibit a strong decrease in $n = n(I)$ when I increases, as predicted for tunnelling. Both multiphoton and Richardson processes have to be excluded.

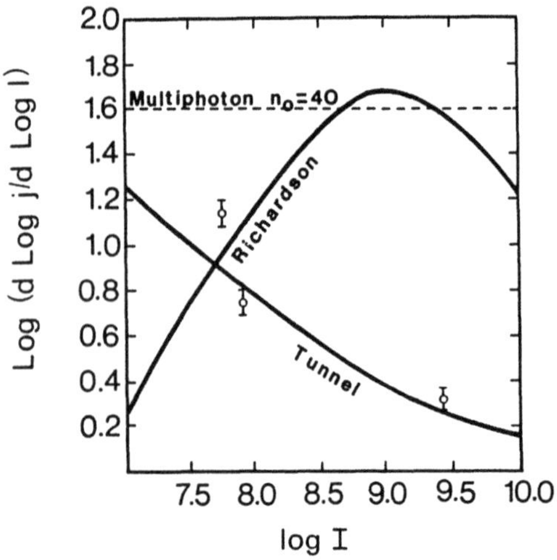


Fig. 8 Experimental (points with error bars) and theoretical (full lines) intensity dependence of slope n for different processes

In a second experiment [38], mode-locked $CO_2$ laser pulse trains containing consecutive 2 nsec pulses separated by 20 nsec periods were used in the 0.3 < I < 1.2 G W/cm$^2$ range. The results obtained confirmed those of the former experiment even numerically. Only bandwidth-limited laser pulses gave evaluable results, furnishing in this way the first new possibility of recognizing such $CO_2$ pulses by a tunnelling detection device [38,39].

As in the case of gas atoms, the observed emission yield (i.e., photocurrent j) was higher than predicted by the idealized nonperturbative formula [13] based on [5] using again short-range model potential, neglecting the image force of the emitted electrons. Again, the elaboration of an improved theory would be needed. In lack of this, a numerical fit of the experimental points to the existing simplified [5,13] theory was carried out. The best fit (continuous curve in Fig. 8) corresponded to a "dynamic" $I_0' = 0.112$ eV and to a "dynamic" $\gamma' = 0.08 < 1 < \gamma = 6.68$ value. The results were interpreted on the basis of the "dynamic" tunnelling discussed previously when the electron in the metallic potential well might acquire dynamically (before the emission) more energy either by fast laser heating or by inverse Bremsstrahlung. We would mention that retarding potential measurement always showed high (up to 50 eV) energy photoelectrons in both the earlier multiphoton [33] and the present experiments.

Although the interpretation presented here certainly needs further confirmation, these experiments have shown that in the case of $CO_2$ laser-induced photoemission both the pure multiphoton and thermionic emission have to be excluded and a new type of interaction occurs.

## 4. Conclusions

We may conclude our survey with the statement that most of the predictions of the intense field QED (either we call them nonperturbative or tunnelling processes) have been verified to some extent experimentally, both for atoms and metals.

## References


1. S.L. Chin and P. Lambropoulos, Eds., Multiphoton Ionization of Atoms (Academic Press, 1984)
2. P. Agostini, F. Fabre, G. Mainfray, G. Petite, N.K. Rahman, Phys. Rev. Lett. 42, 1127 (1979)
3. P. Lambropoulos, S.J. Smith, Eds., Multiphoton Processes (Springer, 1984) and references therein
4. M. Crance, Ed., Multiphoton Ionization of Atoms in Strong Fields, (Conference Digests, Orsay, 28-29 April 1986) and references therein
5. L.V. Keldysh, Sov. Phys. JETP, 47, 1307 (1964)
6. F.V. Bunkin, A.M. Prokhorov, Sov. Phys. JETP, 22, 844 (1964)
7. A.I. Nikishov, V.I. Ritus, Sov. Phys. JETP, 19, 529 (1964)
8. A.M. Perelomov, V.S. Popov, M.V. Terentev, Sov. Phys. JETP, 23, 924 (1966)
9. S. Geltman, M.R. Teague, J. Phys. B7, L 22 (1974)
10. G.J. Pert, J. Phys. B8, L 173 (1975)
11. P. Krstic, M.H. Mittleman, Phys. Rev. A25, 1568 (1982)
12. H.S. Antunes Neto, L. Davidovich, D. Marches, in Coherence and Quntum Optics V, Eds. L. Mandel and E. Wolf, (Plenum Press, New York, 1984) pp. 1109-1116
13. F.V. Bunkin, M.V. Fedorov, Sov. Phys. JETP, 21, 896 (1965)
14. A.P. Silin, Sov. Phys. Solid State, 12, 2886 (1971)
15. S.I. Anisimov, N.A. Inogamov, Y. Petrov, Phys. Lett., 55 A, 449 (1976)
16. H.R. Reiss, Phys. Rev. A22, 1786 (1980)
17. K.G. Müller, A. Tip, M.J. Van der Wiel, J. Phys. B16, L 679 (1983)
18. W. Eberfeld, M. Kleber, W. Becker, M.O. Scully, J. Phys. B19, 2589 (1986)
19. M. Gavrila, in Multiphoton Ionization of Atoms in Strong Fields, Ed. M. Crance, (Conference Digests, Orsay 28-29 April 1986) pp. 1-7
20. F. Ehlotzky, Can. J. Phys. 63, 907 (1985)
21. N.B. Delone, V.P. Krainov, Atoms in Strong Light Fields (Springer, 1985)
22. F.M. Charbonnier, J.P. Barbour, L.F. Garrett, W.P. Dyke, Proc. IEEE 51, 991 (1963)
23. M.V. Ammosov, N.B. Delone, V.P. Krainov, Lebedev preprint, N° 83 (1986)
24. R.V. Karapetyan, Radiophys. Quantum El. 18 (1975)
25. V.P. Krainov, A.V. Masalov, S.S. Todirashku, Izv. Akad. Nauk 45, 2368 (1981)
26. P. Lambropoulos, Phys. Rev. Lett. 55, 2141 (1985)
27. Gy. Farkas, S.L. Chin, in Multiphoton Processes, Eds. P. Lambropoulos and S.J. Smith (Springer, 1984) pp. 191-199

28. N.B. Delone, N.L. Manakov, A.G. Fainshtein, Zh ETF 86, 906 (1984)

29. G. Mainfray, in Electronic and Atomic Collisions, Ed. G. Watel (North-Holland, 178) pp. 699-712

30. J. Bayfield, L. Gardner, P. Koch, Phys. Rev. Lett. 39, 76 (1977)

31. L-A. Lompré, G. Mainfray, C. Manus, C. Repoux, J. Thébault, Phys. Rev. Lett. 36, 949 (1976)

32. L-A. Lompré, A.L'Huillier, G. Mainfray, C. Manus, J. Opt. Soc. Am. B2, 1906 (1985)

33. Gy. Farkas, in Multiphoton Processes, Eds. J.H. Eberly and P. Lambropoulos (Wiley, 1978) pp. 81-100

34. S.L. Chin, Gy. Farkas, F. Yergeau, J. Phys. B16, L 223 (1983)

35. S.L. Chin, F. Yergeau, P. Lavigne, J. Phys. B18, L 213 (1985); S.L. Chin, F. Yergeau, P. Lavigne, to be published

36. U. Johann, T.S. Luk, H. Egger, C.K. Rhodes, Phys. Rev. A34, 1084 (1986)

37. Gy. Farkas, S.L. Chin, P. Galarneau, F. Yergeau, Opt. Comm. 48, 275 (1983)

38. Gy. Farkas, S.L. Chin, Appl. Phys. B37, 141 (1985)

39. Gy. Farkas, Z.Gy. Horváth, Opt. Comm. 12, 392 (1974)

# An $L^2$ Method for Continuum States

*R. Moccia*[1] *and P. Spizzo*[2]

[1]Dipartimento di Chimica e Chimica Industriale dell' Università di Pisa, I-56100 Pisa, Italy

[2]Istituto di Chimica Quantistica ed Energetica Molecolare del C.N.R., Via Risorgimento 35, I-56100 Pisa, Italy

The recent development of laser and synchrotron technology has led to a large amount of detailed experimental information about the continuous spectra of atoms and molecules. The theoretical interpretation of these results often requires a very careful treatment of the continuum, because interesting features like the resonances' shapes and widths arise uniquely from the electron correlation, and hence can be reproduced only by accurate wavefunctions. These calculations are generally performed by close-coupling or related methods, mainly through the numerical integration of the radial equation for the electron wave. These techniques, although quite adequate to the purpose, lie outside the usual framework of the computational quantum chemistry, where several methods have been developed to treat carefully the correlation problem; in fact, pratically all these methods rely on the use of $L^2$ basis functions. It is therefore highly desirable to develop methods for extracting the properties belonging to the continuum from computations performed entirely on $L^2$ bases without the limits of the standardly used techniques, such as the Stieltjes (moment) /1/ and the R-matrix /2/ methods. The Stieltjes technique is a valuable tool for computing slowly varying photoionization cross-sections, but cannot reproduce their features at narrow resonances; moreover it does not yield the states or many other interesting quantities. The R-matrix method, instead, can be usefully employed when the continuum states are fairly approximated by a single ionic configuration coupled to one electron wave. Its multiconfigurational extension, however, leads to intractable bielectronic integrals.

## 1. The $L^2$ Representation of the Continuum

The method used here starts from the assumption that the diagonalization of the Hamiltonian on a proper basis of $L^2$ functions may reproduce accurately, apart from a normalization constant, several continuum eigenstates inside a sphere whose radius $R_s$ depends on the basis nature and dimension. If this sphere is large enough to include a wide outer region in which the electron density is that of the electron wave, the fit of this density to that of shifted Coulomb waves allows to determine both these normalization constants and the phase shifts. The usual STO/GTO bases are, in this respect, quite inadequate, because they require a very large number of orbitals to reproduce relatively few waves inside a rather small sphere, and lead to a very redundant metric. As a matter of fact, computational feasibility requires that the basis orbitals employed in a variational Configuration Interaction (CI) calculation to represent the electron wave meet the following conditions: They must

a) have an oscillating behaviour with non-vanishing amplitude up to large values of r, since otherwise they lead to intractable metrics,
b) have a significant amplitude in the valence shell, because a variational method really optimizes and yields reliable values for the expansion coefficients only for the basis functions which are capable to reproduce the correct behaviour of the true state in the space regions where the potential is large.

Basis sets with these properties have been developed recently /3,4/ by employing, beside the STOs needed to represent the lowest bound states and the ionic or atomic cores, Laguerre orbitals for the Rydberg states and diffuse oscillating orbitals of the form

$$\chi_{klm}(\underline{r}) = N_{lk}\, f_l(r)\, \cos(k\, r)\, Y_{lm}(\vartheta, \varphi) \qquad (1)$$

$$f_l(r) = r^l \sum_j C_{lj} \exp(-A_{lj}\, r) \qquad (2)$$

where $C_{lj}$ and $A_{lj}$ are fitting parameters determined so that $f_l(r)$ simulates, outside the valence shell and up to r about 100 (atomic units are used throughout unless otherwise stated), the proper damping 1/r of a Coulomb wave. These orbitals lead to tractable close-analytic formulae for all the integrals needed in a monocentric expansion /3/ and the k-values can be conveniently chosen equally spaced up to $k_{max}$ according to

$$k_j = \pi(j+1/2) / R_0 \qquad j = j_{min}, \dots j_{max} \qquad (3)$$

so that all these orbitals have a zero in $r=R_0$. The advantages of this choice will be apparent below.

We have verified the basic assumption in a large number of single-particle central-field problems, for quite different wave energies and potential shapes, including delicate cases of narrow-shape resonances /5/. Moreover, they have been successfully employed also for non-spherical potentials /6/ . In particular, there is definitive computational evidence that, provided $\Delta k=\pi/R_0$ is lower than the smallest $A_{lj}$, the variational states closely resemble the true eigenstate of the same wave energy up to $r=R_0$ (that is, $R_s=R_0$) and then vanish rapidly. Although no rigorous mathematical proof can be given, this fact may be heuristically understood if one notes that the k-spread of the variational states may be expected to be about $\Delta k=\pi/R_0$, which is exactly the k-spread for the eigenstates of a particle in a spherical box of radius $R_0$ upon those of a free particle /5/. This behaviour is clearly shown in fig. 1, which reports the radial charge density for a $^1P^o$ CI variational state of Beryllium (wave energy 0.00824) obtained with $R_0$=518.7. This

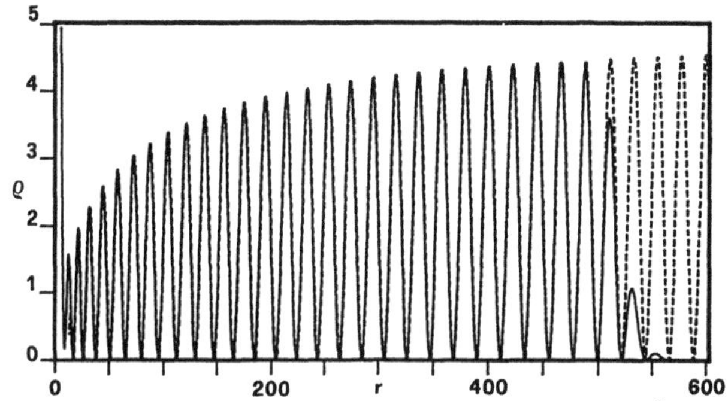


Figure 1 Charge density $\varrho$ for the Beryllium $^1P^o$ variational state of wave energy 0.00824 (full line) and for the shifted Coulomb wave which fits it (broken one).

figure reports also the density for the shifted Coulomb wave that fits the variational state, and the agreement is excellent up to $r=R_0$. We stress that this state represents a critical test for the basis set, since a wave of such low energy is difficult to reproduce by a variational approximation.

Provided that all the STOs and Laguerre orbitals die out well before $R_0$, the choice (3) implies that all the variational states have a node in $R_0$, so the present method selects all the continuum states which vanish in this point up to a wave energy of about $k^2_{max}/2$. If the lowest lying of these states has $n_0$ nodes in the open interval $(0,R_0)$, the basis orbitals (1) with less nodes in it (i.e. with $j<n_0$) appear mainly in the variational bound states, and may cause dangerous redundancies with the basis orbitals included to describe the discrete spectrum, thus the need for a proper choice of $j_{min}$. In other words, the representation of the continuum electron wave is mainly performed by the basis orbitals (1) with $j>n_0$.

The states of different energy may be then obtained by repeating the calculations for different values of $R_0$ (see section 2). In this context, let us note that the present method circumvents the difficulties connected with the bielectronic integrals required by the multiconfigurational extension of the R-matrix method.

Figure 2 refers instead to a state which has a wave energy rather close to that of the broad 2p3s $^1P^o$ resonance, and it is clear that the charge density behaviour resembles that of a shifted Coulomb wave only for large values of r. It should be noted that the charge density for $10<r<20$ does not arise from the $1s^2 2s$ core but is borrowed from the 2p3s $^1P^o$ resonance. It is evident that for higher resonances the shifted Coulomb wave behaviour is encountered at still higher values of r.

## 2. The Treatment of the Resonances

The resonances in the continuous spectra of many-electron systems may be viewed as due to the (near) degeneracy, in the independent particle limit, of the ionization continuum with multiply- or core-excited bound states /8/. Let us denote by $|f\rangle$ the

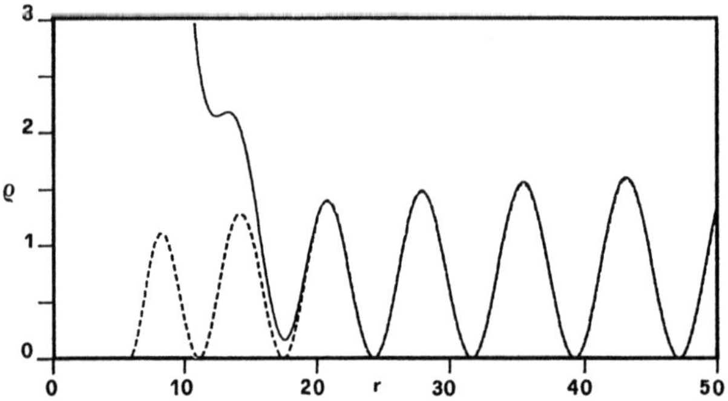


Figure 2 Charge density $\varrho$ for the Beryllium $^1P^o$ variational state of wave energy 0.05737 (full line) and for the shifted Coulomb wave which fits it (broken one)

zeroth order bound state and by $|E\rangle$ the unperturbed continuum function of energy E and assume that the resonance-continuum matrix element $\langle f|H|E\rangle$ is a slowly varying function of the energy E; one can then show that the autoionizing width of the wavepacket $|f\rangle$ is given by

$$\Gamma = 2\pi\, |\langle f|H-E_f|E_f\rangle|^2 \qquad E_f = \langle f|H|f\rangle \tag{4}$$

and that around $E=E_f$ the phaseshift $\delta(E)$ behaves according to

$$\delta(E) = a + b\,(E-E_0) + \tan^{-1}\left[\Gamma/2(E_0-E)\right] + \pi\,\vartheta(E-E_0) \tag{5}$$

where the first two terms account for the slowly varying background contribution of the nonresonant continuum, while the unit step function has been added to make this quantity continuous in $E=E_0$. The resonance position $E_0$ is given at first order by

$$E_0 = E_f + P\int \langle f|H-E_f|E'\rangle^2\, dE' \,/\, (E_f-E') \tag{6}$$

in which P denotes the Cauchy's principal value of the integral. Also the photoionization cross-section $\sigma$ has a typical behaviour which is usually well reproduced by the Fano profile /8/:

$$\sigma = \sigma_0\,(q+\varepsilon)^2/(1+\varepsilon^2) \qquad \varepsilon = 2\,(E-E_0)/\Gamma \tag{7}$$

Equation (5) shows that across a resonance the phaseshift increases of about $\pi$ in an energy interval of $\Gamma$, as is shown in fig. 3 for the 2,3b (or 2,3-) resonance of atomic Helium, together with the photoionization cross-section of the ground state. The points shown in this figure have been obtained performing 21 calculations over bases with different $R_0$'s ranging from 200 to 202 step 0.1. The presence of a resonance, therefore, results in an extra state with a node in $R_0$ and whose energy is almost independent of $R_0$, since the node condition is mainly met by an adjustment in the phaseshift instead of in the wavelength, i.e. there is a persisting eigenvalue.

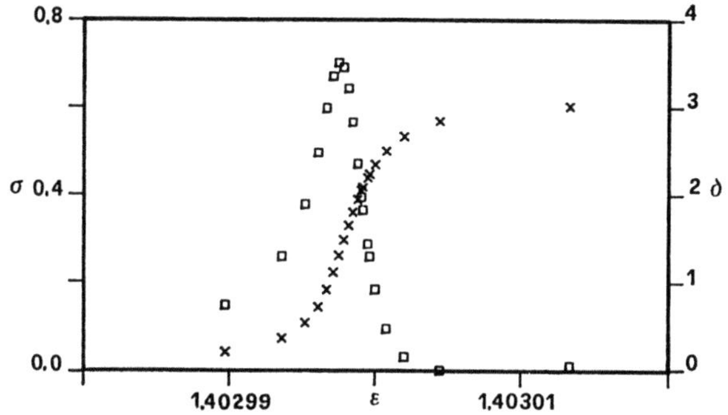


Figure 3 Phaseshift $\delta$ (×) and length gauge photoionization cross-section $\sigma$ from the ground state (□) of the 2,3b resonance in Helium vs. the wave energy $\varepsilon$

A great advantage of the variational method employed with appropriate bases is the fact that a single calculation yields for every resonance, no matter how narrow, at least one easily identifiable corresponding state. This avoids therefore the laborious searching required by the close-coupling and related methods for very narrow resonances.

We have implemented three independent methods for obtaining the positions and the widths of the resonances:

a) the phaseshift analysis: by fitting the computed phaseshift around a resonance to the formula (5), it is possible to obtain the width $\Gamma$ and the position $E_0$ of the resonance.

b) the complex stabilization method: the plot of the eigenvalues against the scale parameter $R_0$ shows avoided crossing due to the presence of the above-mentioned persisting eigenvalues, and near them the resonance complex stabilization condition $dE/dR_0 = 0$ is satisfied for a complex value of $R_0$ at $E=E_0-i\Gamma/2$ /9/. This condition is found by analytically continuing into the complex plane the Hamiltonian matrix elements as functions of $R_0$. This method worked fairly better for Helium /12/ than for Beryllium /7/ and the analysis of the difficulties encountered support the feeling that it is very sensitive to fine numerical details and requires bases with very little redundancy, which for Helium are much easier to realize than for Beryllium.

c) the projectors method: let B indicate a subset of the whole basis including only localized functions adequate to describe localized states (i.e. bound ones and zeroth order resonances), denote by P the projector over the bound states and the resonances reproduced by the subset B and let Q=1-P. The nonresonant background continuum may be obtained by diagonalizing on the whole basis set the projected Hamiltonian QHQ, after which the resonance's widths and positions may be obtained by (4) and (6) respectively. Compared to the two methods described above, this last one has the advantage of requiring the calculation only for a single value of the scale parameter $R_0$.

## 3. Results for Helium and Beryllium

In Helium, the independent particle model predicts three series of $^{1,3}P^o$ bound states below the ionization threshold n=2, which arise from the configurations 2snp, 2pns and 2pnd and are degenerate in the Hydrogenic approximation, so they are expected to mix strongly under the interelectronic repulsion, making questionable their attribution to a dominant configuration. In the photoionization from the ground state, however, only two series of resonances are observed, with quite different intensities and widths. The third series is still unobserved in photoionization and it is very unlikely to be observed in the near future, since the available theoretical calculations predict their lines to be extremely faint and narrow. Several classifications have been proposed (see /10/ for a review); here we follow the "phenomenological" one proposed by LIPSKY and CONNEELY /11/ who label 2,na, 2,nb and 2,nc the resonances of the large, narrow and very narrow series respectively. Indeed, as shown in Table 1, the present calculations /12/ predict a large 2pnd character for all the 2,na (about 20%), 2,nc (about 30%) and 2,(n+1)b (about 50%) resonances (n>2), while only the 2,2a and 2,3b ones have almost pure sp character.

Table 1 Characters, positions E /eV/ over the ground state (upper figures) and widths $\Gamma$ /eV/ (lower ones) for several $^1P^o$ resonances of Helium

| resonance | character % | | | present theoretical results | | | exp. |
|---|---|---|---|---|---|---|---|
| | sp | pd | df | a | b | c | /13/ |
| 2,2a | 91.40 | 8.54 | 0.06 | 60.1542 | 60.1548 | 60.1541 | 60.133 |
| | | | | 3.70(-2) | 3.46(-2) | 3.65(-2) | 3.8(-2) |
| 2,3b | 93.24 | 6.76 | 0.01 | 62.7559 | 62.7560 | 62.7559 | 62.756 |
| | | | | 1.03(-4) | 0.99(-4) | 0.98(-4) | |
| 2,3a | 81.96 | 18.02 | 0.02 | 63.6555 | 63.6555 | 63.6567 | 63.656 |
| | | | | 8.11(-3) | 8.11(-3) | 7.95(-3) | 8.0(-3) |
| 2,3c | 68.91 | 30.91 | 0.18 | 64.1167 | 64.1167 | 64.1167 | |
| | | | | | 4.41(-7) | 5.32(-7) | |
| 2,4b | 49.27 | 50.62 | 0.12 | 64.1324 | 64.1324 | 64.1324 | 64.141 |
| | | | | 4.11(-5) | 4.11(-5) | 4.67(-5) | |
| 2,4a | 80.57 | 19.42 | 0.01 | 64.4622 | | 64.4630 | 64.462 |
| | | | | 3.43(-3) | | 3.33(-3) | |
| 2,4c | 75.36 | 24.56 | 0.07 | | | 64.6459 | |
| | | | | | | 2.28(-7) | |

For a given n>2, there are no doubts that the $^1P^o$ resonance energies lie in the order 2,nb<2,na<2,nc, but the relative ordering of the 2,nc and 2,(n+1)b $^1P^o$ resonances, in absence of unambiguous experimental data, has been a controversial question until recent times. The early close-coupling calculations, in fact, placed the observed 2,(n+1)b resonance below the unobserved 2,nc one and only recent work has revised this prediction, but in the meanwhile some beam-foil lines have been assigned according to these calculations and reported in several compilations /12/.

Another interesting problem is the width of the 2s2p $^3P^o$ resonance. In fact, our calculations agree with all the other available theoretical results in predicting for this resonance a width slightly greater than 8 meV /12/, while the experimental beam-foil results of CEDERQUIST et al. /14/ might suggest a width of only 7 meV. A more direct experimental determination of this width is therefore desirable.

The lowest continuum of atomic Beryllium (i.e. that below the 2p threshold of the ion) is very much structured by the presence of the doubly-excited states which form the so-called "displaced term system" and arise from the configurations 2pnl ($nl \neq 2s$). The $2p^2\ ^1S^e$ resonance has been long considered a bound state /7,15/ and only recently this uncorrect assignment has been revised; this resonance has been finally discovered by CLARK et. al. /16/ in 1985. It lies very close to the ionization limit, only 997 $cm^{-1}$ above the ion ground state and has a width of 74.9 $cm^{-1}$. The present calculations /7/ predict for this resonance a width of about $3.30 \cdot 10^{-4}$ (72.4 $cm^{-1}$) in excellent agreement with the experiment, but locate it at the sligthly greater wave energy of 0.00695 (1525 $cm^{-1}$). This position is, however, in a noticeably better agreement than any other theoretical result /16/. Since our variational wavefunction appears to have recovered all the valence-shell correlation energy, this discrepancy should be ascribed to the non-valence (i.e. inter- plus inner-shell) correlation energy. Indeed, our calculations indicate that this correlation energy is greater than that of the ion ground state in all the levels whose spectroscopic configurations include 2p orbitals, and in some cases (2s2p $^1P^o$, $2p^2\ ^1S^e$, $2p^2\ ^3P^e$) even greater than that of the neutral atom ground state.

Finally, it is worth to remind that the correct handling of the correlation effects, as emphasized and done in this paper, is quite important in order to obtain gauge-invariant transition probabilities. This problem is often overlooked when one tries to assess the quality of the results obtained in a single gauge by comparing them to the experimental data.

## References


1. G.H.F. Diercksen, W.P. Kramer, T.N. Rescigno, C.F. Bender, B.V. McKoy, S.R. Langhoff, P.W. Langhoff: J. Chem. Phys. 76, 1043 (1982)
2. P.G. Burke, W.D. Robb: In Advances in Atomic and Molecular Physics 11, (Academic Press, New York 1975) pag. 143
3. I. Cacelli, R. Moccia, V. Carravetta: Chem. Phys. 90, 313 (1984)
4. I. Cacelli, R. Moccia, V. Carravetta: In Collisions ans Half Collisions with Lasers, ed. by N.K. Rahman and C. Guidotti (Harwood, Chur 1982) pag. 229

5. P. Spizzo: Tesi di Perfezionamento at the Scuola Normale Superiore, Pisa (1986)
6. I. Cacelli, V. Carravetta, R. Moccia: J. Chem. Phys., in press
7. R. Moccia, P. Spizzo: J. Phys. B: At. Mol. Phys. 18, 5537 & 3555 (1985)
8. U. Fano: Phys. Rev. 124, 1866 (1961)
9. C.W. McCurdy, J.F. McNutt: Chem. Phys. Lett. 94, 306 (1983)
10. L. Lipsky, R. Anania, M.J. Conneely: At. Data Nucl. Data Tables 20, 127 (1977)
11. L. Lipsky, M.J. Conneely: Phys. Rev. A 14, 2193 (1976)
12. R. Moccia, P. Spizzo: J. Phys. B: At. Mol. Phys. 19, in press (1986)
13. R.P. Madden, K. Codling: Astrophys. J. 141, 364 (1965)
14. H. Cederquist, M. Kisielinski, S. Mannervik: J. Phys. B: At. Mol. Phys. 16, L479 (1983)
15. S. Bashkin, J.O. Stoner jr.: Atomic Energy Levels and Grotrian Diagrams, vol. 1 (North Holland, Amsterdam 1975) and vol. 1 addenda (North Holland, Amsterdam 1978)
16. C.W. Clark, J.D. Fassett, T.B. Lucatorto, L.J. Moore, W.W. Smith: J. Opt. Soc. Am. B 2, 891 (1985)

# Collective Effects in Multiphoton Ionization: Application to Two-Photon One-Electron Ionization of Xenon

*A. L'Huillier and G. Wendin*

Institute of Theoretical Physics, Chalmers University of Technology, S-412 96 Göteborg, Sweden

## 1. Introduction

The purpose of the present work[1-3] is to study various aspects of multiphoton ionization of heavy atoms[4-11] which are known to show collective behaviour in single photon ionization.[12-14] The ultimate goal is to treat the non-linear collective and single-particle dynamics in cases where the field is strong enough to seriously perturb the atomic states (dynamic Stark shift and broadening; ponderomotive effects; harmonic generation, etc.). However, this is a rather formidable problem for a many-electron atom. Strong-field effects have essentially only been seriously considered within one-electron formulations and model calculations, and many problems remain to be solved.

Here we shall consider the low-intensity limit and discuss 2-photon 1-electron ionization of the 5p-shell of Xe using a zero-field atomic basis. This will demonstrate important ways in which collective effects may enter, and it also represents the first calculation of its kind.[3]

## 2. Linear Response and Collective Effects

An external potential $v_{ext}$ applied to an electronic system will induce a charge density displacement $\delta n$ (polarization) and an induced potential $v_{ind}$,

$$v_{ind} = V\,\delta n \qquad \text{(a)}$$
$$= V\,\chi\,v_{ext} \qquad \text{(b)} \qquad (1)$$
$$= V\,\chi^0\,v_{eff} \qquad \text{(c)}$$

$\chi$ is the full dielectric linear susceptibility, and V is the Coulomb interaction. $\chi^0$ is the susceptibility of the system of independent electrons. In (1b), many-electron polarization is represented by the full response function $\chi$, which can be calculated with the help of e.g. term-dependent "one-electron" wave functions; in (1c), many-electron polarization is represented by the effective (total) perturbing potential.

The effective perturbing potential $v_{eff}$ is given by

$$v_{eff} = v_{ext} + v_{ind} \qquad (2)$$

which, combining (1c) and (2), gives an integral equation for the effective potential

$$v_{eff} = v_{ext} + V\,\chi^0\,v_{eff} \qquad (3)$$

illustrated diagrammatically in Fig. 1. Equation (3) describes the random phase approximation (RPA) and its different variations (RPAE, RRPA, LDRPA, TDLDA, etc.) depending on the precise definition of $\chi^0$. In the case of dipole excitation, we may write the external and effective perturbations according to

$$v_{ext} = -\,\mathbf{E}\cdot\mathbf{r} \qquad \text{(a)}$$
$$v_{eff} = -\,\mathbf{E}\cdot\mathbf{r}(\omega) \qquad \text{(b)} \qquad (4)$$

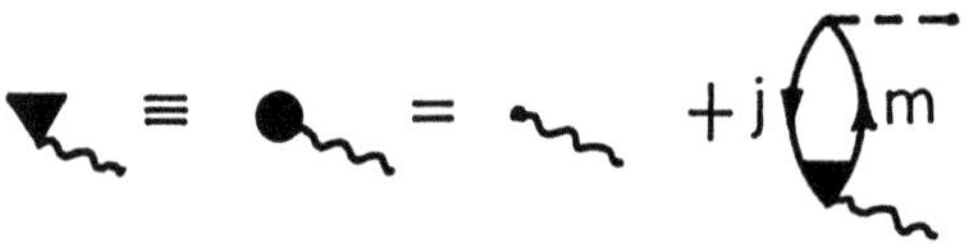


Fig. 1. Diagrammatic representation of the screened electron-photon interaction in the random phase approximation (RPA)

where[1,12-14]

$$\mathbf{r}(\omega) = \mathbf{r} - \sum_{mj} \frac{\langle j|1/r_{12}|m\rangle \ \langle m|\mathbf{r}(\omega)|j\rangle}{(\omega_{mj}^2 - \omega^2)/2\omega_{mj}} \tag{5}$$

## 3. Non-Linear Response

The RPA(E) is equivalent to linearized time-dependent Hartree(-Fock) (linearized TDH(F)) and only allows one real electron-hole pair at any time in the perturbation expansion (incidentally, this is why relaxation effects are not described by the RPAE). However, if one did not linearize the TDH(F) equation, one could in principle obtain an exact description of the time-dependent deformation of the wave functions and charge density of the system under the influence of intense external fields.

To lowest order including non-linear response, the ground state many-electron wave function may be written as

$$|\Psi(t)\rangle = [1 + E\sum_{ni} a_{ni}(t)c_n^+c_i + \text{c.c.} + $$
$$+ E^2 \sum_{nimj} a_{nimj}(t)c_n^+c_i\ c_m^+c_j + \text{c.c.} + \ldots]|0\rangle \tag{6}$$

With this ansatz, we now have two possibilities: two photons may excite one high-energy electron-hole pair by repeated application of the first-order (linear) term, or they may excite two low-energy pairs via the second-order (non-linear) term in (6). In the general case, N photons may excite N electron-hole pairs via the N-th order non-linear term. In a system with large dipole oscillator strength in lowlying discrete or continuum one-electron transitions, it may be advantageous to distribute the photons over as many electrons as possible. This means that the external field may actually drive many electrons (even an entire shell), corresponding to "many quanta excited in a collective oscillation".

## 4. Two-Photon One-Electron Ionization

Figure 2a shows the zeroth-order diagrammatic 2-photon 1-electron amplitude given by the expression

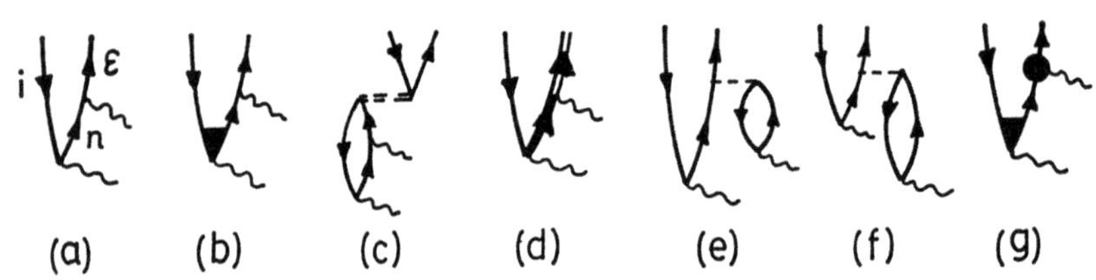


Fig. 2. Diagrammatic representation of 2-photon 1-electron ionization. Black dots represent RPA screening (Fig. 1). See the text for further explanations.

$$t_{\varepsilon i}(\omega) = \sum_n \frac{<\varepsilon|\mathbf{E}\cdot\mathbf{r}|n> <n|\mathbf{E}\cdot\mathbf{r}|i>}{\omega_{ni} - \omega} \tag{7}$$

This independent-electron result cannot be expected to give any accurate description of the true cross-sections. As mentioned above, screening of the external dipole perturbation is extremely important when describing 1-photon ionization of the 5p-shell in Xe.

The question then is how this screening is going to enter the 2-photon amplitude and, in particular, what we mean by RPA in this context. If by RPA we mean linear response, we can only allow single-pair excitations (first-order term in (6)). If in addition we only allow induced dipole polarization, only the first electron-photon operator will be screened at frequency $\omega$ (Fig. 2b), and the 2-photon 1-electron amplitude becomes

$$t_{\varepsilon i}(\omega) = \sum_n \frac{<\varepsilon|\mathbf{E}\cdot\mathbf{r}|n> <n|\mathbf{E}\cdot\mathbf{r}(\omega)|i>}{\omega_{ni} - \omega} \tag{8}$$

If we also allow induced monopole and quadrupole polarization (Fig. 2c), the result may be expressed in terms of screening of the second electron-photon operator at frequency $2\omega$ (ref. 2),

$$t_{\varepsilon i}(\omega) = \sum_n \frac{<\varepsilon|\mathbf{E}\cdot\mathbf{r}(2\omega)|n> <n|\mathbf{E}\cdot\mathbf{r}(\omega)|i>}{\omega_{ni} - \omega} \tag{9}$$

As far as we understand, it is at this level of approximation that many-body calculations have been performed so far[15-17].

If we only consider forward propagating diagrams (Tamm-Dancoff Approximation, TDA), the 2-photon 1-electron amplitude may be written in terms of intermediate- and final-state LS-coupled wave functions (and corresponding energies) and bare (external) electron-photon operators (Fig. 2d),

$$t_{\varepsilon i}(\omega) = \sum_n \frac{<\varepsilon_{eff}|\mathbf{E}\cdot\mathbf{r}|n_{eff}> <n_{eff}|\mathbf{E}\cdot\mathbf{r}|i>}{\omega_{ni} - \omega} \tag{10}$$

There is, however, no reason why the response should be limited to linear response. In fact, it would appear natural to consider the response to at least the same non-linear order as the multi-photon process itself. This means that each incident photon is allowed to polarize the system by exciting electron-hole pairs. To second order, this effect of non-linear response is diagrammatically represented in Figs. 2e,f. It involves double excitations and provides additional contributions to the 2-photon 1-electron ionization amplitude.

Since the two electron-hole pairs in the double excitation in Figs. 2e and 2f are independent, i.e. $\omega_{12}=\omega_1+\omega_2$ ("independent boson" approximation), then adding together the diagrams in Figs. 2e,f leads to the summations over the pairs becoming independent. As a result, the electron-photon vertices become independently screened by the same response function. It follows that the 2-photon 1-electron amplitude takes the form

$$t_{\varepsilon i}(\omega) = \sum_n \frac{<\varepsilon|\mathbf{E}\cdot\mathbf{r}(\omega)|n> <n|\mathbf{E}\cdot\mathbf{r}(\omega)|i>}{\omega_{ni} - \omega} \tag{11}$$

where both effective photon operators are screened in the same way, given by (5). Note that this is an exact result within the "independent boson" approximation.

This form of the amplitude is very appealing from a physical point of view: it says that due to the atomic response there exists a "macroscopic" effective, space-dependent electromagnetic field driving the motion of the atomic shells. In this approximation, every photon operator is screened by the same dielectric function via (5). This includes photons being absorbed and reemitted (or the other way around) in descriptions of e.g. the dynamic Stark effect.

Finally, combining (9) and (11) we obtain

$$t_{\varepsilon i}(\omega) = \sum_n \frac{\langle\varepsilon|\mathbf{E}\cdot\mathbf{r}(\omega,2\omega)|n\rangle \ \langle n|\mathbf{E}\cdot\mathbf{r}(\omega)|i\rangle}{\omega_{ni} - \omega} \tag{12}$$

where the last perturbation ($n\rightarrow\varepsilon$) incorporates also screening at frequency $2\omega$. However, it should be noted that, as usual, there is considerable freedom in accounting for the screening effects. The screening at frequency $2\omega$ could e.g. be represented in terms of an effective (LS-coupled) final state wave function $\langle\varepsilon_{eff}|$ (Eq. (10), Fig. 2c).

## 5. Application to Two-Photon One-electron Ionization of the 5p-Shell of Xenon

In order to illustrate the concepts discussed in Sect. 4, we would like to give a brief presentation of some very recent numerical results.[3]

The 2-photon 1-electron ionization generalized cross-section $\sigma_2$ (in $cm^4$ s) can be written as

$$\sigma_2 = 2\pi\ (\alpha c/a_0)\,(\omega/F_0)^2 \sum_{l_\varepsilon} |t_{\varepsilon i}(\omega)|^2 \tag{13}$$

where $\alpha$ is the fine-structure constant, c is the speed of light, $a_0$ is the Bohr radius, and $\omega$ is the photon energy. $F_0$ is equal to 3.22 $10^{34}$ photons/$cm^2$s. The partial amplitudes $t_{\varepsilon i}$ are defined by

$$t_{\varepsilon i}(\omega) = \sum_n C(\varepsilon,n,i)\ \frac{\langle\varepsilon|r(\omega)|n\rangle \ \langle n|r(\omega)|i\rangle}{\omega_{ni} - \omega} \tag{14}$$

$|i\rangle$, $|n\rangle$, and $|\varepsilon\rangle$ are one-electron wave functions (LDA or HF average). $C(\varepsilon,n,i)$ are spin-angular coefficients which depend on the laser polarization. The screening effects are given by the RPA integral equation for $r(\omega)$ in (5).

The 2-photon ionization cross-section $\sigma_2$ has been calculated over a large energy range, from the 2-photon ionization threshold up to 1.4 Ry. When the photon energy is above the 1-photon ionization threshold, expression (14) describes a so-called ATI (above-threshold ionization) process and requires a special numerical treatment. In particular, one needs to calculate free-free dipole matrix elements.[2]

Figure 3 shows the 2-photon 1-electron ionization cross-section $\sigma_2$ calculated with an LDA basis, for linearly polarized light, as a function of the photon energy. The solid line is the independent-electron (zeroth-order) approximation (Fig. 2a; Eq. (7)), while the dashed line is the LDRPA result (Fig. 2g; Eq. (11)). We have also calculated the 2-photon 1-electron ionization cross-section by only screening the first electron-photon interaction (Fig. 2b; Eq. (8)), an approximation which has been used in several previous calculations.[15-17] This result is indicated by the dotted line in Fig. 3.

The LDA ionization threshold (0.62 Ry) is lower than the experimental threshold ($\cong$0.90 Ry). In the region $2\omega < 0.9$ Ry, the LDA gives an average description of the discrete 2-photon absorption resonances of the real system. The structures in this region of Fig. 3 are associated with 5p→s→p and 5p→d→p transitions. Above $\cong$0.5 Ry, the cross-sections are dominated by 5p→d→f transitions. The pronounced peak in the independent-electron cross-section at 0.68 Ry is related to the 5p→εd LDA oscillator strength distribution, which presents a pronounced maximum in this region. From the

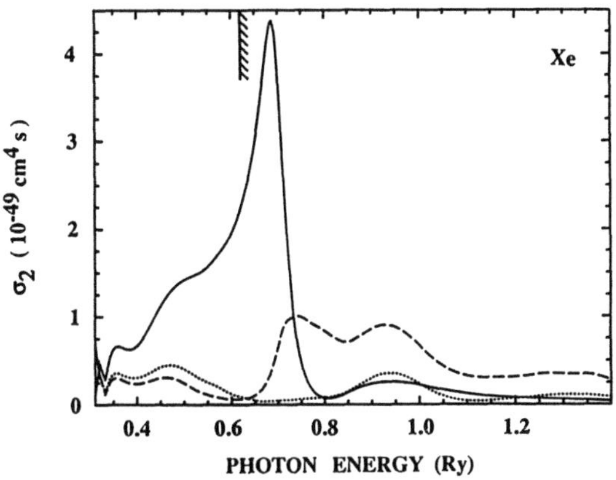


Fig. 3. 2-photon 1-electron ionization cross-section of Xe (linear polarization): (———) independent-electron approximation (LDA); (- - - -) screened electron-photon interactions (LDRPA); (·······) only the first interaction screened (LDRPA).

threshold (0.31 Ry) to ≅0.7 Ry, the RPA screening leads to a large reduction of the oscillator strength compared to the one-electron approximation.

We emphasize that all electron-photon operators should be dynamically screened. As seen in Fig. 3, omission of the screening of the second photon only leads to moderately different results for ω below the LDA 5p-ionization limit but leads to dramatically different results for ω above that limit. The reason is that the present, fully screened, approach accounts for double excitation (ionization) of the 5p-shell (Figs. 2e,f). Above 0.62 Ry, the spectrum is completely dominated by these processes. At photon energies >0.7 Ry, they induce an important enhancement of the 2-photon ionization cross-section compared to the one-electron approximation (7) or the singly screened approximation (8). This is not any anti-screening (field-enhancement) effect since the effective field is still reduced compared to the external field (up to ≅1Ry), but a purely non-linear effect: In the process, two photons are (virtually or really) absorbed by two different electrons.

The results obtained with the HF-basis are presented in Fig. 4, showing the RPAE result over a wide energy region, together with the result of the LDRPA. As shown in

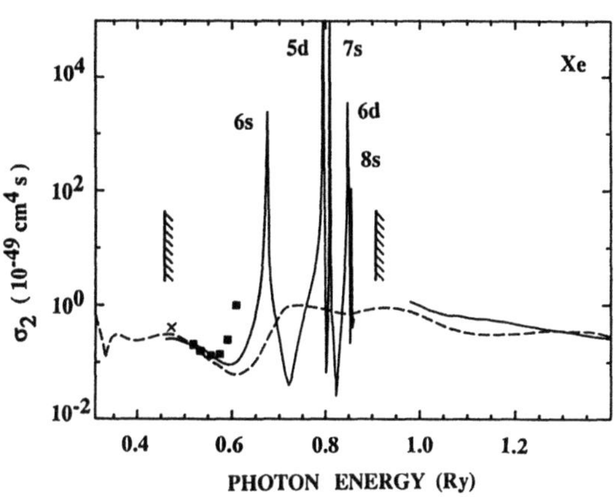


Fig. 4. 2-photon 1-electron ionization cross-section of Xe (linear polarization): (———) RPAE; (-----) LDRPA. The 2-photon (0.46 Ry) and 1-photon (0.92 Ry) HF ionization thresholds are indicated by hatched lines. Black squares indicate the MQDT results by Gangopadhyay et al.[19] The cross is an experimental point (193 nm) by McCown et al.[21]

the figure, we have included the $5p^{-1}ns$, n=6,7,8 and $5p^{-1}nd$, n=5,6 resonances. Apart from the close neighbourhood of the $5p^{-1}ns$ resonances, the dominant channel leading to ionization is 5p→d→f.

Near the 2-photon ionization threshold, our results in Fig. 4 can be compared to calculations of McGuire[18] and Gangopadhyay et al.[19] McGuire's result (Eq. (7) evaluated with a Hartree-Fock-Slater type of basis and experimental excitation energies) is slightly below our independent-electron result[20] (not shown). The multi-channel quantum defect theory (MQDT) treatment of Gangopadhyay et al.[19] leads to good agreement with the present non-linear RPAE and LDRPA results in the 0.5-0.6 Ry range. Inclusion of spin-orbit splitting within the present scheme will probably lead to good agreement also in the vicinity of the $5p_{3/2}$-6s resonance. Finally, the cross in Fig. 4 indicates an experimental point[21] obtained at 0.47 Ry (193 nm). Although the good agreement may be somewhat fortuitous because this point lies in the region of autoionizing resonances (between the two 5p-ionization thresholds $P_{3/2}$, $P_{1/2}$), one can note that the experimental point[21] is much closer to the RPAE, LDRPA, and MQDT results than to the one-electron result. We also note that in this photon-energy region, it makes no great difference whether the last photon operator is screened or not.

Comparing the LDRPA and RPAE results we find that both results are in good agreement at low photon energies (0.46 Ry $< \omega <$ 0.62 Ry) and in the continuum region ($\omega >$ 0.98 Ry). This comparison enables us to point out the respective advantages of the two methods. The RPAE result is, of course, much closer to the real spectrum in the region of the discrete resonances (at least, it has the possibility to be so). However, the LDA greatly simplifies the numerical calculations (because of the absence of discrete states and the short range of the potential), and one obtains a nice overview of the average oscillator strength distribution. The LDA basis might be particularly useful for describing multi-photon ionization of closed-shell negative ions, which have very few bound states. We would also like to suggest that absence of discrete resonances in the LDA basis might provide useful average results also in cases of fairly high intensity with large dynamic Stark broadening.

Finally, in the present formulation[1-3] we neglect the residual Coulomb interaction between individual electron-hole pairs, in which case multiple-excitation energies equal the sum of single-excitation energies. This is a good approximation for a many-electron system with extended states, e.g. a metal. For an atomic system with localized states the Coulomb interaction is much more important. However, in the atomic systems of interest here, the continuum transition strength is dominated by collective (5p→d) excitations describing localized perturbations (small displacements) of the 5p-shell. These collective excitations are only weakly influenced by electrostatic monopole forces, and we feel that our "independent-boson" model could represent a fundamental starting point for describing the gross features of multi-photon ionization processes up to fairly high intensities of the electromagnetic field.

## References


1. G. Wendin, L. Jönsson, and A. L'Huillier: Phys. Rev. Lett 56, 1241 (1986)
2. A. L'Huillier, L. Jönsson, and G. Wendin: Phys. Rev. A 33, 3938 (1986)
3. A. L'Huillier, and G. Wendin: submitted to J. Phys. B
4. A. L'Huillier, L.A. Lompré, G. Mainfray, and C. Manus: Phys. Rev. A 27, 2503 (1983)
5. P. Kruit, J. Kimman, H.G. Muller, and M. van der Wiel: Phys. Rev. A 28, 248 (1983)
6. K. Boyer, H. Egger, T.S. Luk, H. Pummer, and C.K. Rhodes: Phys. Rev. A 32, 214 (1984)
7. C.K. Rhodes: Science 229, 1345 (1985)
8. S.L. Chin, F. Yergeau, and P. Lavigne: J. Phys. B 18, L213 (1985)
9. K. Boyer, and C.K. Rhodes: Phys. Rev. Lett. 54, 1490 (1985)
10. A. Szöke, and C.K. Rhodes: Phys. Rev. Lett. 56, 720 (1986)
11. A.L. Robinson: Science 232, 1193 (1986)
12. G. Wendin: In New Trends in Atomic Physics, Les Houches Summer School, Session XXXVIII, 1982, eds. G. Grynberg, and R. Stora, (Elsevier Science Publishers B.V., 1984) p. 555

13. G. Wendin: Phys. Rev. Lett. 53, 725 (1984)
14. G. Wendin: In Giant Resonances in Atoms, Molecules and Solids, NATO Summer School, Les Houches, 16-26 June, 1986, eds. J.P. Connerade, J.-M. Esteva, and R.C. Karnatak, (Plenum), in press
15. M.S. Pindzola, and H.P. Kelly: Phys. Rev. A 11, 1543 (1975)
16. R. Moccia, N.K. Rahman, and A. Rizzo: J. Phys. B 16, 2734 (1983)
17. A.F. Starace, Proc. 8th Int. Conf. Vacuum Ultraviolet Radiation Physics, 4-8 August, 1986, Lund Sweden; to appear in Phys. Scripta (Sweden)
18. E.J. McGuire: Phys. Rev. A 24, 834 (1981)
19. P. Gangopadhyay, X. Tang, P. Lambropoulos, and R. Shakeshaft: Phys. Rev. A, in press
20. The present independent-electron approximation includes all ladder diagrams and exchange self-interaction. In this basis, the RPA bubble expansion gives accurate RPAE results. The HF configuration-average one-electron result for the 2-photon 1-electron cross-section lies somewhat lower than the result of the present basis set applied to Fig. 2a and Eq. (7), and agrees quite well with McGuire[18].
21. A.W. McCown, M.N. Ediger, and J.G. Eden: Phys. Rev. A 26, 3318 (1982)

# Permanent Quantum Beats from Coupling of Two Atomic Bound States to a Single Autoionizing State

*A. Lami*[1] *and N.K. Rahman*[2]

[1]Istituto di Chimica Quantistica ed Energetica Molecolare del C.N.R., I-56100 Pisa, Italy
[2]Dipartimento di Chimica e Chimica Industriale, Università di Pisa, Via Risorgimento 35, I-56100 Pisa, Italy

The theory of a proposed experiment in which an autoionizing state is coupled by external e.m. fields to two atomic bound states is considered. The crucial process of interest is setting up of permanent oscillations by varying the external parameters such as the two detunings. Some experimental aspects are also discussed.

## 1. Introduction

The original motivation for our theoretical study of double resonance schemes involving a structured continuum has been the following. Suppose one wants to investigate strong field effects on the ionization close to an autoionizing resonance (or on the dissociation close to a predissociating state). It is well known that at sufficiently high field strength, one should in principle observe various unusual effects, like the non monotonic behaviour of the ionization yield as a function of pulse length /1/ or the sharpening of the photoelectron spectrum at a critical detuning, which has been called "confluence" /2/. The experimental observation of such effects is, however, extremely difficult. For example, in order to observe the non monotonic ionization, one should have a field strong enough to make the ionization rate of the same order as the inverse lifetime of the resonance, and one should be able to vary the pulse length. This means working, for a typical resonance ($\tau \sim 10^{-13}$ sec.) in the subpicosecond regime and with a laser power of $10^{13}$ W/cm$^2$ (one-photon ionization). On the other hand the observation of the photoelectron spectrum is practically impossible with the electron resolution available. It is however possible to obtain some interesting effects without going to such high field strengths. The need for limiting field strength arises since, as should be evident from perusal of the articles in this book concerning Above Threshold Ionization, when the field strength is high, the continuum-continuum transitions become increasingly important.

Our suggestion was then /3/ to investigate the above effects by a double resonance experiment, with a strong field to couple the autoionizing resonance to a bound excited state and a weak field to probe the modification of the continuum structure from the ground state (see Fig.1). We followed an effective hamiltonian approach, and generalized in a transparent way the well-known results obtained for three-level system in a $\Lambda$ configuration. In doing this we became aware that the inclusion of a second field in the problem opened new interesting possibilities, like the production and detection of undamped quantum beats in the ionization yield as a function of the pulse length, which has been interpreted by us as due to the creation of a couple of photon-dressed discrete states embedded in the photon--dressed continuum /4/. In the following we give a simple presentation of this theory and discuss some numerical calculations with reasonable values of the parameters involved.

## 2. Theory

The theory has been described in a more formal way in /5/, where we simply illustrate the theoretical approach with the scheme in Fig.1.

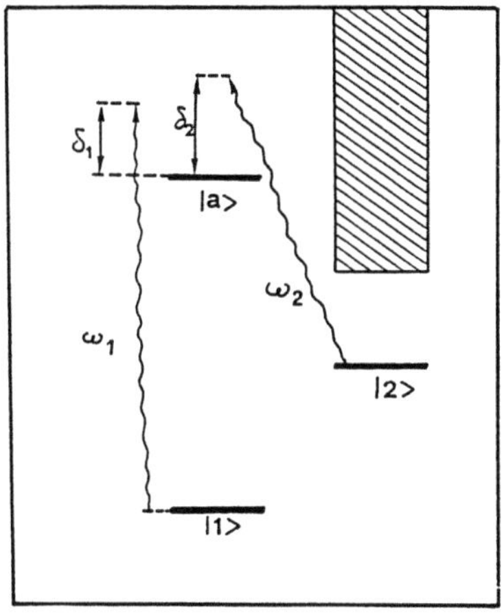


Fig.1

The states $|1\rangle$ and $|2\rangle$ are two bound states, whereas $|a\rangle$ interacting with the degenerate continuum $|E\rangle$, gives rise to the resonance. We use second-quantized fields and adopt the resonant approximation. This means that the transition from $|1\rangle$ to $|a\rangle$ (or to the continuum) is allowed only if a photon $\omega_1$ is absorbed (emitted in the reverse process). The same holds for the transition $|2\rangle \to |a\rangle$ (or $|2\rangle \to |E\rangle$) and its reverse. The problem consists in solving the time-dependent Schrödinger equation involving the following photon $\otimes$ atom states: $|1;\ \omega_1,\omega_2\rangle$; $|a;\ \omega_2\rangle$; $\{|E;\ \omega_2\rangle\}$; $|2;\ 2\omega_2\rangle$. After projecting out the continuum $\{|E;\ \omega_2\rangle\}$, we reduce the problem to a three-level one, in which however the effective hamiltonian matrix is non-hermitian, but complex symmetric. The whole procedure is well justified if the bound-free matrix elements involved (i.e. the transition dipoles $d_{1E}$ and $d_{2E}$ and the internal coupling $V_{aE}$) are slowly varying functions of E. This is why we have explicitly considered $|a\rangle$ without performing the Fano diagonalization.

Hence we have the following effective hamiltonian (here we are supposing that all that goes from $|1\rangle$ to the continuum can be recombined in $|2\rangle$ and viceversa /3,4/ and that free-free transitions can be neglected).

$$H^{eff} = \begin{Bmatrix} -i\gamma_1 + \delta_1 & \pm(\gamma_1\gamma_a)^{\frac{1}{2}}(q_1+i) & \pm(\gamma_1\gamma_2)^{\frac{1}{2}}(q_{12}+i) \\ & -i\gamma_a & \pm(\gamma_2\gamma_a)^{\frac{1}{2}}(q_2+i) \\ & \text{Symmetric} & -i\gamma_2 + \delta_2 \end{Bmatrix} \qquad (1)$$

Notice that the sign of the extradiagonal elements depends on the sign of matrix elements connecting bound states and continuum, but the results that follow are independent of them /5/. The meaning of symbols in (1) is the following. $\gamma_1$ and $\gamma_2$ are the photoionization rates of state $|1\rangle$ and $|2\rangle$, due respectively to field $\omega_1$ and $\omega_2$; $\gamma_a$ is the natural width of the resonance; $q_1$ and $q_2$ are the Fano parameters for the autoionizing resonance as is seen from $|1\rangle$ and $|2\rangle$, respectively; $q_{12}$ is a generalized Fano parameter which gives the ratio of the real to the imaginary part (with the minus sign) of the amplitude for the stimulated continuum Raman scattering

$$|1;\ \omega_1,\omega_2\rangle \xrightarrow{\text{via continuum}} |2;\ 2\omega_2\rangle .$$

Furthermore: $\delta_j = \tilde{\varepsilon}_j + \omega_j - \varepsilon_a$ (j = 1,2), where $\tilde{\varepsilon}_j$ is the (Stark shifted) energy of $|j\rangle$.

In terms of the three complex eigenvalues of $H^{eff}$, ($E_1$, $E_2$, $E_3$), one can easily write the amplitudes for staying in the initial state, i.e. $U_{11}(t)$, and for making a transition to state $|2\rangle$, i.e. $U_{21}(t)$, after a pulse of length t (here $\hbar = 1$):

$$U_{11}(t) = \sum_{\mu=1}^{3} \frac{A_{11}(E_\mu)}{B_\mu} \exp(-iE_\mu t) \tag{2}$$

$$U_{21}(t) = \sum_{\mu=1}^{3} \frac{A_{21}(E_\mu)}{B_\mu} \exp(-iE_\mu t) \tag{3}$$

where $A_{j1}$ ($E_\mu$) is the cofactor of $(E_\mu - H^{eff})^{-1}_{j1}$ and

$$B_\mu = \prod_{\nu\neq\mu} (E_\mu - E_\nu) \tag{4}$$

The populations of $|1\rangle$ and $|2\rangle$ at the end of the pulse are:

$$P_1(t) = |U_{11}(t)|^2 \tag{5}$$

$$P_2(t) = |U_{21}(t)|^2 \tag{6}$$

The total ionization (supposing that one collects also autoionized electrons) is

$$P_C(t) = 1 - P_1(t) - P_2(t) \tag{7}$$

It is clear from Eqs.(2)-(7) that depending on the nature of the $E_\mu$ one has a different behaviour of the populations as a function of t. For example, let us consider the ionization (i.e. the population of the continuum). If one of the $E_\mu$'s, say $E_1$, is real then

$$P_C(t) \xrightarrow{t\to\infty} 1 - \left|\frac{A_{11}(E_1)}{B_1}\right|^2 - \left|\frac{A_{21}(E_1)}{B_1}\right|^2 \tag{8}$$

since the contributions from $E_2$ and $E_3$ go to zero, due to the negative imaginary part. The above result means that in this case the saturation is never reached, no matter how long we make the pulse /6/.

If two of the eigenvalues, say $E_1$ and $E_2$, are real, then the asymptotic behaviour of $P_C(t)$ (as well as of $P_1(t)$ and $P_2(t)$) is oscillatory with frequency $\Omega = |E_1-E_2|$. This permanent oscillation in the ionization is a new phenomenon firstly noted by us /4/ and attributed to the creation of a pair of photon-dressed discrete states embedded in a (photon-dressed) continuum. In /5/ we have extended this result to the case of N bound state and N frequencies, showing that N discrete states can be created, whereas in /7/ the photoelectron spectrum under such conditions has been derived.

The interesting thing is that there is the experimental possibility of varying the $E_\mu$'s. In fact they are functions of the control parameters, i.e. the intensities of the two fields (which are contained in $\gamma_1$ and $\gamma_2$) and the two frequencies (contained in $\delta_1$ and $\delta_2$). A convenient way of representing what is happening consists in

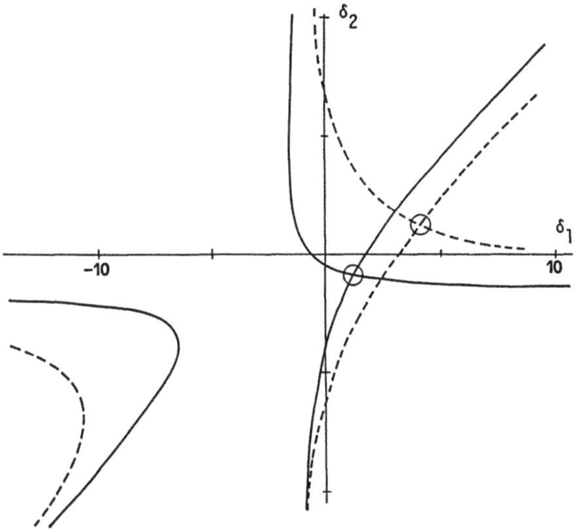


Fig.2

fixing the intensities and varying $\delta_1$ and $\delta_2$. One can search in the $\delta_1$, $\delta_2$ plane the geometrical locus of points where at least one of the $E_\mu$'s is real. This can be achieved by carefully studying the cubic secular equation of $H^{eff}$, by analytical-numerical methods. Typical results are shown in Fig.(2) for $q_1= 2$; $q_2=4$; $\gamma_1/\gamma_a=.4$; $\gamma_2/\gamma_a=.6$ and a) $q_{12}=-5$; b) $q_{12}=0$.

These curves are characteristic in the sense that for various sets of parameters, we have always found three branched curves. Two of the three always intersect and at the point of intersection ($\delta_1^c$, $\delta_2^c$), two of the eigenvalues of the effective Hamiltonian become real. Even though we have no explicit expressions for these curves, $\delta_1^c$ and $\delta_2^c$ can be written down as analytic expressions. They are

$$\delta_1^c = q_1(\gamma_a - \gamma_1) - \gamma_2(q_2 - q_{12}) \tag{9}$$

$$\delta_2^c = q_2(\gamma_a - \gamma_2) - \gamma_1(q_1 - q_{12}) \tag{10}$$

This means that if we can adjust the detunings to satisfy Eq.(9) and Eq.(10), the time evolution of the probability amplitudes of staying in $|1\rangle$ and $|2\rangle$ will be determined by two non-decaying terms (plus a decaying one). The same behaviour may be detected for other observables, which in presence of just one field exhibits damped oscillations, due to ionization, but with two fields may oscillate indefinitely (if conditions in Eq.(9) and Eq.(10) are met). Two of the measurable quantities $P_2$ and $P_C$ are plotted in Fig.3a and Fig.3b as a function of time for the two sets of parameters (Fig.2) at these critical detunings. After some initial irregular behaviour (transients), one observes the locking of these populations at constant frequencies.

## 3. Adiabatic Invariant

We shall now consider an additional aspect of these oscillations which is pertinent for experimental revelation of the same. From what has been discussed and illustrated above, one notes that after the transients have passed, what remain are stable oscillations of the populations with a given frequency when certain conditions are

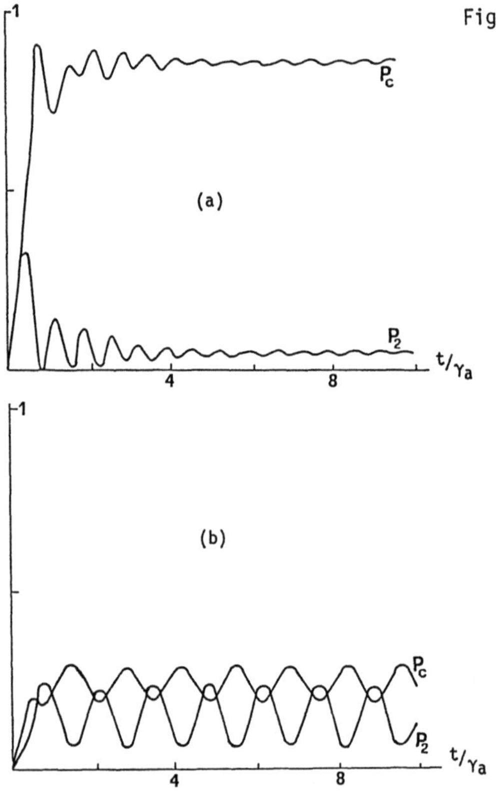


realized. Thus, at long times an "oscillator" has been created with the oscillator frequency given by $\Omega = |E_1 - E_2|$ ($E_1$ and $E_2$ are the real roots). This oscillator can be described by a Hamiltonian

$$H = 1/2\ P^2 + 1/2\ \Omega^2\ Q^2 \qquad (11)$$

where Q refers now to the oscillatory part of one of the populations that we wish to consider. Since the frequency $\Omega$ is the same for all the populations as well as the ionization, the treatment goes through for all the other measurable quantities. It is easy to derive an explicit expression for Q which we do not give here.

The usefulness of introducing such a hamiltonian given by Eq.(11) is the following. In the real world, some of the external parameters can hardly be kept constant and must be treated as slow variables. Introducing the oscillator hamiltonian, we can take into account these slow variables as adiabatic variables. In that case, there exists an adiabatic invariant given by $A = \int PdQ$ over a period of oscillation. This is given by $A = \Omega P_o^2$ where $P_o$ is the amplitude of oscillation in e.g. $P_1$. Experimentally, existence of an adiabatic invariant has the following significance. Consider a reasonable long pulse with which the experiment is feasible. This means that the oscillations will have to be revealed while the pulse is operative on the atomic system. Firstly, there is the rise time till the amplitude goes towards its average value. During this period, the conditions of the oscillation are not met, and therefore we do not expect to see any trace of such. Now, when the field is around the average value, the amplitude of the field cannot be treated to be an absolute constant, but as slowly varying with time. Thus, even though the condition for oscillation is not met exactly, due to the time dependence, the adiabatic invariant A remains a constant and can be checked experimentally.

## 4. Discussion

It may be noted that while we have worked out in detail the principle features of some novel aspects of a double resonance experiment through the continuum containing autoionizing states, no calculations have yet been reported with a specific atom. This is what should be attempted next, for two main reasons. An experiment proposed along the lines discussed here, needs as input the atomic parameters, the q's and the autoionization width, as well as the external field parameters commensurate with these atomic parameters. Now, in a design of an experiment, the atomic parameters must be known with a certain degree of confidence. This may be obtained with some precision from independent experiments or computations. The only exception being $q_{12}$. This generalized and crossed Fano parameter can be obtained only by theoretical calculation, if an experiment has to be designed with known atomic parameters. Otherwise, the double resonance type of experiment is the only method for obtaining it. This means straightforward double resonance experiments, not necessarily in search of oscillations as has been treated here, but those that have been discussed previously in the literature /6/. Measurements from such experiments will yield $q_{12}$ which can then be utilized to design that particular experiment which will exhibit permanent oscillation in the continuum.

Finally, with the technique of the effective hamiltonian, it is possible to generalize our work with N levels, all of which are coupled to an autoionizing state /5/. While it may seem somewhat premature to study such systems in view of possible experimental realization, one obtains a certain number of explicit analytic results. One of these is the explicit condition under which N distinct real eigenvalues are obtained. Needless to say, permanent oscillations originating from more than two eigenvalues will have their own distinctive structure.

## 5. References

1. P. Lampropoulos and P. Zoller: Phys. Rev. A24, 379 (1981)
2. K. Rzazewski and J.H. Eberly: Phys. Rev. Lett. 47, 408 (1981); Phys. Rev.A 27, 2026 (1983)
3. A. Lami and N.K. Rahman: Phys. Rev. A26, 3360 (1982); Opt. Commun. 43, 383 (1982); in Collisions and Half-Collisions with Lasers, ed. by N.K. Rahman and C. Guidotti (Harwood Academic, New York 1984) p.373
4. A. Lami and N.K. Rahman: In Fundamentals of Laser Interactions, ed. by F. Ehlotzky (Springer, Heidelberg 1985) p.299; Phys. Rev. A 33, 782 (1986)
5. A. Lami and N.K. Rahman: Phys. Rev. A 34, XXX (in press)
6. P. Knight: Comments At. Mol. Phys. 15, 193 (1984)
7. E. Kyrölä: J. Phys. B: At. Mol. Phys. 19, 1437 (1986)

# Electric Field Effects on Structure in the Continuum

*D.E. Kelleher, E.B. Saloman, and J.W. Cooper*

National Bureau of Standards, Gaithersburg, MD 20899, USA

We have observed electric fields to have pronounced effects on autoionizing resonances in the continuum region above the first ionization threshold. The types of phenomena come under three categories: I) broadening (and narrowing); II) interference; III) "procreative" Rydberg Stark manifolds.

I) We have reported the observation [1] of a Sr resonance to broaden approximately quadratically in an electric field due to its field-mixing with a nearby much broader resonance of opposite parity. We have also observed [2] the width of a narrow high J resonance in Gd to increase stepwise quite suddenly at a field of less than 2 V/cm. We attribute this to the level's lying near a series of Rydberg levels with n ~ 60. The sudden increase in width is observed at the Ingliss Teller limit for this n, where Stark manifolds of different n levels overlap.

II) In an electric field, a relatively narrow resonance that is nearly degenerate with a much broader resonance of opposite parity has been observed to cause an interference dip in the broad level that deepens as the field is increased, until at higher fields the broad level is split into two components [3]. Also, narrow resonances in the wings of broad resonances lead to field-induced Fano-Beutler type inference patterns whose amplitude increases as the field is increased. Results thus far are in good agreement with a non-perturbative theory developed recently [3].

III) Rydberg levels with $n \gtrsim 5$ exhibit quasi hydrogenic Stark manifolds in which the energies of the nearly degenerate higher-$\ell$ levels with small quantum defect "fan out" linearly in an electric field. These cases [4] correspond to a spherically symmetric core. The manifolds are strikingly different when the ionic core (seen by the Rydberg electron) has angular momentum, which is the case for essentially all autoionizing levels and for most bound levels except those in the first two rows of the atomic table. For example, as the field is increased, Stark-induced levels appear to split into more levels; hence the term "procreative" Stark manifold. We have performed detailed calculations (in jK coupling) of this phenomenon and compare them with our observations of Stark manifolds in barium autoionizing Rydberg levels [5].

A more detailed discussion of these phenomena is presented below. In all cases the observable is photo-ion yield as a function of laser tuning through the indicated final state energies.

Ia. We have observed a relatively narrow autoionizing (AI) resonance to broaden monotonically as an applied dc electric field is increased. We have compared our results to a simple model which predicts that the width should increase quadratically with the field. At sufficiently high fields this effect should be able to broaden a narrow resonance up to widths of several hundred inverse centimeters.

The observed state is broadened when the electric field mixes (dipole couples) it with a broader state of opposite parity. Figure 1 illustrates the case for our experiment. A very broad odd-parity $^1P_1^0$ strontium AI resonance rises above the first ionization threshold. The zero-field width of the nearby $^1D_2$ level is 350 times narrower than the broad $^1P_1^0$ resonance. As an external electric field is applied, the narrow level mixes with the broad one (eigenstates in the field will have mixed

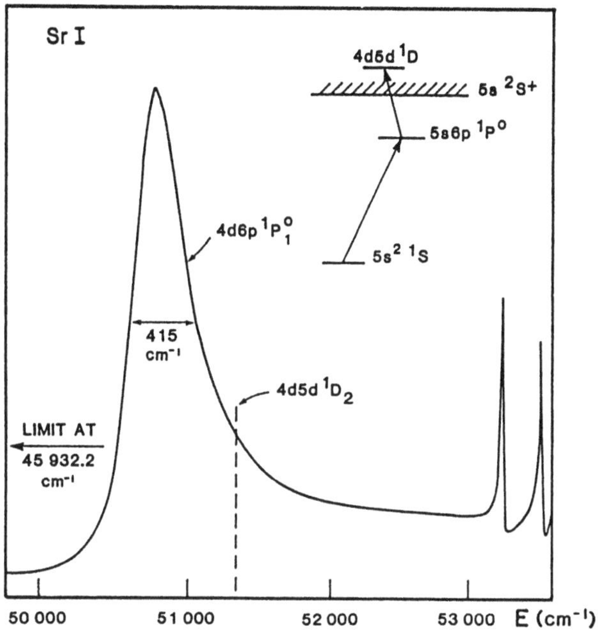


Fig. 1. Spectrum above the ionization threshold of neutral strontium. The solid line indicates the odd-parity spectrum. The dashed line indicates the position of the narrow even-parity 4d5d $^1D_2$ resonance. The inset depicts the laser excitation scheme

parity), thus sharing its AI decay channels with the broader level and broadening accordingly.

A simple perturbative model for this phenomenon, presented by JACOBS and DAVIS [6], states that the field-induced width in an electric field F of a state "i" which is AI metastable in zero field is

$$\Gamma_i = \sum_j \frac{|p_{ij}|^2 F^2}{|E_i - E_j|^2 + \frac{1}{4}\Gamma_j^2} \Gamma_j \quad , \tag{1}$$

plus interference terms, where the $\Gamma_j$ are the zero-field AI rates (widths) of the j levels which are coupled to the opposite-parity i level by the field. The dipole coupling is given by $p_{ij} = e\langle i|r \cos\theta|j\rangle$. We show in [1] that if the narrow level i has a non-negligible width $\Gamma_0$ in zero field, then one can replace $\Gamma_i \rightarrow \Delta\Gamma_i = \Gamma_i - \Gamma_0$ in the above expression.

This is analogous to the line broadening by field mixing of bound levels. For example, in hydrogen the bound metastable 2s level field mixes with the relatively broad 2p levels [7,8]. As the electric field is increased, the 2s will increasingly mix with the 2p levels; its radiative decay rate, and thus its width, will increase quadratically with the field. Indeed, (1) is equivalent to (76) of LAMB and RETHERFORD [8] for the field quenching of the 2s by field mixing with the 2p. The mixing effect, however, will begin to saturate when the field is high enough that the width of the narrow level becomes comparable to that of the broad one (i.e., when complete mixing is approached). Then a nonperturbative theory must be used.

A validity criterion for using perturbation theory for the effect of discrete-discrete field mixing on the narrow level can be shown to be roughly given by $2pF < \Delta E$ where $\Delta E$ is the energy separation between the levels. If this condition is satisfied, the results predicted by the simple perturbative model for both the shift and the broadening of the narrow level are reasonably close to the "exact" theory's predictions, even when the energy difference is small compared to the width of the broad level. We emphasize that this is not true at all for the broad level,

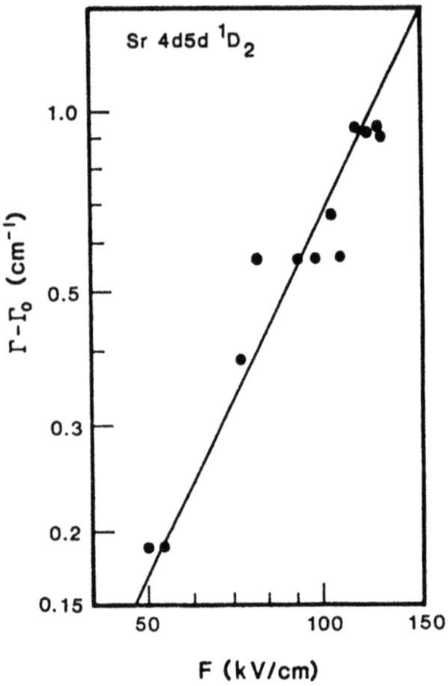


Fig. 2. Field-induced width (with zero-field subtracted) of the 4d5d $^1D_2$ line of Sr. The solid line represents the best-fit quadratic field dependence

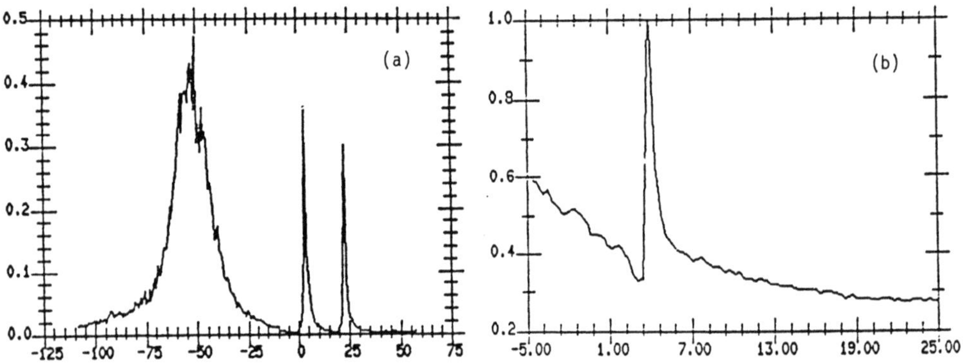


Fig. 3. (a) Field mixing of the Ba 5d5f $^1P_1$ resonance with 5d5g resonances at a field strength of 36 kV/cm. Profile of the lower Ba 5d5g resonance at a field strength of 10 kV/cm. The energy scale is cm$^{-1}$ minus 43300

for which interference effects predominate and the perturbative model is not even qualitatively correct [3], as is also the case with the narrow level outside the above validity criterion. In the present experiment $2pF/\Delta E = 0.08$, so the above condition is satisfied.

We observed the width of the Sr 4d5d $^1D$ level to increase monotonically with the electric field. After subtracting the measured zero-field width, $\Gamma_0 = 1.2$ cm$^{-1}$, a least-squares fit yielded $(\Gamma-\Gamma_0) = 6.5 \times 10^{-10}$ $F^m$, where F is the electric field in V/cm and m = 1.8(4). (Quoted uncertainties correspond to two standard deviations.) Due to technical limitations, shift measurements were not made. The best-fit quadratic field dependence (m=2) is indicated in Fig. 2. Comparing our data with (1), we obtained p = 5.8(9) $ea_0$ for the dipole matrix element between the two AI resonances in Fig. 1.

The impact of a sharp resonance in the wing of a broad, opposite parity resonance is seen in Figs. 3a and 3b. The sharp "satellites" on the high energy side of the broad resonance in Fig. 3a are field induced. Figure 3b shows a detail of one of the satellites, indicating a Fano-type interference pattern. The theoretical treatment in Ref. 3 is in excellent agreement with the observed one.

Ib. Several years ago [9], a narrow even parity autoionizing state was observed in gadolinium lying approximately 230 $cm^{-1}$ above the first ionization limit. Two interesting features of this state were noted, namely its narrow width (0.075 $cm^{-1}$, corresponding to a decay time of $4.4 \times 10^{-10}$ sec) and the fact that the observed line broadened to ~0.35 $cm^{-1}$ when a field of 100 V/cm was applied to the interaction region.

We have observed the width of a narrow Gd AI level to remain roughly constant up to a field strength of about 2 V/cm, and then increase rapidly to a new plateau as shown in Fig. 4. The field-enhanced width is about four times larger than the zero-field width. The effects we observe are at a much lower field strength than that required for classical field ionization ("forced" autoionization). An exploration of the opposite parity in the same spectral range was made in order to determine whether there were odd parity states in the same spectral range which might account for the weak field broadening. Rydberg series converging to the first excited state of the ion were found and their field dependence was also studied.

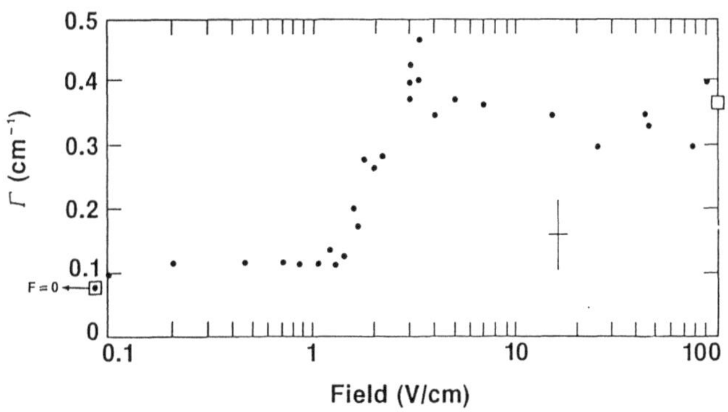


Fig. 4. Field dependence of the Gd J=5 even parity level at 49836.0 $cm^{-1}$. The sudden increase in width occurs in the region where the Stark manifolds of adjacent underlying Rydberg levels overlap

In zero field, the observed even parity resonance is narrow because of its high angular momentum J. In the absence of a field, the configuration mixing with the nearby Rydberg levels is very small, and it can mix only with levels of the same J, which are also very narrow. Autoionization, which is due to configuration interaction in the atomic core, scales roughly as $\ell^{-5}$ (the centrifugal barrier hinders the high $\ell$ level from penetrating the core). An electric field F mixes the $\ell$-levels of a high n Rydberg level at very low fields. At F(a.u.) $\cong 1/3n^5$, the Ingliss-Teller limit, adjacent n levels begin to overlap, and even the lowest $\ell$-levels, which in general have the largest quantum defects and the largest autoionization widths, become mixed with the other $\ell$ levels. Thus for this field region and higher, they form a virtual continuum of overlapping mixed-$\ell$ states. Their widths are intermediate between the broader low-$\ell$ states and the narrow high-$\ell$ states.

In the above sense, the Ingliss Teller limit corresponds not only to the onset of the n-mixing region, but to the full $\ell$-mixing region as well, at least for high-n Rydberg levels. Due to this strong $\ell$-mixing (and thus J-mixing) in fields beyond the Ingliss-Teller limit, virtually any of these Rydberg states can configuration mix with "isolated" states via the $r_{12}^{-1}$ interaction including levels of high J. If the Rydberg levels are broader than the isolated level, this field-induced mixing will result in the width of the isolated level increasing to a value comparable to that of the mixed Rydberg levels.

While the above model is a qualitative one, it does lead to two quantitative predictions consistent with our observations. The isolated J = 5 level at 49836 $cm^{-1}$ broadens in the region 1.5 to 3.0 V/cm; the Ingliss-Teller limit for n = 63 (see Fig. 2) is 1.7 V/cm. The field-induced width of the level is about 0.35 $cm^{-1}$, comparable to widths of the Rydberg/perturber levels.

It appears that the width of narrow, high-angular-momentum, autoionizing levels can be significantly altered at electric fields near to or greater than F(a.u.) = $1/3n^5$ (the Ingliss-Teller limit), where n is the principal quantum number of nearby Rydberg levels. The presence of such Rydberg levels is a common situation for autoionizing levels of complex atoms like Gd, due to the high density of low-lying excited states of the ion.

II. We have studied the various stages of the effect of field-induced coupling of two levels of quite different widths — from the embryonic stages of interference to the ultimate complete splitting into two separate components [3]. The zero-field spectrum is shown in Fig. 5. Analysis of the profile of the 5d9p $^3P_1^0$ resonance at 43027.4 $cm^{-1}$ yielded a width of 23±2 $cm^{-1}$ in agreement with HUDSON, CARTER, and YOUNG [10], and a Fano q parameter of 4.5±1.0. The zero-field even spectrum, obtained with use of two-step pumping with π polarization of both lasers, is also shown in Fig. 5. We measured the width of the narrow 5d8d J=2 resonance [11] at 43030.61 $cm^{-1}$ to be 1.5±0.1 $cm^{-1}$.

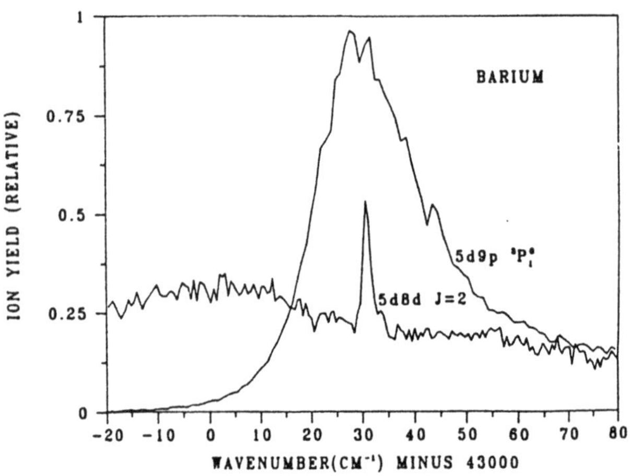


Fig. 5. Zero-field spectrum of both odd and even parities with π polarization of the lasers

In Fig. 6 we see that in the presence of an electric field the sharp even level causes an interference dip in the spectral profile of the broad odd level. The dip grows deeper and broader as the field is increased, and eventually the broad level splits into two components. Before the components are separated, the width of the interference dip is approximately equal to the width of the field-broadened even level causing the dip. The field couples the two opposite parity levels, leading to the interference pattern in the broad level when a sharp level is nearly degenerate with it. At large fields $F \gtrsim \Gamma_>/d$, the eigenstates are split ino separate eigenstates which are ± admixtures of the zero field states which repel one another as the field increases. A quantitative treatment of this is given in Ref. [3].

The observed effect of field mixing on the sharp even-parity level (not shown) is completely different. It broadens with increasing field, until the broadened level becomes indistinguishable from the background at field strengths around 15 kV/cm.

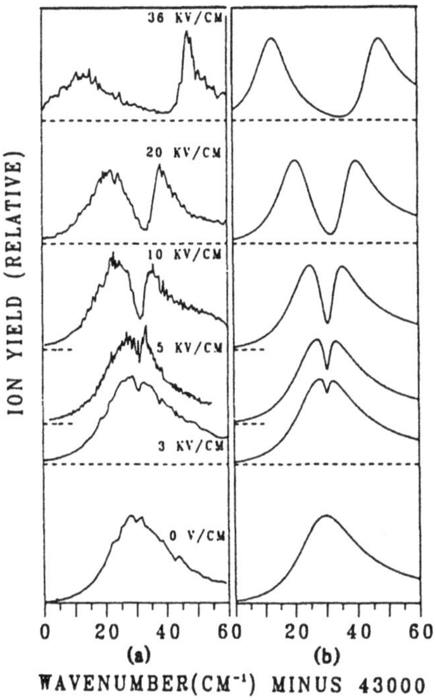


Fig. 6. Stark spectra of the barium 5d9p $^3P_1^0$ resonance for six different values of the electric field (kV/cm). (a) Observed spectra. (b) Computed spectra. Dashed lines indicate the zero level for each spectrum, spaced according to $F^{1/2}$

The theory predicts that the width of the sharp even level increases quadratically in the field (as predicted by perturbative models [6]), up to a field of 15-20 kV/cm, at which point a "forbidden" component becomes significant. At still higher fields this component is predicted to enlarge and split from the "allowed" component, forming two separate components similar to that observed in the odd-parity spectra. However, this latter phenomenon was not observable because the broadened line became indistinguishable from the background.

III. We have measured the Stark spectra of $5d_{3/2}$ $8\ell$ autoionizing states in barium. Due to the anisotropic 5d ion core, the Stark manifolds at higher fields are considerably more complex than for the analogous 6s $n\ell$ bound states. The electrostatic coupling of the Rydberg electron with the anisotropic core gives rise to relatively large fine structure splittings. For nonpenetrating orbitals ($\ell_{Ryd} > \ell_{core}$), jK coupling is a useful representation and the number of fine structure components due to this interaction with the core increases the number of eigenstates ($2\ j_{core} + 1$) fold. Results of our theoretical model for barium are in quantitative agreement with the experimental observations for both $\pi$ and $\sigma$ polarizations up to fields where there is extensive overlap between adjacent-n manifolds. These fine structure effects occur for any states with non-isotropic cores ($j_{core} > 1/2$), i.e., the states of most atoms except for bound singly excited states of alkali and alkaline earths.

The fine structure has two principal effects on the Stark manifold: 1) It breaks the quasidegeneracy of the higher $\ell$ levels at low fields. The field dependence of the manifold does not become linear until a field is reached where the Stark shifts exceed the fine structure splittings. 2) The most dramatic effect of the fine structure is that it gives rise to many more components in the Stark manifold. These come from the different possible couplings of the angular momenta associated with the electrostatic interaction between the Rydberg electron and the anisotropic core.

From a given lower level, most of the fine structure levels are observable only at higher fields. Thus, as the field is increased, more and more components appear.

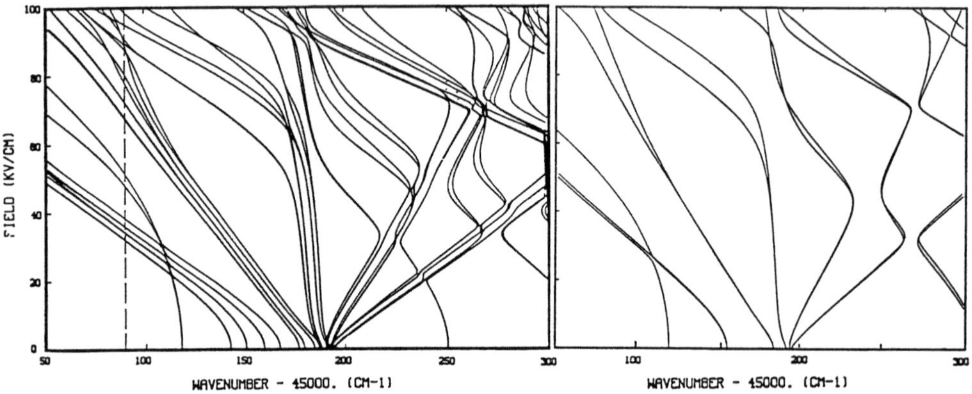


Fig. 7a. (left) Theoretical M = 0 Stark map in the region of the $5d_{3/2}$ $8\ell$ autoionizing levels. Dashed line indicates the range of experimental observation. Fig. 7b. (right) Theoretical M = 0 Stark map in the region of the $6s_{1/2}$ $8\ell$ autoionizing levels of BaI. This spectrum, which lacks the fine structure due to an anisotropic core, should be contrasted with the much more complex spectrum in Fig. 7(b). For purposes of comparison, the energies are fictitiously scaled up to the values they would have if the $6s_{1/2}$ limit was equal to the $5d_{3/2}$ limit

Two or more components appear to grow or split off from a single component at lower field giving rise to "procreative" Stark manifolds. The effect of this can be seen by comparing Fig. 7a [n = 8 Rydberg electron; $j_c$ = 3/2 (anisotropic core)] with Fig. 7b [n = 8 Rydberg electron; $j_c$ = 1/2 (isotropic core)]. A detailed theoretical model for the effects of core anisotropy on Stark spectra is presented in Ref. [5]. Good agreement is obtained with the experimental spectra shown in Fig. 8.

In summary, we have observed a number of different electric field effects on autoionizing resonances in the continuum. Broadening quadratic in the field was observed for a narrow line in the wing of a much broader line of opposite parity. "Sudden" broadening of a narrow high angular momentum resonance was observed near the Ingliss-Teller limit for Rydberg levels lying near the resonance. In the spectrum of broad resonances, narrow lines (of opposite parity) in their wings give rise to satellites which have a Fano-type interference shape; narrow lines near the center of a broad resonance cause interference dips which at sufficiently large fields

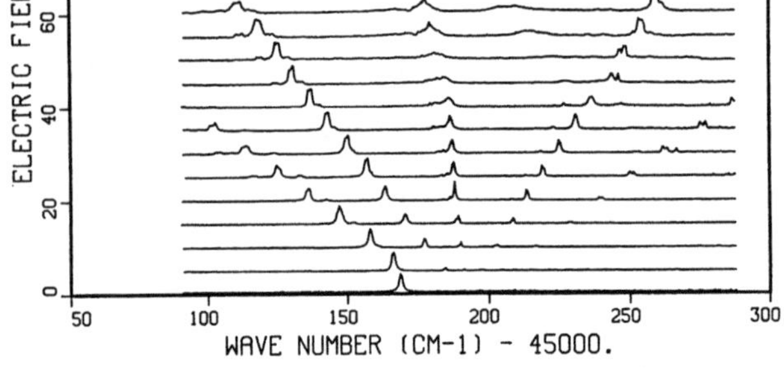


Fig. 8. Experimental photo-ion spectra from the $5d^2$ $^1S_0$ state of BaI at different electric fields, with $\pi$ polarization (laser field parallel to applied dc field), in the region of the $5d_{3/2}$ $8\ell$ autoionizing levels

split the resonance into ± admixtures of the zero-field states. Finally, we have observed that for Rydberg autoionizing resonances the fine structure due to the noncentral electrostatic interaction with the anisotropic core considerably complicates the Stark spectra and gives rise to "procreative" Stark manifolds.

In this brief overview of our work, we have not had space to mention the important contributions of others in this field, including those at UVa, Stony Brook, Orsay, and New Mexico.

This work was supported in part by the AFOSR and SDIO. One of us (D.E.K.) is grateful to the Joint Institute for Laboratory Astrophysics for hospitality during his exchange fellowship and for help in preparing this manuscript.

References

1. D. E. Kelleher, J. F. Delpech and J. Weiner, Phys. Rev. A 32, 2230 (1985).
2. V. I. Mishin, G. G. Lombardi, J. W. Cooper and D. E. Kelleher, Phys. Rev. A, in press.
3. E. B. Saloman, J. W. Cooper and D. E. Kelleher, Phys. Rev. Lett. 55, 193 (1985).
4. M. L. Zimmerman, M. G. Littman, M. K. Kash and D. Kleppner, Phys. Rev. A 20, 2251 (1979).
5. D. E. Kelleher and E. B. Saloman, submitted to Phys. Rev. A.
6. J. Davis and V. L. Jacobs, Phys. Rev. A 12, 2017 (1975).
7. G. Lüders, Z. Naturforsch. 50, 608 (1950); Ann. Phys. (Leipzig) 8, 301 (1951).
8. W. E. Lamb and R. C. Retherford, Phys. Rev. 79, 549 (1950).
9. G. I. Bekov, V. S. Lotokhov, O. I. Mateev and V. I. Mishin, JETP Letters 28, 283 (1978).
10. R. D. Hudson, V. L. Carter and P. A. Young, Phys. Rev. A 2, 643 (1970).
11. P. Camus, M. Dieulin, A. El Himdy and M. Aymar, Phys. Scr. 27, 125 (1983).

# Part II

# Above-Threshold Ionization

# Above-Threshold Ionization with Linear and Circular Polarizations

*F. Yergeau, G. Petite, and P. Agostini*

Service de Physique des Atomes et des Surfaces, CEN Saclay, F-91191 Gif-sur-Yvette, France

## 1. INTRODUCTION

Ionization in the field of a strong electromagnetic wave had been studied for more than one decade when it was observed that electron spectra were not as predicted by lowest non-vanishing order perturbation theory (LOPT) [1]. Instead of one peak at $E_0 = N\hbar\omega - E_I$ (where N is the minimum number of photons $\hbar\omega$ required to overcome the ionization potential $E_I$), the spectra consist of a number of peaks separated by $\hbar\omega$, with significant amplitudes. Since the electrons are produced in a regime totally free of collisions, inverse Bremsstrahlung as a possible mechanism is ruled out. The only remaining possibility is an ensemble of direct transitions involving (N+S) photons and coupling the ground state to higher ionization continua. The computation of the corresponding generalized cross-sections within the framework of LOPT is more difficult than in the case of Multiphoton Ionization, but the problem has been solved by several authors [2],[3],[4],[5] and it can be said that up to intensities of $10^{12}$ W.cm$^{-2}$ and hydrogenic atoms, the agreement between theory and experiment is quite reasonable [5]. It may happen that, even at such intensities, higher-order corrections (AC-Stark shifts of intermediate bound states) must be taken into account [5]. However, the general features of Above-Threshold Ionization (ATI), as it was named [3] remain essentially perturbative in the sense that the spectra consist of a few peaks with rapidly decreasing amplitudes, each of them following the usual power law $I^{N+S}$ [6]. This is not surprising since, at such intensities, the perturbation of continuum states, given by the "ponderomotive potential" $e^2F^2/4m\omega^2$ (where F and $\omega$ are the amplitude and frequency of the electric field of the wave, e and m the electron charge and mass), is negligible, at least at optical frequencies (1.03 eV at 1064 nm and $10^{13}$ W.cm$^{-2}$). A totally different situation was found [7] at higher intensities which demonstrated features quite unexpected on the ground of LOPT: the spectra show a very large number of peaks which do not follow the "normal" power law and the first few peaks can be relatively suppressed. This result has attracted a lot of interest and both theories and experiments have blossomed in the last two years [8],[10],[16]. Although there is a large consensus that the most important effect of the strong field is to be found on the continuum part of the spectrum rather than on the bound states, there is still a long way to go before quantitative agreement is found between the two. The aim of this paper is to discuss a few recent observations. In Section 2 we discuss briefly the question of space-charge effects. In Section 3 we summarize results about the intensity dependence of the spectra. In Section 4 we debate about the influence of circular polarization in ATI.

## 2. SPACE-CHARGE EFFECTS IN ABOVE-THRESHOLD IONIZATION

One of the most intriguing effects observed in ATI is the suppression of low-energy peaks from the spectra. Several physical mechanisms, related to the high intensity of the electromagnetic wave, have been proposed to explain this effect. The one to be considered in this section is more trivial and well known. After ionization, there is in the interaction volume a mix of positive ions and electrons of different energies. The fast ones get away rapidly, leaving behind a net positive charge that could trap the slow electrons [9]. Space charge could also modify the electron trajectories and deviate them from the detector in the case of a small collection angle [9]. The method to get rid of these effects in the experiment is quite obvious. The parameters must be arranged in order to produce only a small number of charges for each laser shot (typically less than 100). For example for an 11-photon ionization (Xenon 1064 nm), if one uses a pressure of $5 \times 10^{-5}$ Torr, the maximum intensity which can be used is $0.5 I_s$ where $I_s$ is the saturation intensity. In fact the restriction is even more severe if one takes into account the ionization of the background vacuum. In order to comply with these conditions it is essential to use a spectrometer with a very good collection efficiency and counting techniques [10].

## 3. INTENSITY DEPENDENCE OF ATI SPECTRA IN LINEAR POLARIZATION

Fig. 1 shows typical ATI spectra in Xenon at 1064 nm, with 200 psec pulses. The main characteristics of the intensity dependence of the spectra clearly appear on this picture and are discussed in the next sub-sections.

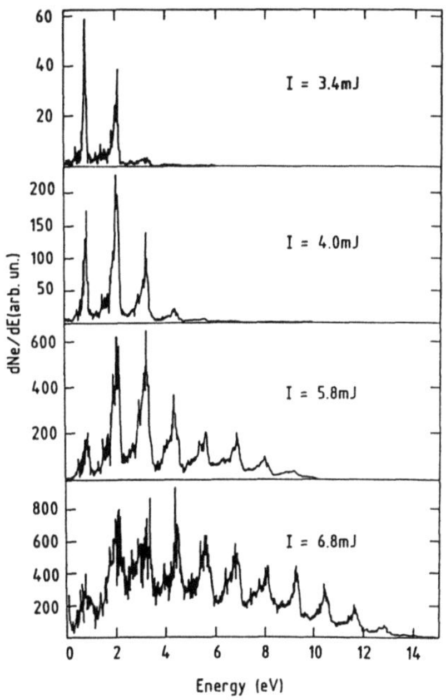


Figure 1 - Typical electron spectra of eleven-photon MPI at 1064 nm for different pulse energies (18 mJ/pulse correspond to $1.3 \times 10^{13}$ W.cm$^{-2}$). The maximum count rate in these spectra was 10 per laser shot.

### 3.1 Slopes

First, the number of ATI peaks increases rapidly with intensity: this implies that the power law differs from peak to peak with the high-order peaks growing faster than the low-order ones. This is confirmed by the dependences shown on Fig. 2 for the first three peaks. The exponents which can be determined on this figure are respectively 6.6, 8 and 9.6. These values are in good agreement with, on the one hand, previous measurements [2],[8], on the other hand, with theoretical prediction [11]. (Note that [11] is the only calculation, to our knowledge, which predicts a value of N-4 for the slope of the first peak). However, previous measurements, as well as several theories [11],[12] report that the exponents for higher peaks would be all equal to N (here 11). These have not been measured in this experiment so we cannot refute or confirm this important point, but it must be stressed that if this was true, it would imply a difficult problem: since the total ionization signal scales like $I^N$, if all the peaks behave also $I^N$ then the spectra would be invariant when intensity is increased. This question is still open and will require more precise experimental data to be resolved.

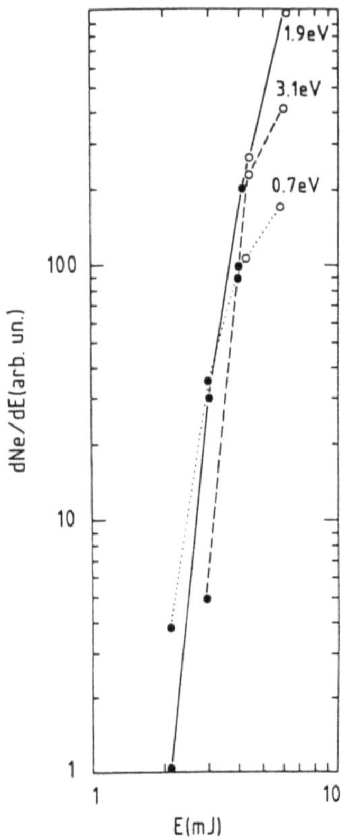


Figure 2 - Amplitudes of the first three ATI peaks plotted against pulse energy.

3.2 Peak suppression

Second, the first peak is relatively suppressed when the intensity is increased. Theoretical explanations of this fact are to be found elsewhere in this book. From the experimental viewpoint, it must be pointed out that this suppression is only relative, since, as shown on Fig. 2 the absolute value of the peaks amplitudes are increasing functions of I. This is in agreement with other findings but contradicts theories [7],[11] which predict an absolute peak suppression. However, on one hand, the explored intensity range is quite limited, due to other experimental constraints, (extension of this range can be expected in the near future), on the other hand the theories do not take into account the spatial distribution of intensity. It is therefore possible that the contradiction is less serious than it seems now.

3.3 Saturation intensity

Third, it appears clear that the peak suppression occurs at an intensity much lower than the saturation intensity, which, as a consequence, does not play any particular role in the process. Using known values of the saturation intensity [13], gives an intensity of $5x10^{12}$ $W.cm^{-2}$ in agreement with the value predicted on the basis of the ponderomotive potential ($6.6x10^{12}$ $W.cm^{2}$). However, other results [13], are not in agreement with the corresponding value (especially for Neon [14] and the question is still an open one. So far, to our knowledge, there is no theoretical framework able to reproduce all the existing experimental data.

3.4 Peak broadening

Fourth, another very clear result of this measurement is the peak broadening quite apparent on Fig. 2. Part of the width measured at the lowest intensity could be due to the fine structure splitting of the ground state of the ion. At the highest intensity the observed width is more than 300 meV. This broadening is also quite a puzzle: it is much larger than the ionization width of the ground state since, even at the saturation intensity, it would be just the inverse of the pulse duration ($2x10^{-5}$ eV !). It has been suggested that the width would be related to the saturation of continuum-continuum transitions [15], and calculations of this effect are currently in progress. Other explanations, based on the ponderomotive potential, have been proposed but the subject of the ponderomotive acceleration is still somewhat controversial [11] and this point of view will not be developed here.

Otherwise, in this experiment, as in all the work published so far, the different electron peaks have always been found centered at the expected energies. This seems to suggest that there is no shift of the ionization potential or, if there is one, that it is compensated by an equal energy variation due to the ponderomotive potential [7].

## 4. CIRCULAR POLARIZATION IN ABOVE-THRESHOLD IONIZATION

Recently [16] an interesting observation was reported about the influence of circular polarization in ATI spectra. The experimental result was as follows: in 11-photon ionization of Xenon, at saturation intensity, and above, the first 5 peaks are suppressed from the ATI spectra. The interpretation which has been proposed is the following: with circular polarization, the final continuum states angular momentum are restricted to $L > N$. To explain the consequence of this on the peak suppression, the authors calculate the transition matrix element, using legitimate approximations, and find that, for that intensity range, the matrix elements corresponding to the first few continua are much smaller than for continua with $L > 15$. They interpret this effect by the bad overlap between the ground state wavefunction and the final state wavefunction due to the centrifugal barrier term in the atomic Hamiltonian. To summarize, it is emphasised that the peak suppression is essentially an atomic effect, perhaps enhanced by the intensity. The authors also conclude that this observation supports the one-step (perturbation theory) model of ATI. We have repeated this experiment [10] at a much lower intensity. This was possible in our setup due to the high collection efficiency of the spectrometer. The result is shown on Fig. 3, where all the low-order peaks which where not seen in the previous experiment [16] are now detected. When the intensity is increased, though kept well under saturation, one peak is suppressed, as in linear polarization, but at a lower intensity value. This result does not contradict but complements the result of Bucksbaum et al. since it is taken in a different intensity range. However, it somehow shifts the emphasis on the intensity rather than on the atomic part of the matrix element as the main factor for peak suppression. The result of a numerical calculation of the generalized cross-sections, reported in Table 1, confirms this view. The calculation [17] has been performed using the implicit summation technique, and within the limitations of LOPT, is exact. The cross-sections are regularly decreasing when the order is increased. However, as seen from the values of the rates, the relative amplitudes of the peaks are, as expected, intensity-dependent. At sufficiently high intensities the rates are no longer monotonically decreasing functions but first increase then decrease. Therefore, it is quite clear that, if the peak suppression is to be explained within the framework of perturbation theory, in which the interpretation given in [16] takes place, then the intensity is the most important ingredient. Two more aspects can be considered: first, it is apparent in both result that circularly polarized light is more efficient to suppress the low-energy peaks than the linearly polarized light. This is in agreement with theoretical calculations of the generalized cross-sections [17] and is probably due to the reason mentioned above. However, it must be pointed out that other factors, like the angular distributions, may seriously influence the collection of low-energy electrons in a spectrometer with a small collecting angle, since the number of electrons emitted in a given solid angle is much smaller when using circular polarization than when using linear polarization and detecting along the polarization vector. As a consequence, space-charge effects could be more important in this situation. Note that

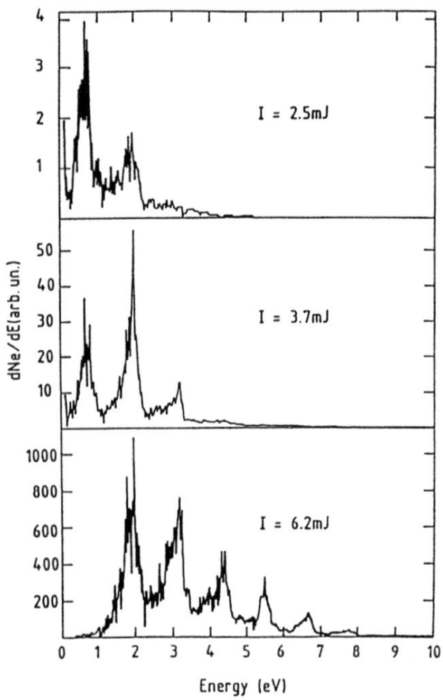


Figure 3 - Electron spectra of eleven-photon MPI of Xenon at 1064 nm with circular polarization for different pulse energies.

| S | $\sigma_{6+S}$ | $W_{6+S}W_6$ | $W_{6+S}/W_6$ |
|---|---|---|---|
| 0 | $4.7x10^{-72}$ | 1 | 1 |
| 1 | $4.1x10^{-85}$ | 0.872 | 1.31 |
| 2 | $5.3x10^{-99}$ | 0.113 | 0.254 |
| 3 | $3.4x10^{-113}$ | 0.00723 | 0.0243 |
| 4 | $1.5x10^{-127}$ | 0.000319 | 0.00161 |

Table 1 - (6+S)-photon ionization of Hydrogen at 532 nm in circular polarization. Second column: generalized cross-sections. Third and fourth column: transition rates at $1x10^{13}$ and $1.5x10^{13}$ $W.cm^{-2}$ respectively, normalized to that of the S=0 peak.

non perturbative approaches [18] can also account, apparently very well, for some of the observations.

## 5. CONCLUSION

Above-threshold ionization is an interesting example of non-resonant multiphoton process which cannot be described by LOPT even at not too high intensity ($10^{12}$ $W.cm^{-2}$). This might be surprising since the total ionization rate was found to obey the "normal" power law up to much higher intensities ($10^{15}$ $W.cm^{-2}$). It readily follows, since ATI is an intrinsic part of MPI processes at such intensities, that LOPT does not correctly account for MPI at intensities above $10^{12}$ $W.cm^{-2}$. Since phenomena such as Multiple Multiphoton Ionization are observed at such intensities only, they cannot be fully understood without taking ATI into account.

Some of the theoretical attempts to describe ATI are presented in this book. Obviously, a considerable effort is still necessary to achieve the kind of understanding which has been reached in the case of other MPI experiments (resonance or coherence effect, for instance). The experimental part of it certainly lies in providing more precise quantitative data on points like the peak-amplitude intensity-dependence, peak shifts and broadening, as well as in new investigations on the effect of other laser characteristics such as the wavelength or the pulse duration.

## REFERENCES

1. P. Agostini, F. Fabre, G. Mainfray, G. Petite and N. K. Rahman: Phys. Rev. lett. 42, 1127 (1979).

2. W. Zernick and R.W. Klopfenstein: J. Math. Phys., N.Y. 6, 262 (1965); E. Karule: J. Phys. B 11 441 (1978).

3. Y. Gontier and M. Trahin: J. Phys. B 13 4383 (1980).

4. R. Shakeshaft: this book.

5. G. Petite, F. Fabre, P. Agostini, M. Crance and M. Aymar: Phys. Rev. A 29 2677 (1984).

6. G. Petite, F. Fabre, P. Agostini: in "Collisions and Half-collisions with lasers", ed. by K. Rahman and C. Guidotti Harwood Academic Publishers Chur (1984).

7. M. Van Der Wiel in "Collisions and Half-collisions with lasers" ed. by N. K. Rahman and C. Guidotti Harwood Academic Publishers Chur (1984).

8. H. J. Humpert, R. Hippler, H. Schwoer and H. O. Lutz: in Fundamental Processes in Atomic Collision Physics ed. by H. Kleippoppen, J. S Briggs and H. O. Lutz, Pelnum N.Y. (1984).

9. M. Crance: J. Phys. B 1986, (to be published) and this book.

10. F. Yergeau, G. Petite and P. Agostini: J. Phys. B 19 to be published (1986).

11. J. Kupersztych (to be published).

12. Z. Deng and J. H. Eberly: J. Opt. Soc. Am. B 2 486 (1985).

13. L. A. Lompre, A. L'Huiller, G. Mainfray and C. Manus: J. Opt. Soc. Am. B 2 1906 (1985).

14. L. A. Lompre: private communication.

15. M. Trahin: private communication

16. P. H. Bucksbaum, M. Bashkansly, R. R. Freeman, T. J. Mc.Ilrath and L. F. Dimuro: Phys. Rev. Lett. 56 2590 (1986).

17. Trahin (unpublished).

18. H. Reiss this book.

# Dressing of Continuum States and Manipulation of Intermediate Resonances in Multiphoton Ionisation

*H.B. van Linden van den Heuvell, H.G. Muller, and M.J. van der Wiel*

FOM Institute for Atomic and Molecular Physics, Kruislaan 407,
NL-1098 SJ Amsterdam, The Netherlands

## 1. Introduction

A particle such as an atom, that is ionised by a strong electromagnetic radiation field, can give rise to the production of electrons of different energies. If the kinetic energy of such a photoelectron is larger than that of the largest photon in the radiation field, we will call the production process of this electron excess-photon ionisation (EPI).

In a typical EPI experiment (Agostini et al. [1], Kruit et al. [2]) a number of peaks is observed in the electron energy spectrum, all corresponding to the same initial state of the atom and the same final state of the ion, but differing in the number of absorbed photons. A convenient way to label those peaks is by their excess photon number (EPN), i.e. the amount by which the number of absorbed photons exceeds that minimally required for ionisation. In this way the ordinary multiphoton ionisation (MPI) process has an EPN of zero, while all EPI peaks have non-zero EPN.

EPI is interesting because in a perturbative description of the ionisation process the virtual intermediate state is embedded in a continuum. Such a continuum state has the property of decaying very rapidly, in the sense that within a very short time the electron will be so far from the remaining ion that absorption of additional photons from the field is no longer possible. (For an electron in a homogeneous radiation field, absorption of energy only can happen if there is also a potential present with a nonvanishing second derivative, and this derivative decays rapidly even for a Coulomb-like potential.) A consequence is that the EPI process, although formally resonant because the intermediate state is a quantum-mechanically allowed one, behaves more like non-resonant multiphoton ionisation. Nevertheless a very strong field can saturate the transitions between these continuum states (Deng and Eberly [3]) in a way very similar to transitions between bound states.

The main question which we try to answer in this paper, can be formulated as follows: 'Under what circumstances do the excess photon transitions behave like resonances, and when like virtual intermediate states?'.

Indirect evidence for saturation in EPI processes has come from the behaviour of the order of non-linearity for the peaks in an EPI experiment corresponding to a different excess-photon number (Kruit et al. [2]), as well as from measurements on the total cross-section for MPI (Lompré et al. [4]). Such measurements on the total cross-section are usually performed on ions. These experiments strongly indicated that although many EPI peaks of similar size are present (indicating the breakdown of perturbation theory), the total ionisation cross-section nevertheless obeys the same power-law dependence as at low field strength. This suggested that the EPI process could be decomposed into a rate-limiting ionisation step (to the lowest reachable continuum state), followed by a redistribution of the photoelectrons over the various EPI states by saturated transitions between continuum states. Here 'saturated' means that the transitions are fast compared to the total ionisation rate, and has nothing to do with depopulation of the ground state, an effect that is sometimes also referred to as saturation. Note that in the preceeding sentences the verb "followed" should not be taken to imply any time-ordering of the EPI process, but is used in the sense of perturbation theory, where the initial state is connected to the final state through a sequence of successive actions of the perturbing potential. From a quantummechanical point of view the question of time ordering during the process is meaningless.

This explanation implies that the rate for the fast transitions is such that an electron is driven many times up and down the transition during the ionisation process. Until recently any direct evidence for this mechanism was lacking, since it is experimentally impossible to discern an electron that has made many transitions by absorbing photons and then re-emitting them, from an electron created by only absorption of the net number of photons. A two-colour experiment, however, does not have this disadvantage, since the emitted photon can be of a different energy from the ones so far absorbed.

## 2. Two-colour experiment

We therefore performed a two-color experiment, (Muller et al. [5]), where we used the fundamental of a Nd:YAG laser ($\omega_2$, corresponding with a wavelength of 1.064 μm) to induce transitions between the continuum states of xenon atoms. From the experiment of Kruit et al. [2] it is known that at this wavelength these transitions are already strongly driven at intensities where the ionisation of xenon is still negligibly small. Its intensity could therefore be chosen such that there is no ionisation due to the $\omega_2$ beam, but that there are many transitions between continuum states when these states are present. A second beam of a much higher frequency ($\omega_1$) and lower intensity was then used to cause ionisation of the atoms. The frequency was such that three photons bring xenon in close vicinity of its continuum limit, and four photons suffice for ionisation.

With only the UBV beam present, and just above threshold for three-photon ionisation, all electrons produced have the same (very low) energy, corresponding to EPN zero. EPI is negligible under these circumstances, due to the relatively low intensity of the UV light, as well as the high frequency. Note that matrix elements between continuum states typically grow as $\omega^{-2}$.

If $\omega_1$ is decreased to below the threshold for three-photon ionisation, the only detectable electrons come from a four-photon process. In this case the $Xe^+$ ion remaining after ionisation could be in two states, $P_{3/2}$ or $P_{1/2}$, which differ in energy by 1.3 eV due to spin-orbit interaction. All excited states of xenon in the energy region around 3 $\hbar\omega_1$ correspond to states with a 'core' that looks like one of these ion states, but with an extra electron loosely bound to it in a much larger orbital. Ionisation to a $P_{1/2}$ state of $Xe^+$ is strongly disfavoured: the only pathways leading to this state (again in terms of perturbation theory) go at the three-photon level either through an excited state with a different (i.e. $P_{3/2}$) core configuration (with a very low dipole coupling to the final state), or through a state with the same core configuration, which is far off-resonant. Because the $P_{1/2}$ threshold lies much higher up, states with a core resembling this ion state are still sparse in this energy region. Although the Rydberg states converging to the $P_{3/2}$ limit can not be considered proper resonant intermediate states, since their spacing is less than the laser linewidth, they still form a pathway that can be used for ionisation with relative ease.

Switching on the $\omega_2$ beam dramatically alters the electron spectrum. Without $\omega_2$ beam the spectrum consists of a single peak at $4\omega_1$. With $\omega_2$ beam the spectrum can be seen in figure 1. The ionisation with four $\omega_1$ photons becomes strongly

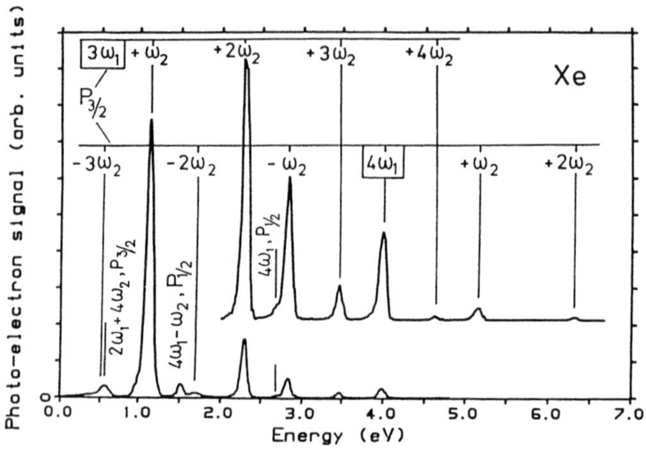


Figure 1. Energy spectrum of photoelectrons, by ionising xenon in a radiation field $\omega_1 \simeq 4$ eV, intensity $\approx 10^{10}$ W/cm$^2$ and field $\omega_2 \simeq 1$ eV, intensity $10^{12}$ W/cm$^2$.

suppressed, and now the dominant process becomes ionisation with $3\omega_1 + 1\omega_2$ photons. This is not surprising, since the $\omega_2$ intensity is much larger than that of the $\omega_1$, and for every pathway of ionisation with $4\omega_1$ there is a corresponding one that goes with $3\omega_1$ through the same (virtual) intermediate states, but in the final 'step' absorbs an $\omega_2$ photon. This last path has a stronger dipole matrix element, as well that it profits from the higher $\omega_2$ intensity.

A lot of other peaks are present as well, mostly corresponding to absorption of an excess of $\omega_2$ photons. This gives rise to the series $3\omega_1 + n\omega_2$ with n ranging from one to four, and $4\omega_1 + n\omega_2$, in which n also takes negative values. The observation of a number of EPI peaks suggests that we are indeed in the region where saturation starts to become an important effect. Indeed downward transitions (i.e. emission of $\omega_2$ photons) now can be observed in the members of the second series with $n < 0$. The fact that the $n = -1$ member has a higher probability than the $n = +1$ one, and, indeed, than the $n = 0$ master peak, seems somewhat odd. This is not what one would expect for a saturated transition.

To explain this, one has to realise that ionisation with photons of various frequencies introduces a whole new aspect to the perturbation expression for the matrix elements associated (through Fermi's golden rule) with these processes. This is the fact that there are now many pathways leading to the same final state, differing in the order in which the photons are absorbed. (Again this is not meant as temporal order!) All these pathways have virtual intermediate states at different energies, and therefore can behave quite differently.

A detailed comparison of all the possible intermediate energies in the various pathways with the energy levels of xenon (Moore [6]), reveals that there is a pathway to the $n = -1$ final state through absorption of three $\omega_1$ photons, subsequent emission of one $\omega_2$ photon, and then again absorption of an $\omega_1$ photon. This pathway benefits from the presence of the state $5p^5(^2P^o_{3/2})7p$ in the discrete spectrum, which lies very closely to the last virtual intermediate state. Since no continuum embedded intermediate states appear in this pathway, it can be viewed as ordinary MPI, and indeed the $4\omega_1 - \omega_2$ signal is no EPI according to the definition given at the beginning of this paper.

The other peaks in figure 1 all can be explained in terms of coincidental resonances that enhance these signals. Since there are so many ways to combine the different photons, such peaks are almost always present. An example is the peak due to ionisation to the $P_{1/2}$ ion state. This is the only process in which the competition between the two ion states does not end completely in favour of the $P_{3/2}$ state, a fact which is probably due to the presence of a $5p^5(^2P_{1/2})6p$ state.

To conclude the description of the two-colour experiment we remark that it is indeed possible to observe downward transitions due to emission of photons, a process in which the electron comes out with an energy corresponding to a net number of absorbed photons smaller than the minimum number required for ionisation. One could call such a process deficit-photon ionisation (DPI).

## 3. Threshold experiment

Another way to study the nature of the intermediate states in EPI, is to observe what happens if we scan the energy of this state over a threshold, so that it changes from a continuous state into a discrete one. This can be done by scanning the frequency of the ionising photons. In our experiment the same $\omega_1$ beam as in the two-colour experiment was used for this purpose, with $\hbar\omega_1 \simeq 4.04$ eV. Because the ionisation potential of xenon is 12.127 eV, the atom is, after absorption of three photons, either in the continuum, or slightly below it.

As discussed before, above the threshold, we can observe electrons of two different energies coming out of the atom. They correspond to absorption of three and four photons, leaving the ion in the $P_{3/2}$ state. Under these circumstances the four-photon peak is an EPI signal with EPN = 1. Below threshold, three photons bring the atom in a state somewhere in the Rydberg series converging to this threshold. Therefore the three-photon process leads to excitation instead of ionisation, and the associated signal can no longer be observed. The four-photon signal is now formally a REMPI signal, the Rydberg states acting as resonances (REMPI is Resonance Enhanced MPI). Theory predicts in this case a strongly peaked behaviour of the cross-section as a function of photon frequency.

In practice the resonant behaviour does not set in until the photon frequency is well below the threshold, and the EPI behaviour extends continuously over the threshold. This can be seen in figure 2, where the three- and four-photon ionisation signal are plotted as a function of the photon energy. The three-photon threshold is around 306.63 nm and the resonant behaviour of the four-photon sinal starts below 306.85 nm. We thus seem to have a situation in which the Rydberg series of resonances acts as a continuum, and we call this process Quasi EPI.

We claim that this observed continuity between QEPI and EPI, instead of the expected discontinuity in going from EPI to REMPI, is caused by the finite duration of the laser pulse. Above it was already remarked that the interaction between an electron and a homogeneous radiation field can only lead to the absorption of energy (and therefore of photons) if the electron is sufficiently close to the remaining ion. Therefore its energy is essentially determined as soon as it is a few Bohr-radii away from the nucleus. In this small neighbourhood the wavefunctions of low-lying continuum states and the Rydberg states all look the same, and therefore the rate with which probability amplitude populates these states will be essentially equal. Above as well as below the threshold, a wavepacket is formed by superposition of all the states that are coupled by the field to the ground state.

A wavepacket of Rydberg states eventually has to return to its starting point, whereas a continuum wavepacket escapes to infinity. The first time the Rydberg packet returns, is after the elapse of the classical orbit time of the Rydberg electron. If this time is long compared to the duration of the laser pulse, no more interaction with the field is possible. The electron can therefore only absorb the energy of four photons immediately during its emission from the atom. This explains why QEPI is so much like EPI.

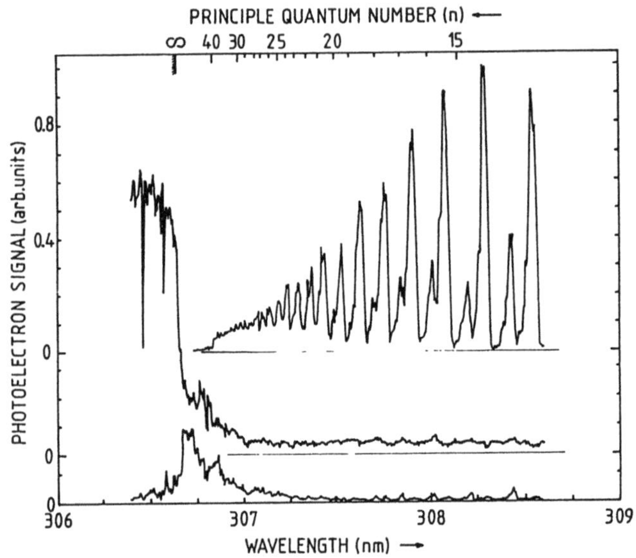


Figure 2. Ionisation of xenon by four photons (upper trace), three photons combined with ionisation not caused by the laser (middle trace) and ionisation that is not caused by laserlight isolated (lower trace). The lower trace can be seen as a background in the three-photon ionisation signal (middle trace). The indicated n-states are calculated hydrogenic values with the limit of the series at the three-photon ionisation limit.

If the electron returns before the pulse has finished, we get into the interesting situation where the returning wavepacket interferes with the one that is currently being produced. This interference can be constructive, if we are exactly on the energy of a Rydberg state, or destructive, if we are somewhere in between. This not only modulates the probability for excitation to the Rydberg states (through absorption of three photons), but also that for the absorption of four photons, leading to the characteristic REMPI spectrum, as can be seen below.

For a more quantitative description we can argue along the above lines as follows: After the radiation field has been on for N orbits, there is a current traveling around the classical trajectory of the orbit. This current is at any point the superposition of the N waves generated by three-photon absorption at the N times at which that point on the orbit passed near the nucleus. Note that usually we can couple only to very low l quantum numbers. On the average (i.e. if we assume the phases to be random) this will enhance the circulating current by a factor of N compared to the originally emitted current. If we are on resonance this factor will be $N^2$, because the current is proportional with the square of the wavefunction, but for other frequencies the interference will be destructive. The total probability excited to the Rydberg states can be obtained by observing how much charge passes along a certain point of the orbit during a single orbit time. On the average it is thus $T_{pulse} \cdot R_{em}$, where $R_{em}$ is the rate for emission of an electron (which would be called ionisation rate if we were above threshold). This three-photon absorption cross-section is thus independent of N, and indeed independent from the fact whether we have to do with ionisation (N = 1) or with averaged excitation (N > 1) as Fano and Cooper [7] already pointed out.

The four-photon absorption (= ionisation) rate, however, can be obtained by multiplying the circulating current by the probability that a (fourth) photon will be absorbed during a single passage near the nucleus. It therefore is proportional to N, and increases if either the pulse duration is increased, or the intermediate state lies more below threshold (shortening the orbit time).

Although the experimental data qualitatively show the predicted effects, the wavelength at which the four photon peak starts to increase does not correspond to an orbit time equal to the pulse duration (which was 6 ns), but rather to 50 ps. We assume that this is caused by the magnetic field of 1 T in which the measurement had to be performed, because of the constuction of our electron spectrometer (Kruit and Read [8]). High-lying Rydberg states are rather easily perturbed by a field of this magnitude, and a calculation of classical trajectories in such a field revealed that an orbiting electron starting exactly at the nucleus (an $l = 0$ electron) has a distance of closest approach on its first return which differs from zero.

In fact this "perinucleon" distance increases for a given B field very sharply with the principal quantum number, proportional with n . In our case this means that from n = 41 on, this distance is more than a few Bohr radii, and therefore absorption of a fourth photon on this and subsequent returns becomes impossible even if the duration of the pulse normally would allow for it. For a 'pulse' of infinite duration the B-field would not influence the cross-section, since eventually the electron would have to return to the nucleus because of ergodic considerations. In this case the number of orbits is only limited by the lifetime of the Rydberg state with respect to spontaneous photon emission. In order to test this interpretation measurements at different values of B are now in preparation in our Institute.

## 4. Conclusion

What we have learned from this can be resumed as follows: Three cases of MPI can be distinguished. In the REMPI case the intermediate state is exactly on an existing level, leading to ever-lasting constructive interference and consequently strong population of the intermediate state, provided the pulse duration and lifetime of the state are long enough to profit from this interference.

In the case of EPI/QEPI the useful life of the intermediate state or the pulse is such that there is no build up due to interference. This means that if absorption of still more photons is going to happen at all, this has to occur in one go. This then leads to much smaller cross-sections for these processes.

The final case is non-resonant MPI, where all states are in energy far from the intermediate state. This leads to a very rapidly emerging phase difference between probability amplitudes brought in states coupled to the initial state at different times. Therefore the population of all real states stays very small at all times, leading to an even smaller cross-section.


## Acknowledgement

This work is part of the research programme of the Stichting voor Fundamenteel Onderzoek der Materie (Foundation for Fundamental Research on Matter) and was made possible by financial support from the Nederlandse Organisatie voor Zuiver-Wetenschappelijk Onderzoek (Netherlands Organisation for Advancement of Pure Research).


## References


1. P.Agostini, F.Fabre, G.Mainfray, G.Petite and N.K.Rahman, Phys. Rev. Lett. 42 1127 (1979)
2. P.Kruit, J.Kimman, H.G.Muller and M.J. van der Wiel, Phys. Rev. A 28 249 (1983)
3. Z.Deng and J.H.Eberly, Phys. Rev. Lett. 53 1810 (1984)

4. L.A.Lompré, G.Mainfray, C.Manus and J.Thebault, Phys. Rev. A 15 1604 (1977)
5. H.G.Muller, H.B.van Linden van den Heuvell and M.J. van der Wiel, J.Phys.B, in print, (1986)
6. C.E.Moore, Atomic Energy Levels, NBS Circular No 467 (Washington, DC: US Govt. Printing Office,1958)
7. U.Fano and J.W.Cooper, Rev. Mod. Phys. 40 441 (1968)
8. P.Kruit and F.H.Read, J. Phys. E 16 313 (1983)

# Theoretical Models for Above-Threshold Ionization in Atomic Hydrogen

*Y. Gontier and M. Trahin*

Service de Physique des Atomes et des Surfaces, Centre d'Etudes Nucléaires de Saclay, F-91191 Gif-sur-Yvette Cedex, France

We compare the behaviour of A.T.I. spectra resulting from lowest-order perturbation theory to those obtained within a non-perturbative model.

It is shown that both of them predict a lowering of the amplitudes of the first few peaks with respect to the others. In the all-order model considered, it appears that the effect of higher-order terms is to produce a lowering of intensity values at which the inversions in the amplitudes of the peaks occur.

## 1. INTRODUCTION

The problems related to the creation of high-energy electrons are not very new. They have been studied since the early seventies [1]. Nevertheless, during the last few years there has been an experimental [2-5] and a theoretical [6-12] renewal of interest for processes called Above-Threshold Ionization (A.T.I.) [6]. Briefly, in ATI process a bound electron absorbs more photons than the minimum number required to ionize the atom. It is a one-step process where the electron absorbs N photons in the discrete spectrum of the atom and 1, 2,...,S additional photons in the potential of its parent ion. As a consequence, the electron yield has a linear pressure dependence. In a vapor since the N, N+1, ..., (N+S)-photoabsorptions occur simultaneously, the electron spectrum consists in equally spaced peaks whose spacing is equal to the photon energy.

## 2. LOWEST-ORDER THEORY

In the most general case, the total ionization probability is

$$P_{tot} = P_N + P_{N+1} + \dots + P_{N+S} + \dots \tag{1}$$

From perturbation theory, each ATI probability $P_{N+S}$ can be expanded in the form of a power series in intensity as

$$P_{N+S} = P^{(0)}_{N+S} + P^{(2)}_{N+S+2} + \cdots + P^{(2k)}_{N+S+2k} + \ldots \quad (2)$$

which contribute through 0, 2, ..., 2k extra pairs of absorption-emissions of photons. Calculations of ATI probabilities have been initiated within lowest-order perturbation theory (LOPT) [6-8] by retaining only the first term $P^{(0)}_{N.S}$. The main difficulty is the calculation of high-order bound-free matrix elements. The method we have used to calculate them has been discussed extensively in previous accounts [6,7] and will not be presented here. We only mention that these matrix elements have been calculated exactly in Hydrogen by using an implicit summation technique. Since the calculations were done within LOPT, the intensity at which this approximation was expected to fail was calculated for each peak. It is the intensity at which the probabilities corresponding to two successive peaks become equal. We have checked that all these intensity values were greater than the saturation intensities for which the probabilities are equal to unity. Thus in every case the series of Eq. (2) converges. The overall electron populations in the peaks are calculated by averaging the probabilities over a spatio-temporal intensity distribution (realistic pulse shape). The results of the calculations for Hydrogen are recalled in Fig.(1) where the normalized spectra corresponding to N=10 and N=12 are displayed.

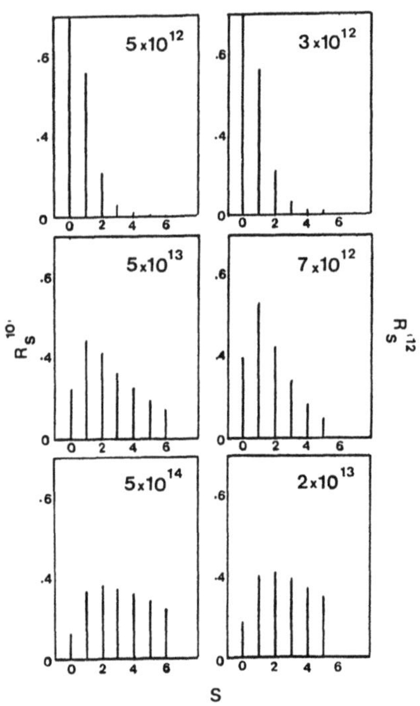


Fig.(1) - Population rates [7] corresponding to N=10 and N=12 for different values of the intensity, as a function of S

Each line represents the rate of the electron population in the $S^{th}$ peak over the total number of electrons. In both cases the spectrum shows the same features. At low intensity the amplitudes of the peaks decrease with increasing S. At higher intensities the amplitude of the first peak decreases with regard to the others, while those of high-order peaks increase. The intensity at which the "switching" of the peaks occurs decreases when N, the order of the coupling between the ground state and lowest continuum, increases. In most cases this behaviour is observed and indicates that ATI processes are favoured for large N.

Although the series (1) and (2) converge, one observes that the amplitudes of the first terms begin to increase before decreasing at high intensity. Such a behaviour indicates that a lowering of the first few peaks may happen within LOPT provided that the intensity values lie just below those for which LOPT fails. The average over intensity distributions amplifies the phenomenon principally when saturation occurs. In spite of some lacks, the merit of this model was to predict the lowering of the first few ATI peaks which was observed few years later in experiments [4]. Nevertheless, the anomalous intensity dependence of the peaks which appears far from saturation cannot be interpreted within LOPT. This effect outlines the importance of describing A.T.I. processes by non-perturbative theories.

## 3. ALL-ORDER THEORY

We focus our attention on two equivalent methods which enable to calculate transition probabilities up to arbitrary order in the field. One of them involves exact re-summed expression for the transition probabilities [12] while in the other the Schrödinger equation is solved by using a finite basis of states. In the latter case one is led to solve a set of algebraic equations [9-11]. This is the method which is utilized here for Hydrogen, together with a particular technique for the evaluation of the principal parts of the integrals over the continuum.

One chooses a set of discrete states $|j\rangle$ and continuum states $|E;j\rangle$ both labeled $|j\rangle$ for shortness. The state vector $\phi(t)$ is expanded on this basis according to

$$\phi(t) = \sum_j C_j(t)|j\rangle + \sum_J \int dE\, C_{Ej}(t)|E,J\rangle \tag{3}$$

where the C's are the probability amplitudes to find the electron in the state $|j\rangle$ at time t if it was in the state $|0\rangle$ at time t = 0

$$C_j(t) = \langle j|\phi(t)\rangle = \langle j|U(t)0\rangle \tag{4}$$

We assume that the light is circularly polarized, and we consider only couplings between "circular" states [11], i.e., couplings between states for which both $\ell$ and m differ by $\pm$ 1.

By solving the Schrödinger equation for $\phi(t)$ and owing to Eq. (2), one obtains a set of first-order differential equations for the matrix elements of U(t). From the inversion integral

$$U(t) = \frac{1}{2\pi i} \oint e^{-izt} G(z)dz \tag{5}$$

the set of equations becomes time-independent and enables one to calculate the matrix elements of the resolvent operator $G(z) = (z-Ho-V)^{-1}$. Despite the considerable simplification provided by working within time-independent scheme, the equations are not solved for G(z) but rather R(z). With the help of this operator one expresses the "on-space" and "off-space" parts of G(z) as

$$PG(z)P = \frac{P}{z-Ho-PRP} \quad \text{and} \tag{6a}$$

$$QGP = \frac{1}{z-Ho} QRP \frac{1}{z-Ho-PRP} \tag{6b}$$

respectively. In Eqs.(6)

$$R = V + V\,QG_0V + VQG_0V\,QG_0V + \ldots \quad \text{and} \tag{7}$$

$$G_0 = (z-Ho)^{-1} \quad , \quad P+Q = 1 \tag{8}$$

P being the projector onto a subspace spanned by particular states of Eq. (1). In the case we consider only the ground state lies in the subspace, i.e., $P \equiv |0\rangle\langle 0|$.

For an A.T.I. process involving the absorption of N photons in the discrete spectrum and S photons above the lowest ionization threshold, R(z) is obtained from the following set of equations

$$R_{0,0} = V_{0,1}\,\bar{R}_{1,0}$$

$$R_{1,0} = V_{1,0} + V_{1,2}\,\bar{R}_{2,0}$$

$$R_{2,0} = V_{2,1}\,\bar{R}_{1,0} + V_{2,3}\,\bar{R}_{3,0}$$

— — — — — — — — — — — — — — — —

$$R_{N-1,0} = V_{N-1,N-2}\,\bar{R}_{N-2,0} + \int dE'\, V_{N-1,NF'}\,\bar{R}_{NE',0}$$

$$R_{NE,0} = V_{NE,N-1}\,\bar{R}_{N-1,0} + \int dE'\, V_{NE,(N+1)E'}\,\bar{R}_{(N+1)E',0} \tag{9}$$

$$R_{(N+1)E,0} = \int dE'\, V_{(N+1)E,NE'}\,\bar{R}_{NE',0} + \int dE' V_{(N+1)E,(N+2)E'}\,\bar{R}_{(N+2)E'0}$$

— — — — — — — — — — — — — — — — — — — — — — — — — — — — — — — — — —

$$R_{(N+S)E,0} = \int dE'\, V_{(N+S)E,(N+S-1)E'}\,\bar{R}_{(N+S-1)E,0}$$

where

$$\bar{R}_{j,0} = \frac{1}{z+jE_p-\varepsilon_j} R_{j,0} \quad : \quad \bar{R}_{jE,0} = \frac{1}{z+jE_p-E'} R_{jE,0} \tag{10}$$

The probability amplitude of populating the j-th ATI peak is given in terms of R(z) as:

$$U_{j,0}(t) = \frac{1}{2\pi_i} \oint \frac{1}{\bar{z}-E+(N+j)E_p} R(z)_{(N+j)E,0} \frac{1}{z-E_o-R_{o,o}} e^{-izt} dz \tag{11}$$

where the integral is calculated by using the residue theorem. Because of the integrals over energy, Eqs. (9) are not algebraic equations. To solve the set of Eqs. (9) we have carried out a method which does not seem to have been used in ATI computations. In order to make the problem more tractable, we transform Eqs. (9) by using the following relation

$$\int dE' \, V_{kE,jE'} \bar{R}_{jE,0} = P \int dE' \frac{V_{kE,jE'} R_{jE,0}}{z+jE_p-E'} - i\pi V_{kE,jE} R_{jE,0} \tag{12}$$

where P denotes the Cauchy principal value and the substitution $z+jE_p \rightarrow E$ has been done in the complex term.
We put $V_{kE,jE'} R_{jE',0} \equiv f(E')$ and $z+jE_p \equiv z_j$. Then by adding to and subtracting from the integrant the quantity $f(E)/\exp((z_j-E')/\alpha)^2$ where $\alpha$ is a parameter, the principal part integral reads

$$P \int_0^\infty dE' \frac{f(E')}{z_j-E'} = I_\alpha - ½ f(z_j) E_1[( \frac{z_j}{\alpha} )^2 ] , \tag{13}$$

where $\alpha$ is a parameter to be determined, $I_\alpha$ is an unknown integral and $E_1$ is the exponential integral function. Since $E_1$ increases when its argument decreases, the second term in the right-hand side of Eq. (13) must increase with respect to $I_\alpha$ because the right-hand side of Eq. (13) must be finite and independent of $\alpha$. By varying the parameter $\alpha$ one minimizes the contribution of $I_\alpha$. One finds numerically an optimal value of $\alpha$ for which the main contribution to the integral of Eq. (13) comes from the term containing the function $E_1$. The probabilities for A.T.I. processes up to fifth-order are summed and plotted against the energy E. A typical spectrum is displayed in Fig.(2) for different increasing values of the intensity ($1.4 \times 10^{12}$, $7 \times 10^{12}$ and $2.8 \times 10^{13}$ W/cm$^2$). The behaviour is very similar to that shown in Fig.(1). In considering the probabilities obtained in

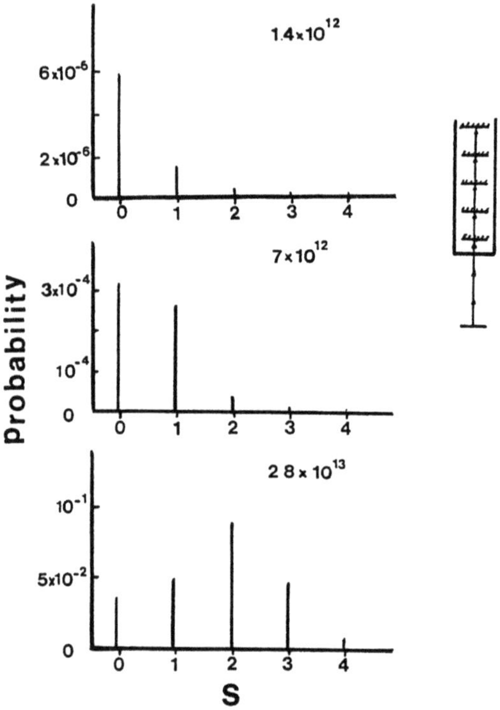


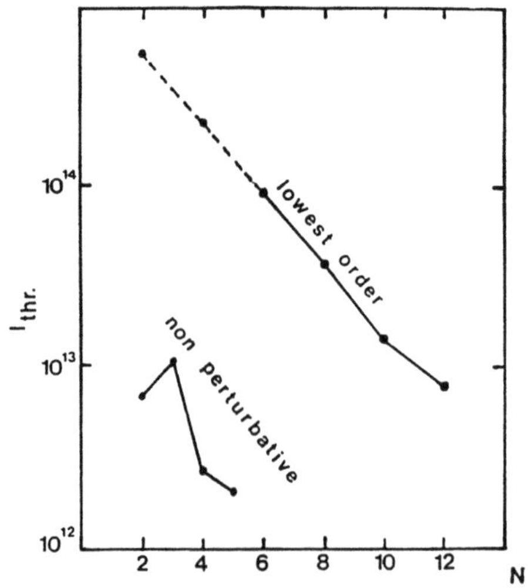


Fig.(2)- A.T.I. probabilities (N = 3) in H for different intensity values.

Fig.(3) - Intensities at which the first A.T.I. peak is lower than the remaining others as functions of N

the two models, one determines in each of them the intensities $I_{th}$ for which the first peak becomes lower than the second one. Such an intensity is plotted in Fig.(3) as a function of N (the minimum number of photons required to ionize the atom).

By comparing the (extrapolated) lowest-order and the non-perturbative results, it is clear that in all-order theory the lowering of the first peak occurs at much lower intensities than those predicted by perturbative theory, principally for small N.

This spectacular effect due to higher-order terms is to be confirmed by the more expanded calculations of all-order A.T.I. processes, we are doing within a somewhat different formalism [13].

**REFERENCES**

1. E.A. Martin and L. Mandel, Appl. Opt. 15, 2378 (1976).
2. P. Agostini, F. Fabre, G. Mainfray, G. Petite and N.K. Rahman, Phys. Rev. Lett. 42, 1127 (1979).

3. F. Yergeau, P. Agostini and G. Petite, J. Phys. B, in press.
4. P. Kruit, J. Kimman, M.G. Muller and M.J. Van der Wiel, Phys. Rev. A28, 248 (1983).
5. R. Hippler, H. Schwier and H.O. Lutz, in Fundamental processes in Atomic collision physics edited by H. Kleinpoppen, J.S. Briggs and H.O. Lutz (Plenum Press, New York, 1985).
6. Y. Gontier, M. Poirier and M. Trahin, J. Phys. B 13, 1381 (1980).
7. Y. Gontier and M. Trahin, J. Phys. B 13, 4383 (1980).
8. M. Aymar and M. Crance, J. Phys. B 14, 3585 (1981).
9. Z. Bialynicka-Birula, J. Phys. B 16, 4351 (1983).
10. Z. Deng and J.H. Eberly, J. Opt. Soc. Am. B2, Vol. 3 (1985).
11. M. Edwards, L. Pan and L. Armstron Jr., J. Phys. B 18, 1927, 1985.
12. Y. Gontier, N.K. Rahman and M. Trahin, Phys. Rev. A 34, 1112 (1986).
13. Y. Gontier, C. Manus and M. Trahin, in preparation.

# Non-Perturbative Theory of Above-Threshold Ionization

*H.R. Reiss*

Physics Department, The American University, Washington, DC 20016, USA
Arizona Research Laboratories, University of Arizona, Tucson, Arizona, USA

## 1. Introduction

Within the past few years, the phenomenon called above-threshold ionization (ATI) has been observed[1-9], in which the spectrum of the photoelectrons from atomic ionization exhibits a series of peaks corresponding to the absorption of more than the minimum number of photons required for ionization. Starting with Kruit et al.[4], it was then discovered that the lowest order electron energy peaks were suppressed when continuing increases were made in the laser intensity. Most recently, observations[7-9] with circularly polarized light have revealed major qualitative and quantitative differences arising from the state of polarization.

The occurrence of ATI and the disappearance of low-order peaks are predicted by the non-perturbative theories of Reiss[10] and of Muller et al.[11,12]. Both theories have the feature in common that they are formulated for short-range potentials, and so they would not be expected to provide accurate descriptions of atomic photoionization until the intensity of the ionizing field is such that it is competitive with, or exceeds, the strength of the Coulomb field exerted by the residual ion on the detached electron. This is generally true, however, for the ATI experiments that have been reported.

An important fact is that the 1980 work[10] predicted not only ATI and low-order peak suppression, but it also examined the differences between circular and linear polarization and exhibited all the qualitative phenomena that have recently been observed. That is, the theory showed that low-order peak suppression sets in at lower intensities for circular than for linear polarization, but that higher-than-lowest order peaks will normally be dominant in the circular case when the intensity is sufficient for peak suppression. The envelope of the peaks in the electron spectrum becomes bell-shaped for circular polarization, whereas this is not true in the linear case, even for very high intensities. The transition probability in the circular case is much less than for linear in general, but this distinction erodes at higher intensities and for the higher order peaks in the spectrum.

The 1980 theory[10] preceded the experiments, and so the examples given therein did not reflect the conditions that were employed in the experiments, which have used xenon exposed to 1064 nm radiation. The 1980 theory will here be applied to model calculations using very simple wave functions, but otherwise employing physical parameters matching the experiments. The principal oucome is that all the essential features of the recent experiments are given correctly.

## 2. Applicability of the Theory

The theory as set forth in [10] is rigorously applicable for photodetachment of negative ions. Detailed bounds on the error involved in using the method for negative ions are given in that paper, and the errors are shown to be very small, even for very high field intensity. Such rigor cannot be established for the problem of atomic photoionization, and no such claims are made in [10]. Nevertheless, the method of [10] should be quite reliable even for atomic photoionization when the magnitude of the interaction term of the electron with the applied field is of the same order of magnitude as the Coulomb potential exerted by the residual ion on the detached electron. Consider, for example, experimental conditions as reported by the Bell Labs group[8]. The intensity parameter z (defined below) has approximately the value 2 in their work. The field interaction term has the magnitude given approximately by the product of eA (in units with $\hbar = c = 1$) with the velocity of the detached electron. This is to be compared with the Coulomb potential exerted on the electron by the ion at a distance of a Bohr radius. The ratio depends on the degree to which the electron is above threshold. For example, for the peak at n = 18 in the data[8], this ratio is about 0.4. It is larger for higher ATI peaks. Even for lower peaks, the theory [10] should be qualitatively reliable. By contrast, one can examine the status of perturbation theory for this same situation. The radius of convergence of perturbation theory for the case of circularly polarized light was explored in [10], and for the conditions of the experiment[8], this radius is z = 0.59. Under no circumstances will it ever be greater than z = 1, and this statement can be made for atomic photoionization as well as for photodetachment of negative ions. This is to be compared to the experimental value of z of about 2. Thus perturbation theory is not convergent for this problem, and the meaningfulness of higher order terms and corrections should be questioned very sharply.

## 3. Formulation of the Theory

The non-perturbative theory of Reiss[10] is a refinement and extension of the theory given by Keldysh[13]. The starting point is the S matrix (using units with $\hbar = 1$ and $c = 1$)

$$S_{fi} = -i\int dt\, (\Psi_f,\ (-e\mathbf{A}\cdot\mathbf{p}/m + e^2\mathbf{A}^2/2m)\ \Phi_i) \qquad (1)$$

where the final state wave function is the complete solution for the detached electron in the field, the initial state wave function is the no-field solution for the initial bound state, and the full interaction Hamiltonian is included. When a Volkov solution is inserted for the field-interacting wave function in (1), this constitutes the Keldysh approximation. From (1), the differential transition rate can be expressed in closed analytical form. The result for circular polarization is[10]

$$dW/d\Omega = (2m^3\omega^5)^{1/2}(2\pi)^{-2}\sum_{n=n_0}^{\infty}(n-z)^2(n-z-\varepsilon_B)^{1/2}\,|\hat{\phi}_i(\mathbf{p})|^2 J_n^2(z^{1/2}\gamma), \qquad (2)$$

where

$$\mathbf{A} = \tfrac{1}{2}a(\hat{\varepsilon}e^{i\omega t}+\hat{\varepsilon}^* e^{-i\omega t}), \qquad (3)$$

$$z = e^2a^2/4m\omega, \qquad (4)$$

$$\varepsilon_B = E_B/\omega, \qquad (5)$$

$$n_0 = [z+\varepsilon_B], \qquad (6)$$

$$\gamma = 2(n-z-\varepsilon_B)^{1/2}\sin\theta, \qquad (7)$$

$$p = (2m\omega)^{1/2}(n-z-\varepsilon_B)^{1/2}. \qquad (8)$$

The quantity z in (4) is the basic intensity parameter, (5) gives the initial binding energy of the ionized electron in units of the energy of a single photon, and (8) is the outcome of the energy-conserving delta function for the electron. Equation (6) is particularly interesting because it expresses an increase in the minimum allowable photon order as the intensity increases. This corresponds to the phenomenon initially observed by Kruit et al.[4]. The square brackets in (6) signify the smallest integer containing the quantity in the brackets. The function contained within the absolute value signs in (2) is the momentum space inital state wave function. The J functions in (2) are the ordinary Bessel functions.

The simple Yukawa wave function can be used as in [10], and shows all the basic features of photoionization by a circularly polarized field. The squared momentum space wave function is

$$|\phi_i(\mathbf{p})|^2 = 8\pi(2mE_B)^{1/2}/(n-z)^2(2m\omega)^2. \tag{9}$$

An equally simple alternative is the basic exponential wave function

$$\phi_i(\mathbf{r}) = (\pi a_0^3)^{-1/2}\exp(-r/a_0) \tag{10}$$

which leads to

$$|\phi_i(\mathbf{p})|^2 = 2^6\pi a_0^3/[1+2m\omega a_0^2(n-z-\varepsilon_B)]^2. \tag{11}$$

The expression (9) is used in the following numerical examples, although the results of (11) are qualitatively very similar, albeit somewhat larger in magnitude. One can also use more realistic wave functions (a model in which a single-particle 5p function is used to represent an outer-shell electron in xenon has been employed[14]), but the basic qualitative features remain unchanged.

It is notable that Faisal[15] was aware of (2) at an early date, arriving at it by a related but independent technique. He did not, however, exhibit (6) with all its consequences, nor did he point out the great qualitative differences between circular and linear polarization. The importance of higher-than-lowest order processes (as in ATI) has been known since the earliest days of non-perturbative electrodynamics[16].

## 4. Results of the Theory

Equation (2) is applied to models of photoionization in which parameters analogous to those of the circular-polarization experiments[7-9] with xenon are used. The wavelength assumed is 1064 nm, and there are two ionization channels with ionization energies of 12.13 eV and 13.43 eV. These have threshold photon orders of 11 and 12, respectively. Figures 1 and 2 show the total transition rate integrated over solid angle corresponding to intensities of $z = 0.1$ and 1.0. Even the result at $z = 0.1$ shows a strong suppression of the peak at $n = 11$. It is not totally absent, but it is very small. At $z = 1.0$, the lowest apparent order is 13, with $n = 15$ the dominant channel. Figure 3 gives results for $z = 10$, where it is seen that both ionization channels contribute significantly. For Figs. 1 and 2, the 12.13 eV channel is totally dominant. Figure 3 also illustrates quite dramatically the strong growth of the qualitative intensity features that made their appearance in Figs. 1 and 2. Photon orders 11 through 24 are suppressed, and the spectrum extends out through $n = 46$. The contribution of the ionization channel with threshold value of 12 is never dominant, but it becomes important for the tallest peaks in the spectrum, and recedes slowly thereafter.

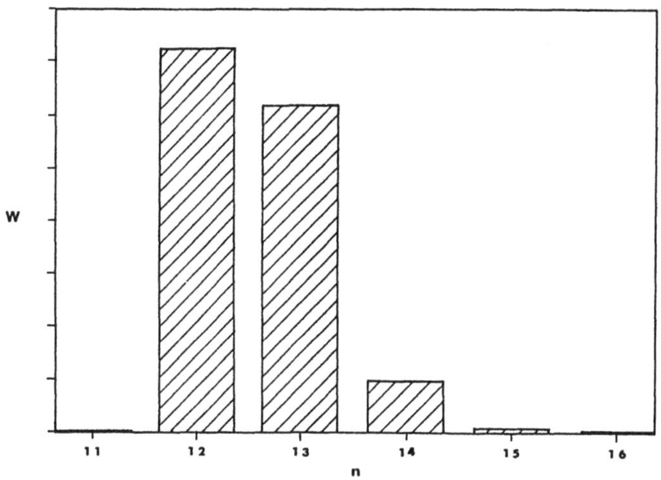


Figure 1. Electron spectrum for z = 0.1. W is the total transition rate and n is the photon number. Conditions are for photoionization of xenon by 1064 nm radiation

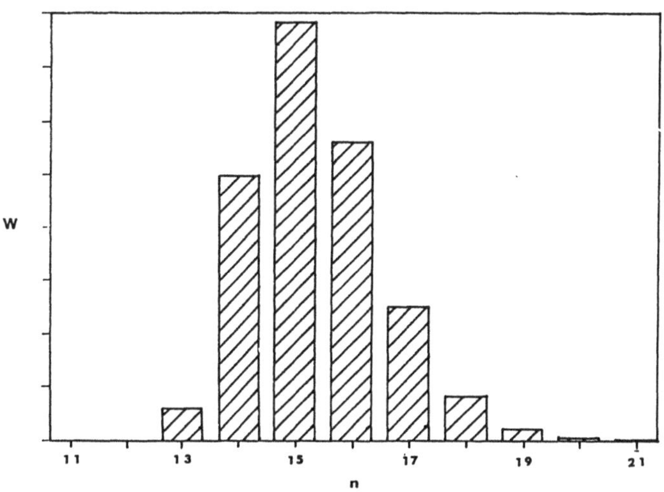


Figure 2. Electron spectrum for z = 1.0. W is the total transition probability per unit time and n is the photon number. Calculated for conditions appropriate to photoionization of xenon by 1064 nm radiation

A direct comparison of theory with experiment involves the ambiguity that experimental field intensities are known only very approximately. Furthermore, even if the maximum intensity were known explicitly, the measured spectra include contributions from regions of lower field intensity. Nevertheless, transition rates rise so rapidly with intensity that the highest intensity regions should dominate. The theory as constituted[10] is applied for a fixed field intensity parameter, although the formalism does permit extension to the treatment of wave packets. Comparison is here made between the theory and the experiments of Bucksbaum et al.[8], since their intensity is

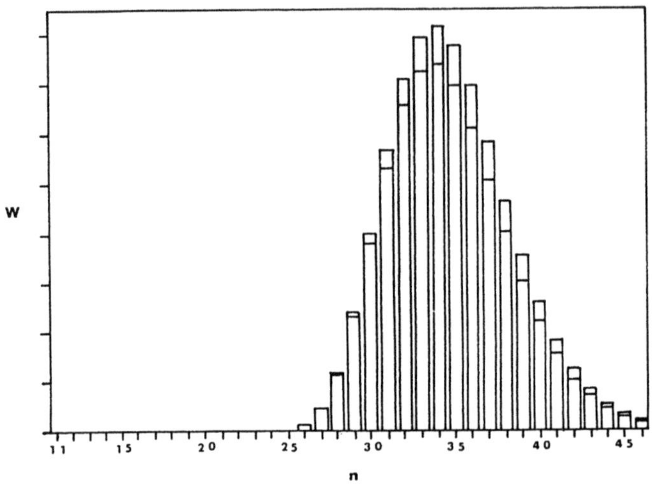


Figure 3. Electron spectrum for z = 10. W is the total transition rate and n is the photon number. The lower part of each bar is the contribution of the ionization channel with a low intensity threshold order n = 11. The upper part of each bar is for the channel with low intensity threshold order n = 12. Conditions are as for ionization of xenon by 1064 nm radiation

the highest of the circular polarization experiments[7-9]. Bucksbaum et al. state their intensity as about 1.5x10E13 W/cm2, corresponding to a z value of about 1.4. However, the theoretical fit is much better if z = 1.9 is used, and the experimenters agree that this is an entirely possible choice[17]. Figure 4 is a reproduction of the Bucksbaum et al. data[8] with the theoretical prediction from (2) (after integration over solid angle) superimposed. The theory is normalized to match the peak at n = 18 since this is quite a prominent peak, while at the same time it lies fairly far above threshold, where the applicability of the theory is relatively reliable. As can be seen from Fig. 4, agreement is very good.

Table 1 gives a summary of the qualitative features of the calculations. These features are virtually independent of the assumed wave function, even when a more detailed representation of xenon is used[14] than is given by (9) or (11). For comparison, Table 1 also includes entries for the Saclay[7] and Bell Labs[8] experiments,

Table 1. Comparison of theory and experiment for the main qualitative features of photoionization by intense circularly polarized radiation

| Intensity Parameter z | Peaks Missing n | Highest Peak n | Number of Peaks | Source |
|---|---|---|---|---|
| 0.1 | 11 | 12 | 4 | Theory |
| ≈ 0.5 | 11 | 12 | 6 | Saclay exper. |
| 1.0 | 11,12 | 15 | 7 | Theory |
| ≈ 2.0 | 11-13 | 17 | 9 | Bell Lab exp. |
| 10.0 | 11-25 | 34 | 21 | Theory |

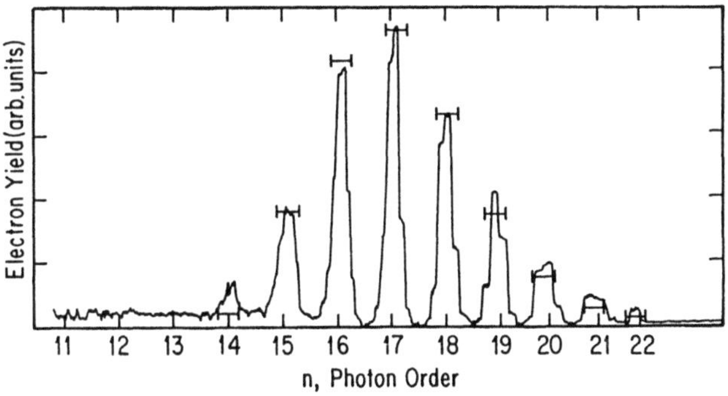


Figure 4. Experimental data of Bucksbaum et al. reproduced from [8]. For comparison, the present theory is shown superimposed on the data in the form of the short horizontal bars drawn on the peaks. The value z = 1.9 is assumed.

although the intensities are not very precisely known. A very steady progression is shown in Table 1 in the loss of low-order peaks with increasing intensity, the increasing order of the highest peak, and the total number of peaks visible. In this progression, one cannot distinguish the experimental entries from the theoretical entries. They all follow the same pattern.

## 5. Literature


1 P. Agostini, F. Fabre, G. Mainfray, G. Petite, N. K. Rahman: Phys. Rev. Lett. 42, 1127 (1979)

2. P. Agostini, M. Clement, F. Fabre, G. Petite: J. Phys. B: At. Mol. Phys. 14, L491 (1981)

3. P. Kruit, J. Kimman, M. J. van der Wiel: J. Phys. B: At. Mol. Phys. 14, L597 (1981)

4. P. Kruit, J. Kimman, H. G. Muller, M. J. van der Wiel: Phys. Rev. A 28, 248 (1983)

5. H. J. Humpert, R. Hippler, H. Schwier, H. O. Lutz: In Fundamental Processes in Atomic Collision Physics, ed. by H. Kleinpoppen, J. S. Briggs, H. O. Lutz (Plenum, New York 1985) p.649

6. L. A. Lompré, A. L'Huillier, G. Mainfray, C. Manus: J. Opt. Soc. Am. B 2, 1906 (1985)

7. F. Yergeau, P. Agostini: To be published

8. P. H. Bucksbaum, M. Bashkansky, R. R. Freeman, T. J. McIlrath, L. F. DiMauro: Phys. Rev. Lett. 56, 2590 (1986)

9. R. Hippler, H. Schwier, H. J. Humpert, H. O. Lutz: To be published

10. H. R. Reiss: Phys. Rev. A 22, 1786 (1980)

11. H. G. Muller, A. Tip, M. J. van der Wiel: J. Phys. B: At. Mol. Phys. 16, L679 (1983)

12. M. J. van der Wiel, H. G. Muller: In Electronic and Atomic Collisions, ed. by D. C. Lorents, W. E. Meyerhof, J. R. Peterson (North-Holland, Amsterdam 1986) p.735

13. L. V. Keldysh: Sov. Phys. JETP 20, 1307 (1965)

14. H. R. Reiss: To be published

15. F. H. M. Faisal: J. Phys. B:At. Mol. Phys. 6, L89 (1973)

16. H. R. Reiss: J. Math. Phys. 3, 59 (1962)

17. T. J. McIlrath: Private communication

# Multiphoton Ionisation Towards Multiple Continua: The Role of Space Charge

*M. Crance*

Laboratoire Aimé Cotton, C.N.R.S. II, bat. 505,
F-91405 Orsay Cedex, France

## 1. Introduction

When an atom with ionisation potential $E_o$ is irradiated by a strong electromagnetic field of frequency w, it ionises. Energy balance requires that at least (n+1) photons have been absorbed (n is defined by $n\hbar w < E_o < (n+1)\hbar w$ ). The energy spectrum of emitted electrons consists of several peaks, evenly spaced. Peak number k corresponds to electrons of energy

$$E_k = (n+k)\hbar w - E_o \tag{1}$$

ejected when the atom has absorbed n+k photons. The latter description is consistent with a perturbative interpretation of multiphoton absorption. In such a frame, the height of k-th peak, $N_k$, is predicted to vary as the (n+k)-th power of light intensity I. A perturbative description is justified as long as (n+k)-photon process is more efficient than (n+k+1)-photon process, and thus the heights of peaks are expected to decrease when k increases. These features have been observed for moderate intensities such as the ones required to ionise alkali atoms [1,2,3].

The energy spectrum of electrons emitted after multiphoton ionisation of noble gases seems to be much more difficult to interpret. The most puzzling point is that different experiments, carried out in similar conditions, have given rather different results. It is not surprising that electron energy spectrum depends strongly on the wavelength of the light exciting the medium. A good example is provided by the comparison of the results obtained for irradiation at 1.06μ and 0.53μ [4,5] (Nd-YAG laser frequency and its first harmonic): two or three peaks are observed in the latter case, while ten or more are observed in the former case. Electron energy spectrum observed for Xenon irradiated at 193nm [6] contains only two peaks at intensity $10^{14}$ W $cm^{-2}$ (for higher intensities, Xenon ejects more than one electron and the electron energy spectrum cannot be described by Eq.1 any longer). We shall focus on a particularly interesting situation: Xenon irradiated with a Nd -YAG laser. We describe the differences between experimental results obtained in various conditions. The trend is that first peaks vanish when the intensity is increased. The striking question is how many peaks vanish and why. We shall discuss possible mechanisms and insist particularly on the effect of space charge.

Multiphoton processes are often governed by power laws reflecting the number of photons involved. Numerous studies of multiphoton processes have shown that it is fruitful to attempt a perturbative interpretation and thus search for power laws. A good way to test the validity of perturbation theory is to study the effective order of non linearity. If perturbation theory is valid, the probability P for a process involving absorption of p photons varies as $I^p$, the p-th power of light intensity I. The same power law holds for the number N(I) of events observed. Experimentally, the effective order of non linearity is defined as the slope of the curve N(I) versus I in a Log-log plot. i.e.

$$\pi = d(\mathrm{Ln}N)/d(\mathrm{Ln}I) \qquad (2)$$

Observation of $\pi = p$ does not imply the validity of perturbative theory, but $\pi \neq p$ means clearly that a perturbative treatment is not justified.

A necessary condition to observe several peaks in electron energy spectrum is to shine the atoms at a light intensity of the order of saturation intensity. Let us recall that saturation intensity $I_s$ is defined, for pulsed excitation, as the intensity above which the probability for an atom to be ionised at the end of the pulse is one. Obviously, $I_s$ depends on the pulse shape and especially on its duration. When maximum light intensity is larger than $I_s$, the number of events is equal to the number of atoms available. Then for a light pulse with maximum intensity larger than saturation intensity, the number of electrons emitted is proportional to the volume where intensity is larger than $I_s$. Intensity dependence depends on the shape of the light beam and corresponds to effective order of non linearity much smaller that the number of photons absorbed.

## 2. Above-Threshold Ionisation of Xenon

Above-Threshold Ionisation of Xenon irradiated at 1.06μ was first studied by two groups [4,5]. Fabre and coworkers used a monomode laser delivering 20ns pulses of intensity $10^{12}$ W cm$^{-2}$. They observed eleven peaks and for all of them they found an effective order of non linearity of 3.4. The first peak was lower than the second one but could be clearly identified. The effective order of non linearity much lower than the number of photons absorbed suggests that maximum intensity was larger than saturation intensity. Kruit and coworkers used a multimode laser delivering nanosecond pulses. They detected nine peaks. Intensity was varied from 2.4 $10^{13}$ to 6.4 $10^{13}$ W cm$^{-2}$. For all the peaks but the first one, the effective order of non linearity was measured to be of the order of 10 (see Table 1). The value of $\pi$ suggests that the experiment was carried out at an intensity lower than saturation intensity, moreover, the results cannot be interpreted by perturbation theory at lowest non vanishing order. At intensity 6.4 $10^{13}$ W cm$^{-2}$, second and third peaks had about the same height. First peak was too low to be distinguished from the background.

Table 1: Comparison between the number of photons required, n+k, and the effective order of non linearity, π, from reference [5].

| n+k | 11 | 12 | 13 | 14 | 15 | 16 | 17 | 18 | 19 |
|---|---|---|---|---|---|---|---|---|---|
| π | 7.7 | 10.0 | 10.2 | 10.4 | 10.8 | 10.5 | 10.6 | 10.5 | 10.8 |

Effective order of non linearity has been studied indirectly in an experiment carried out with picosecond pulses with intensity of the order of $10^{13}$ W $cm^{-2}$ [7]. Electrons were detected by using the Retarding Potential method. A DC electric field is applied in a region between the interaction volume and the detector so that electrons have to cross a potential barrier of intensity V. By this method, one measures the number of electrons emitted with a kinetic energy larger than V. By varying V it is thus possible to study the number of electrons $N_p$ which have absorbed p photons or more. If the number of electrons $N_p$ is governed by a power law, the slope of $N_p$ versus I should be at least p. We compare in table 2 the effective order of non linearity π of $N_{n+k}$ for various k-values.

Table 2: Comparison between the effective order of non linearity π and the minimum number of photons required n+k.

| n+k | 11 | 19 | 28 |
|---|---|---|---|
| π | 11.3±0.3 | 19.0±1.5 | 30.0±2 |

The latter result is in contradiction with data of reference 5, although the intensity is about the same. One essential difference is the collection set-up. The French group observed electrons emitted in the direction of the light polarisation, while the Dutch group detected electrons in a 2π solid angle.

Still using the Retarding Potential Method, Lompré and coworkers [8] have studied the evolution of electron energy spectrum for intensity up to 3.6 $10^{13}$ W $cm^{-2}$. They observe a complete vanishing of first peak. Recently three groups have begun experiments which could shed some light on the discrepancies between earlier results. For an intensity of the order of $10^{13}$ W $cm^{-2}$, Humpert and coworkers [9] studied the angular distribution of electrons in the first five peaks. Qualitatively, observed angular distributions are more and more peaked along the polarisation direction when the number of the peaks increases. Bucksbaum and coworkers [10] compared the effect of linear and circular polarisation. Yergeau and coworkers [11] using a picosecond laser have begun to investigate the effect of pressure on the shape of electron energy spectrum.

## 3. Calculation of Electron Energy Spectrum

Several models have been proposed to interpret the experimental results described above. Before comparing these models, we shall recall the procedure one generally follows to calculate measured quantities. A first step is to calculate for an atom located at point $\mathbf{r}$, which is irradiated by a light pulse of maximum intensity $I(\mathbf{r})$, the probability of ejecting an electron with energy $E_k$. Note that the atom is treated as if it were isolated. A second step is to integrate the $P_k(I(\mathbf{r}))$'s over the spatial distribution of intensity to obtain the electron energy spectrum described by $N_k(I_M)$ when maximum intensity in the interaction volume is $I_M$.

Models invoking ponderomotive forces predict the vanishing of the first peaks in $P_1(I(\mathbf{r}))$, $P_2(I(\mathbf{r}))$...for large intensity $I(\mathbf{r})$ [11-13]. This effect has been treated in various formalisms but can be described qualitatively in a simple way. AC Stark shift affects all states bound and continuum. For strong intensity, continuum threshold would be shifted away from the ground state by more than $\hbar\omega$. As a result, absorption of additional photons would be necessary to reach ionisation. An important point is that the threshold shift is exactly compensated by the effect of ponderomotive forces and kinetic energy of electrons reaching the detector are still given by Eq. 1. However, the condition on k depends on the intensity: for weak field, k has to be positive while for strong field, the first allowed value of k can be several units. Some treatments using particular model atoms predict an inversion of peaks' height for large intensity $I(\mathbf{r})$ [14-16]. The height of peak k should grow with k, reach a maximum and then decrease for large k. None of the previous models predict a complete vanishing of first peaks in electron energy spectrum. In the latter models, it is assumed that ionisation occurs, electron escapes and reaches the detector as if only one ion had been created for each laser shot. However, in most experiments and particularly in [8] a large number of atoms are ionised (up to $10^5$) and the attractive potential well created by ions is far from negligible. In some cases, it could be deep enough to trap some electrons belonging to the first peaks. The potential well could also scatter electrons and thus alter the angular distribution. More generally, the large number of ions or electrons in a restricted region could be responsible for space-charge effects.

## 4. The Effect of Space Charge

It is rather difficult to calculate accurately the effect of space charge. One should first assume realistic values for the probability $P_k$ to emit an electron with energy $E_k$ and then follow the trajectory of each electron until either it is trapped or it has left the interaction volume and is far enough from any other charge to have a linear trajectory. It is not possible to "correct electron energy spectrum for space charge effect". If electrons are trapped or do not reach the detector because they have been scattered, some information on the $P_k$'s is lost. First peaks are strongly attenuated or missing and it is not possible to rebuild the $P_k$'s from the $N_k$'s. It is more fruitful to build tractable models for ionisation and calculate space charge effect for such systems. This procedure gives orders of magnitude which can be used to delimit the range of parameters where space charge is negligible.

## 4.1 Model for multiphoton ionisation

Intensity distribution can be written as a function of time t and position **r**

$$I(\mathbf{r},t) = I_M\, g(\mathbf{r})\, f(t) \qquad (3)$$

Max(g)=Max(f)=1. f(t) describes the pulse shape. $I_M g(\mathbf{r})$ is the maximum intensity seen at point **r**.

When maximum intensity $I_M$ is larger than saturation intensity $I_S$, every atom is ionised in the part of the light beam where $I_M g(\mathbf{r})$ is larger than $I_S$. The number of ions created out of this volume is negligible. We assume that the spatial distribution has a spherical symmetry and we replace g(**r**) and f(t) by Gaussian functions of **r** and t respectively. r = |**r**| is the distance to the best focus in the light beam. For each point of the interaction volume, we assume that ionisation occurs once the intensity reaches the saturation value $I_S$. The probability of an electron being ejected with energy $E_k$ is assumed to follow a power law:

$$P_k = (1-X)\, X^{k-1} \qquad (4)$$

We have carried out two calculations. One assuming an isotropic distribution of velocity [18], one assuming that the electron velocities are parallel to the polarisation of the light field [19].

## 4.2 Isotropic Distribution of Velocity

After ionisation has begun, electrons leave the ions creating a distribution of charges. We calculate the potential U(r,t) created at time t for a distance r from the best focus of the beam, as if no electron were trapped. Then we calculate the trajectories of slow electrons in potential U(r,t). We have applied this method to experimental situation of Lompré and coworkers [8]. X= 0.83 has been deduced from the $N_k$'s for $k > 8$. We found that all electrons of energy $E_1$, 70% of electrons of energy $E_2$, 30% of electrons of energy $E_3$ remain trapped by ions.

## 4.3 velocity along the polarisation axis

It has been shown that angular distribution of electrons is strongly peaked along the polarisation axis. We modelise this feature by assuming that electrons leave either backwards or forwards with a velocity parallel to the light field polarisation.

Moreover, the previous calculation shows that the trajectory of slow electrons escaping the interaction volume deviates strongly from a straight line. This means that another effect of space charge should be to alter the angular distribution of emitted electrons. After ionisation has occurred, electrons leave the cloud of ions and form two cylinders, one forward, one backward, and would stay together if they did not interact. If the number of electrons created is large enough, the electron cloud expands because of electrostatic interaction between electrons. Electrons gain a transverse component of velocity and the angular distribution of electrons sprays. This process should not affect the result of experiments in which electrons are collected in a 2-π angle, however, it could be responsible for the vanishing of first peaks when the collection angle is small, as in experiments reported in [7,8]. It could also explain the trend of angular distributions recorded by Humpert and coworkers [9].

## 4.4 Energy Balance

Space-charge effect can be discussed from another point of view: energy balance between the situations right after ionisation and at the time of electron detection. Each atom has been ionised as if it were isolated. An electron is ejected with a probability $P_k$ to have energy $E_k$. Some electrons escape, probably the fastest, some electrons remain trapped, essentially the slowest. Trapping of slow electrons is due to the potential well created by ions. Right after ionisation has occurred, there is no potential well and electrons begin to escape freely. Fast electrons leave first, ions in excess create an attractive potential. Let us compare the situation right after ionisation has occurred and later, at a time where all electrons which have left the interaction volume are far from any other charge, but ions are still where they have been created (the velocity of ions is at least hundred times smaller than the velocity of slow electrons). The cloud of ions plus trapped electrons has a potential energy due to electrostatic interaction. Schematically, we can describe the process in the following way: "fast electrons" have escaped keeping their kinetic energy, "slow electrons" have been trapped and their kinetic energy has been converted into potential energy of the charge cloud. This "hit or miss" assumption is not realistic but gives an order of magnitude of the number of peaks missing in the electron energy spectrum detected.

We analyse the experimental situation of reference 8 using the energy balance argument. Potential energy of the ions plus trapped electrons cloud is

$$W_p = \frac{e^2}{4\pi\varepsilon_0} \left( N_I - \sum_{1}^{m} N_p \right)^2 \tag{5}$$

$N_I$ is the number of ions. We assume that electrons of the first m peaks have been trapped having lost kinetic energy

$$W_K = \sum_{1}^{m} E_k N_k \tag{6}$$

p is determined by equating $W_P$ and $W_K$. For experimental conditions of [8], the energy balance argument predicts m = 3. This result means, in fact, that most electrons belonging to the first m peaks have been trapped . This is in qualitative agreement with experimental results.

## 5. Conclusion

Above-Threshold Ionisation of noble gases is far from a complete understanding. A first task is to explain the differences between experiments carried out in similar conditions. In this perspective, space-charge effect may partly explain the discrepancies between experiments in which the number of ions created is very different (a few units in [11], up to $10^5$ in [8]). It is clear that the trend of first peaks to vanish when intensity is increased is not only due to space charge. However, model calculations show that the conditions for avoiding space-charge effects are more severe than usually believed. It is thus important to control both the number of charges and their density before attempting an interpretation based on the assumption of "isolated atoms".

1. G.Petite, F.Fabre , P.Agostini, M.Crance, M.Aymar Phys. Rev.A29, 2677 (1984)
2. R.N.Compton, J.A.D.Stockdale, C.D.Cooper, X.Tang, P.Lambropoulos Phys. Rev.A30, 1766 (1984)
3. A.Dodhy, R.N.Compton, J.A.D.Stockdale Phys. Rev. Lett. 54, 422 (1985)
4. F.Fabre, G.Petite, P.Agostini, M.Clement J.Phys.B15, 1353 (1982)
5. P.Kruit, J.Kimman, H.G.Muller, M.J.Van der Wiel Phys.Rev.A28, 248 (1983)
6. C.K.Rhodes Fundamentals of laser interaction Lecture Notes in Physics 229, 111 (Springer Verlag 1985)
7. L.A.Lompré, A.L'Huillier, G.Mainfray, J.Y.Fan J.Phys.B17, L817 (1984)
8. L.A.Lompré, A.L'Huillier, G.Mainfray, C.Manus J.Opt.Soc.Am. B2 1906 (1986)
9. H.J.Humpert, H.Schwier, R.Hippler, H.O.Lutz Phys.Rev.A32, 3787. (1985)
10. P.H.Bucksbaum, M.Bashkansky, R.R.Freeman, T.J.McIlrath, L.F.DiMauro Phys.Rev. Lett.56 2590 (1986)
11. F.Yergeau, G.Petite, P.Agostini J.Phys.B19(1986) to be published
12. H.G.Muller, A.Tip, M.J.Van der Wiel J.Phys.B16, L679 (1983)
13. M.H.Mittleman Prog. Quantum Electron.8, 165 (1984)
14. S.I.Chu, J.Cooper Phys.Rev.A32, 2769 (1985)
15. Z.Bialynicka-Birula J.Phys.B17, 3091 (1984)
16. Z.Deng, J.H.Eberly J.Opt.Soc.Am.B2, 486 (1985)
17. M.Edwards, L.Pan, L.Armstrong Jr J.Phys.B18, 1927 (1985)
18. M.Crance J.Phys.B19 L267 (1986)
19. M.Crance J.Phys.B19 (1986)to be published

# Two-Photon Above-Threshold Ionization: Study of a Schematic Model Atom

*G.P. Arrighini*[1], *N. Durante*[2], *and C. Guidotti*[1]

[1]Dipartimento di Chimica e Chimica Industriale, Università di Pisa, Via Risorgimento 35, I-56100 Pisa, Italy

[2]Istituto di Chimica Quantistica ed Energetica Molecolare del C.N.R., Via Risorgimento 35, I-56100 Pisa, Italy

Two-photon ionization of a schematic one-electron model atom described in terms of Fermi pseudopotential is investigated in the frequency domain corresponding to channels already open after absorption of a single photon.

## 1. Introduction

The advent of powerful laser-sources able to produce within the shined atom/molecule an electric field amounting to a considerable fraction of the coulombic field acting between pairs of charged particles of the target has determined a strong stimulus to developing theoretical techniques which are appropriate for a non-perturbative description of the interaction between electromagnetic radiation and matter /1-3/.

Under milder conditions, however, especially when resonant intermediate processes are absent, a perturbation-theory approach to the problem appears quite justified, so that a N-photon transition is admittedly well described at the lowest order by the N-th term of a perturbative expansion of the time-dependent Schrödinger equation. Although in these conditions the problem is very considerably simplified with respect to the case of an electromagnetic perturbation of arbitrary intensity (and frequency) the evaluation of N-photon absorption cross-sections remains, even for the simplest systems, a very difficult task /4-9/ that has given rise to a remarkable development of clever and tricky techniques in the purpose of making the problem more tractable /6,7/.

We have recently suggested an alternative approach to the calculation of photoionization cross-sections /10-12/, which avails itself of a reformulation of those quantities in terms of Fourier transforms of appropriate time-dependent correlation functions. The development of a time-dependent picture according to which to reconsider various dynamical processes that from the experimental point of view are usually interpreted in the frequency domain is to be pointed out as a surely remarkable result, having noticeable applicative consequences especially whenever an implementation of classical or semiclassical techniques is meaningful /13-16/. Although the value of a time-dependent approach to photoionization at present is not too well established at a computational level, we recognize its validity from both formal and conceptual point of view, because one's attention is turned to consider few really fundamental dynamical quantities in terms of which the problem can be analyzed /10,16/.

In the present paper we propose to extend the evaluation of the 2-photon ionization cross-section, for a particularly simple one-electron model atom (see sect. 3), to the situation where the ionization channel is already open after absorption of a single photon ("above-threshold ionization" region) /5/.

## 2. Formalism

In the so-called "length" form /10-12/, the ionization cross-section $\sigma^{[2]}$ corresponding to the absorption of two (identical) photons from a linearly polarized radiation field of frequency $\omega$ is expressed in the following form (atomic units with $e = m_e = \hbar = 1$ are used throughout the whole paper):

$$\sigma^{[2]}(\omega) = (8\pi^3\omega^2/c^2)\sum_{L=0,2}\int dE_f\,\delta(E_f-E_b-2\omega)|\langle E_f^- L0|\,zG^+(E)z|\,E_b00\rangle|^2, \quad E = E_b+\omega \qquad (1)$$

$|E_f^- L0\rangle$ denotes the (energy-normalized) final state, with incoming-wave boundary conditions, appropriate to the channel of angular momentum ($L, M_L=0$), while $|E_b00\rangle$ represents the (spherically symmetric) ground state of the atom. $G^+(E) \equiv [H-(E+i0^+)]^{-1}$ is the propagator in the energy-domain, i.e. the Green's operator of the (unperturbed) atom. It should be observed that we are assuming $E = E_b + \omega > 0$, corresponding to ionization channels already open after absorption of a single photon.

It is rather straightforward /10-12/ to cast $\sigma^{[2]}$ into an equivalent, alternative form which exhibits it as the expectation value of an appropriate operator quantity. From the completeness of the set of states supported by the model atom, we find

$$\sigma^{[2]}(\omega) = (8\pi^3\omega^2/c^2)\langle\Phi(E)|z\delta[H-(E_b+2\omega)]z|\Phi(E)\rangle \qquad (2)$$

where we have defined the state ket

$$|\Phi(E)\rangle = G^+(E)z|E_b00\rangle, \qquad E = E_b+\omega \qquad (3)$$

The expression of $\sigma^{[2]}$, eq. (2), is easily transformed into the following one

$$\sigma^{[2]}(\omega) = (2\pi\omega/c)^2\int_{-\infty}^{\infty} dt\; e^{2i\omega t}\langle E_b00|\xi_z^\dagger(t)\,\xi_z(0)|E_b00\rangle \qquad (4)$$

where

$$\xi_z(0) \equiv z\,G^+(E)z, \quad \xi_z(t) = e^{iHt}\,\xi_z(0)\,e^{-iHt} \qquad (5)$$

Through eq. (4) we arrive at the searched cross-section by Fourier transforming an appropriate time-dependent correlation function /10/. For the present purposes, however, we prefer to evaluate $\sigma^{[2]}(\omega)$ according to eq. (2), which involves the expectation value in the "doorway" state $z|\Phi(E)\rangle$ /17/ of the "microcanonical" density operator $\delta[H-(E_b+\omega)]$. By using simple manipulations, eq. (2) can be set into the following final form /11/

$$\sigma^{[2]}(\omega) = \frac{1}{2}(4\pi\omega/c)^2 \langle\Phi(E)|z[\mathrm{Im}\, G^+(E_b+2\omega)]z|\Phi(E)\rangle. \tag{6}$$

## 3. Application to a Schematic Model Atom

As a simple example of the formalism of the previous section, we re visit the zero-range potential model, most frequently adopted for modeling the behaviour of negative ions /1,18-20/. Each outer-shell electron is assumed to move in an effective short-range potential, that we have chosen in the form of the so-called Fermi pseudo-potential /1,21/

$$V_{eff}(\vec{r}) = 2\pi a\,\delta(\vec{r})\,\frac{\partial}{\partial r}r \tag{7}$$

where $r = |\vec{r}|$ and a is a real constant [an equivalent potential has been considered by MULLER et al. /2/ and MULLER and TIP /3/ in their non-perturbative approach to the photoionization of model atoms in arbitrarily intense electromagnetic fields]. Such a potential supports, in addition to the continuum spectrum, a single S-symmetry bound state

$$u_b(r) \equiv \langle\vec{r}|E_b 00\rangle = \mathcal{N} e^{-(r/a)}/r \tag{8}$$

with detachment energy $-E_b = 1/2a^2$ /11/. All that we need in the following is contained in the Green's function $G(\vec{r},\vec{r}\,';u) \equiv \langle\vec{r}|[H-u]^{-1}|\vec{r}\,'\rangle$, u being an arbitrary complex quantity /21/

$$G(\vec{r},\vec{r}\,';u) = G_0(\vec{r},\vec{r}\,';u) + G_1(\vec{r},\vec{r}\,';u) \tag{9}$$

$$G_0(\vec{r},\vec{r}\,';u) = \frac{\exp[ik|\vec{r}-\vec{r}\,'|]}{2\pi|\vec{r}-\vec{r}\,'|}, \qquad G_1(\vec{r},\vec{r}\,';u) = -\frac{a\exp[ik(r+r')]}{2\pi(1+ika)r\,r'} \tag{10}$$

where $k = (2u)^{\frac{1}{2}}$. The contribution $G_0(\vec{r},\vec{r}\,';u)$ to $G(\vec{r},\vec{r}\,';u)$ is immediately seen to be the free-particle Green's function, while $G_1(\vec{r},\vec{r}\,';u)$, which involves only the L=0 term in a partial wave analysis, takes into account the whole effect of the potential $V_{eff}(\vec{r})$, eq. (7).

If we express the matrix element of eq. (6) in the coordinate representation, we find

$$\sigma^{[2]}(\omega) = \frac{1}{2}(4\pi\omega/c)^2 \lim_{\eta\to 0}\iint d\vec{r}\,d\vec{r}\,'\; z\,z'\;\Phi^*(\vec{r};E)\;\Phi(\vec{r}\,';E)\;\mathrm{Im}G(\vec{r},\vec{r}\,';E_b+2\omega+i\eta) \tag{11}$$

where

$$\Phi(r;E) \equiv \langle\vec{r}|\Phi(E)\rangle = \lim_{\varepsilon\to 0}\int d\vec{r}\,''\; z''\;u_b(r'')\;G_0(\vec{r},\vec{r}\,'';E_b+\omega+i\varepsilon) \tag{12}$$

(the contribution to $\Phi(\vec{r};E)$ from $G_1(\vec{r},\vec{r}';u)$ vanishes owing to symmetry reasons). On resorting to a standard analysis in partial waves of the quantity $\exp[ik|\vec{r}-\vec{r}''|]/|\vec{r}-\vec{r}''|$ we are led after some algebra to the following explicit result for $\Phi(\vec{r};E)$:

$$\Phi(\vec{r};E) = \mathcal{N}(4\pi/3)^{\frac{1}{2}} \frac{1}{(\omega r^2)} \left[e^{i\beta r}(1-i\beta r)-e^{-\gamma r}(1+\gamma r+\omega r^2)\right] Y_{10}(\hat{r}) \equiv f(r;E)Y_{10}(\hat{r}) \quad (13)$$

where we have put

$$\gamma = 1/a\ , \quad \beta = (2\omega - \gamma^2 + i\,0^+)^{\frac{1}{2}}. \quad (14)$$

As expected, the "doorway" state $z|\Phi(E)\rangle$ has angular momentum components L=2 and L=0; both of these are determined by the motion in the absence of effects arising from the presence of the potential $V_{eff}$.

If we come back to eq. (11), from eqs. (9), (10), (12) we are led to write

$$\sigma^{[2]} = \sigma^{[2]}_{Born}(\omega) + \Delta\sigma^{[2]}(\omega) \quad (15)$$

where $\sigma^{[2]}_{Born}$ is the contribution to $\sigma^{[2]}(\omega)$ arising from the free-particle Green's function $G_0(\vec{r},\vec{r}';u)$, which consists, as expected, of S- and D-symmetry partial wave components. It is rather straightforward to verify that

$$\sigma^{[2]}_{Born}(\omega) = (16\pi^2\omega^2/15\,c^2)\,k\left[4|S_2(k)|^2 + 5|S_0(k)|^2\right] \quad (16)$$

where

$$S_L(k) = \int_0^\infty dr\, r^3\, j_L(kr)\, f(r;E) \quad (17)$$

with $j_L(z)$ a spherical Bessel function of the first kind, of argument z and

$$k = (4\omega - \gamma^2)^{\frac{1}{2}} \quad (18)$$

The change $\Delta\sigma^{[2]}(\omega)$ caused by the presence of the (pseudo)potential is related entirely to $G_1(\vec{r},\vec{r}';u)$. Because of the exclusive S-symmetry of the corresponding contribution, we have a single-term expression,

$$\Delta\sigma^{[2]}(\omega) = -(4\pi^2\omega^2/3c^2)\,k^2\,\mathrm{Im}\left[(\gamma - ik)\,T(k)\,T^*(-k)\right] \quad (19)$$

with

$$T(k) = \int_0^\infty dr\, r^3\, f(r;E)\, h_0^+(kr) \quad (20)$$

$h_0^+(z)$ being a Hankel function of first kind. Eq.(19) puts in evidence how $\Delta\sigma^{[2]}(\omega)$ is not necessarily a positive quantity.

Table 1 - Contributions to the 2-photon ionization cross-section in the "normal" region ($\gamma^2/4<\omega<\gamma^2/2$) and "above threshold" ($\omega^2>\gamma^2/2$) for the Fermi pseudopotential model atom.

| 2-photon ionization cross-section (a.u.) | Normal 2-photon ionization ($\gamma^2/4<\omega<\gamma^2/2$) | Above threshold 2-photon ionization ($\omega>\gamma^2/2$) |
|---|---|---|
| $\sigma^{[2]}_{Born,L=0}$ | $\pi^3 N^2 k(2\gamma^2-5\omega)^2/(9\omega^6c^2)$ | $\pi^3 N^2 k(2\gamma^2-5\omega)^2/(9\omega^6c^2)$ |
| $\sigma^{[2]}_{Born,L=2}$ | $16\pi^3 N^2 k^5/(45\omega^6c^2)$ | $16\pi^3 N^2 k^5/(45\omega^6c^2)$ |
| $\Delta\sigma^{[2]}$ | $\pi^3 N^2 k\left[4\beta^4(2\gamma^2-3\omega-2\gamma\beta)+ + \omega^2(4\gamma^2-9\omega)\right]/(9\omega^7c^2)$ | $\pi^3 N^2 k(4\gamma^4-12\gamma^2\omega+7\omega^2)/(9\omega^6c^2)$ |
| $\sigma^{[2]}_S = \sigma^{[2]}_{Born,L=0}+\Delta\sigma^{[2]}$ | $8\pi^3 N^2\beta^4 k(\gamma^2-\omega-\gamma\beta)/(9\omega^7c^2)$ | $8\pi^3 N^2\beta^4 k/(9\omega^6c^2)$ |
| $\sigma^{[2]}$ | $8\pi^3 N^2 k\left[\frac{2}{5}\omega k^4+\beta^4(\gamma^2 + -\omega-\gamma\beta)\right]/(9\omega^7c^2)$ | $8\pi^3 N^2 k(\beta^4+\frac{2}{5}k^4)/(9\omega^6c^2)$ |
| | $\beta=(\gamma^2-2\omega)^{\frac{1}{2}}$, $k=(4\omega-\gamma^2)^{\frac{1}{2}}$ | $\beta=(2\omega-\gamma^2)^{\frac{1}{2}}$, $k=(4\omega-\gamma^2)^{\frac{1}{2}}$ |

The model considered in the present paper is susceptible to an entirely analytical treatment. In Table 1 we report the explicit expressions for various contributions in terms of which $\sigma^{[2]}(\omega)$ can be partitioned, either in the "normal" 2-photon ionization domain (i.e. as $\gamma^2/4<\omega<\gamma^2/2$) /10-12/ or in "above-threshold ionization" regime (i.e., as $\omega>\gamma^2/2$). The entries collected in the Table allow to investigate the relative prominence of S- and D-symmetry partial wave contributions to $\sigma^{[2]}(\omega)$ over the whole spectral domain as well as the degree of accuracy of possible simplifications founded on the approximation $G(\vec{r},\vec{r}';u) \simeq G_0(\vec{r},\vec{r}';u)$ /22,23/, which leads to the quantities labeled by us $\sigma^{[2]}_{Born,L}(\omega)$. An identical formula valid below and above the threshold corresponding to ionization induced by absorption of a single photon is found to hold for both $\sigma^{[2]}_{Born,L=0}$ and $\sigma^{[2]}_{Born,L=2}$, while the same behaviour is not exhibited by $\Delta\sigma^{[2]}$, for which different expressions below and above $\omega=\gamma^2/2$ are to be pointed out. As a further element worth of mention we remark that $\sigma^{[2]}_S$ vanishes at the threshold $\omega=\gamma^2/2$.

In order to better appreciate the behaviour of the quantities collected in Table 1 three curves for $\sigma^{[2]}$ (full line), $\sigma^{[2]}_S$ (dashed line) and $\sigma^{[2]}_D \equiv \sigma^{[2]}_{Born,L=2}$ (dotted line) respectively are represented as a function of $Log_{10}(k)$ in Fig. 1. The calculations refer to the choice $\gamma = 0.235588$ a.u., which mimics the case of the negative hydrogen ion $H^-$. The vertical line at $Log_{10}(\gamma) = -0.6278$ marks the beginning of the

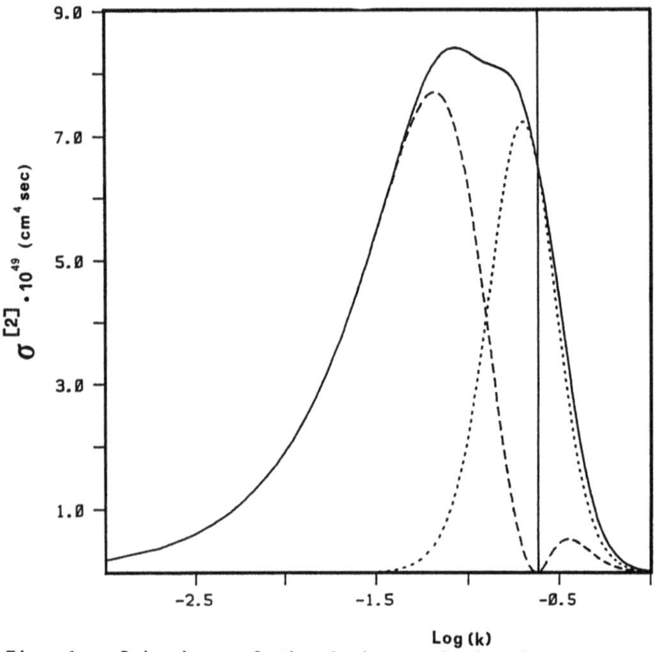


Fig. 1 - Behaviour of the 2-photon ionization cross-section versus the decimal logarithm of the momentum k of the ejected electron. The full line represents the quantity $\sigma^{[2]}$, the dashed line refers to the quantity $\sigma_S^{[2]}$ and the dotted line corresponds to the quantity $\sigma^{[2]}_{Born,L=2}$. The vertical line at $Log_{10}(k)=Log_{10}(\gamma)$ marks the beginning of the "above-threshold ionization" region.

"above-threshold ionization" region. At low values of the momentum $k = (4\omega - \gamma^2)^{\frac{1}{2}}$ of the ejected electron one sees that $\sigma^{[2]} \simeq \sigma_S^{[2]}$ , that involves an essentially isotropic distribution of the emitted electrons. Near the threshold $k = \gamma$, however, $\sigma^{[2]} \simeq \sigma_D^{[2]}$ , with a corresponding strongly changed type of distribution of the emitted electrons, which display to a prevailing extent forward scattering. This latter kind of behaviour is also evident after entering the "above-threshold ionization" region; from Table 1, it is finally recognized that at large k values $(\sigma_D/\sigma_S) \to 8/45$, so that the distribution of the emitted electrons again becomes largely ($\sim$ 85%) isotropic.

Acknowledgments

This work was supported by Ministero della Pubblica Istruzione.

1. I.J. Berson: J. Phys. B: Atom.Molec. Phys. 8, 3078 (1975)
2. H.G. Muller, A. Tip, M.J. Van der Wiel: J. Phys. B: Atom. Molec. Phys. 16, L679 (1983)
3. H.G. Muller, A. Tip: Phys. Rev. A 30, 3039 (1984)

4. P. Lambropoulos: In Advances in Atomic and Molecular Physics, ed. by D.R. Bates and B. Bederson, Vol. 12 (Academic Press, New York 1976)
5. J. Morellec, D. Normand, G. Petite: In Advances in Atomic and Molecular Physics, ed. by D.R. Bates and B. Bederson, Vol. 18 (Academic Press, New York 1981)
6. M.G.J. Fink, P. Zoller: J. Phys. B: Atom. Molec. Phys. 18, L373 (1985)
7. M. Crance, M. Aymar: J. Phys. B: Atom. Molec. Phys. 18, 3529 (1985)
8. W.M. Huo, R.L. Jaffe: Chem. Phys. Letters 101, 462 (1983)
9. R. Moccia, A. Rizzo: J. Phys. B:Atom. Molec. Phys. 18, 3319 (1985)
10. G.P. Arrighini, F. Biondi, C. Guidotti: J. Phys. B: Atom. Molec. Phys. 19, 1297 (1986)
11. G.P. Arrighini, N. Durante, C. Guidotti: Nuovo Cimento D7, 559 (1986)
12. G.P. Arrighini, N. Durante, C. Guidotti: Phys. Rev. A (accepted for publication)
13. E.J. Heller: J. Chem. Phys. 68, 3891 (1977)
14. K.C. Kulander, E.J. Heller: J. Chem. Phys. 69, 2439 (1978)
15. E.J. Heller: Acc. Chem. Res. 14, 368 (1981)
16. E.J. Heller: In Potential Energy Surfaces and Dynamics Calculations, ed. by D.G. Truhlar (Plenum Press, New York 1981)
17. W. Rhodes: Chem. Phys. 22, 95 (1977)
18. S.A. Adelman: J. Phys. B: Atom. Molec. Phys. 6, 1986 (1972)
19. Y.N. Demkov: In Electronic and Atomic Collisions, ed. by N. Oda, K. Takayanagi (North Holland, Amsterdam 1980)
20. G. Drukarev: In Advances in Quantum Chemistry, ed. by P.O. Löwdin, Vol. 11 (Academic Press, New York 1978)
21. T.T. Wu: In Resonances-Models and Phenomena, ed. by S. Albeverio, L. S. Ferreira, L. Streit, Lecture Notes in Phys. Vol. 211 (Springer, Berlin, Heidelberg 1984)
22. S. Geltman: Phys. Letters 4, 168 (1963)
23. F.H.M. Faisal: Nuovo Cimento 33 B, 775 (1976)

# Multiphoton Absorption Above the Ionization Threshold

*R. Shakeshaft*

Physics Department, University of Southern California,
Los Angeles, CA 90089, USA

## 1. INTRODUCTION

The topic of this paper is above-threshold processes, where the number, N, of photons absorbed by an atom exceeds the minimum number, $N_0$, required to ionize the atom. In contrast to the case $N=N_0$, one encounters in the treatment of above-threshold processes continuum-continuum matrix elements, whose evaluation presents a numerical problem far more challenging than the evaluation of bound-bound and bound-continuum matrix elements. This extra complexity is, however, somewhat artificial in the sense that nonnormalizable functions appear only as an artifact of the particular theoretical approach; the emergent electron is, after all, initially bound and can in principle be described by a normalizable wavepacket throughout the ionization process, even if the electron continues to absorb photons after it has sufficient energy to escape. The root of the difficulty lies in the (unphysical) requirement that the electron depart with a definite energy--for then it must be described by a nonnormalizable (continuum) wavefunction. Furthermore, the (physical) requirement that this wavefunction be regular at the origin implies that it has the character of a stationary wave, comprised of outgoing and ingoing parts. The ingoing part, as it stands, is unphysical; only when stationary waves of different energies, spread over a narrow bandwidth, are superposed does the wavefunction acquire the correct asymptotic character of an outgoing wave, which describes the electron leaving the atom.

When the coupling of the electron to the radiation field can be treated as a perturbation, the aforementioned difficulty can be circumvented by a procedure which starts by splitting the stationary wavefunction of the emergent electron into its outgoing (+) and ingoing (-) parts. This decomposition results in a decomposition of the multiphoton matrix element, $M^{(N)}$, into two parts, $M^{+(N)}$ and $M^{-(N)}$. The matrix element $M^{+(N)}$, which contains only the outgoing part of the final wavefunction, can be handled easily by rotating the integration contour into the upper right quadrant of the complex plane. This rotation transforms the outgoing part of the final wavefunction into a normalizable function, and the evaluation of $M^{+(N)}$ becomes equivalent to the evaluation of a bound-bound matrix element. A similar trick--rotating the integration contour into the lower right quadrant--cannot be used to evaluate $M^{-(N)}$ since its integrand contains outgoing Green's functions which, under this transformation, would increase exponentially at large distances. However, a simple recurrence relation connects $M^{-(N)}$ to $M^{+(N)}$, so that it is unnecessary to calculate $M^{-(N)}$ directly. This technique has, in particular, been used to transform the Sturmian expansion of $M^{(N)}$, which diverges when $N>N_0$, into an expansion which is convergent for $N \geq N_0$.

In the next section we describe this technique in more detail, in the context of perturbation theory. We assume that the atom can be treated as a one-electron atom, with the active electron experiencing a local potential. In section 3 we briefly consider ionization by a field of sufficiently high intensity that the coupling to the field can no longer be regarded as a perturbation. Above the ionization threshold, absorption is favoured over emission and as a result a large number of photons can be absorbed, in strong fields.

## 2. WEAK FIELDS

### 2.1 Introduction

The basic quantity required to calculate the generalized cross-section or the angular distribution for the absorption of N photons, each of frequency $\omega$, is, in lowest order, perturbation theory,

$$M^{(N)} = \int_0^\infty dr_N \dots \int_0^\infty dr_1 \; u_{\nu_N}(r_N) \; r_N \; g^+_{\nu_{N-1}}(r_N, r_{N-1}) \dots r_1 \; u_g(r_1) \tag{2.1}$$

where $\nu_n = (E_n, \ell_n)$ and $E_n = E_0 + n\hbar\omega$; $E_0$ is the energy of the atomic ground state and $\ell_n$ is the angular momentum of the set of intermediate states reached by the absorption of n photons. In Eq. (2.1) $u_{\nu_N}(r)$ and $u_g(r)$ are the radial parts of the final continuum and ground state wave-functions respectively. The quantity $g^+_\nu(r,r')$ is the radial Green's function which satisfies the following equation:

$$(E+i\epsilon-H_\ell)g^+_\nu(r,r') = \delta(r-r') \qquad (\nu=(E,\ell)) \tag{2.2a}$$

where $\epsilon$ is positive but infinitesimal and where

$$H_\ell = -\frac{1}{2}\frac{d^2}{dr^2} + \frac{\ell(\ell+1)}{2r^2} - \frac{Z}{r} + \bar{V}(r) \quad . \tag{2.2b}$$

We use atomic units throughout this section. The potential $\bar{V}(r)$ is the non-Coulombic part of the potential seen by the electron. We assume here that $\bar{V}(r)=0$ for r greater than some cutoff distance $r_c$ from the nucleus. The solution of Eq. (2.2a) automatically satisfies the boundary conditions

$$g^+_\nu(0,r') = 0 \quad , \tag{2.2c}$$

$$g^+_\nu(r,r') \sim e^{ikr} \quad , \quad r \sim \infty \quad , \tag{2.2d}$$

where $k = \sqrt{2E}$; we draw a branch cut along the positive energy axis and take the branch of the square root which is positive when E is on the upper edge of this cut. The solution of Eqs. (2.2) can be written as

$$g_\nu^+(r,r') = -2\pi\hat{R}_\nu(r_<)\hat{I}_\nu^+(r_>) \quad , \tag{2.3a}$$

where $\hat{R}_\nu(r)$ and $\hat{I}_\nu^+(r)$ are, respectively, regular and irregular solutions of $(E+i\epsilon-H_\ell)\phi=0$, which for $r>r_c$ have the form

$$\hat{R}_\nu(r) = i^{\ell+1}(2/\pi k)^{\frac{1}{2}} e^{-i\delta_\ell} e^{\pi\gamma/2} \left\{ \frac{1}{2} \frac{|\Gamma(\ell+1-i\gamma)|}{(2\ell+1)!} M_{i\gamma,\ell+\frac{1}{2}}(-2ikr) + \frac{(-1)^\ell e^{i\eta_\ell-\pi\gamma}}{1+i\cot\delta_\ell} W_{i\gamma,\ell+\frac{1}{2}}(-2ikr)\right\} \quad , \tag{2.3b}$$

$$\hat{I}_\nu^+(r) = \frac{(-i)^\ell}{(2\pi k)^{\frac{1}{2}}} e^{-(\pi\gamma/2)+i\delta_\ell+i\eta_\ell} W_{i\gamma,\ell+\frac{1}{2}}(-2ikr), \tag{2.3c}$$

where $M_{a,b}(z)$ and $W_{a,b}(z)$ are the regular and irregular Whittaker functions. We introduced above the Coulomb phase shift $\eta_\ell=\arg[\Gamma(\ell+1-i\gamma)]$, where $\gamma=Z/k$, and the phase shift, $\delta_\ell$, due to $\bar{V}(r)$. In Eq. (2.3a), $r_<$ and $r_>$ are, of course, the lesser and greater, respectively, of r and r'. The value of the non-Coulombic phase shift $\delta_\ell(E)$ $(=\delta_\ell)$ can be obtained from the relationship [i,2]

$$\cot\left(\delta_\ell(E)\right) = \left(1-e^{-2\pi\gamma}\right) \left\{ \cot\left(\pi\mu_\ell(E)\right) + \frac{iH(-E)}{1-e^{2\pi\gamma}} \right\} \quad , \tag{2.4}$$

where H(E) is the Heaviside step function and $\mu_\ell(E)$ is the quantum defect. A value of $\mu_\ell(E)$ can be found for any E by fitting the measured values of the quantum defects (for a given $\ell$) to a linear function of the energy. This linearity is assumed to hold for both positive and negative values of the energy.

We decompose $u_{\nu_N}(r)$ into its outgoing and ingoing parts, $u_{\nu_N}^+(r)$ and $u_{\nu_N}^-(r)$, respectively, and express $M^{(N)}$ as $M^{+(N)} + M^{-(N)}$, where $M^{\pm(N)}$ is obtained by replacing $u_{\nu_N}(r)$ by $u_{\nu_N}^\pm(r)$ on the right of Eq. (2.1). Using the identity (in operator notation)

$$g_\nu^+ - [g_\nu^+]^\dagger = -2\pi i\delta(E-H_\ell) \quad , \tag{2.5}$$

we can establish the following recurrence relationship [3]:

$$M^{(N)} = 2\mathrm{Re}[M^{+(N)}] - 2\pi i \sum_{n=1}^{N-N_0} J_n^{(N)} M^{(N-n)} \quad , \tag{2.6}$$

where the integrals $J_n^{(N)}$ are defined by

$$[J_n^{(N)}]^* = \int_0^\infty dr_N \ldots \int_0^\infty dr_{N-n+1}\, u^+_{\nu_N}(r_N) r_N g^+_{\nu_{N-1}}(r_N, r_{N-1}) \ldots$$
$$\ldots g^+_{\nu_{N-n+1}}(r_{N-n+2}, r_{N-n+1}) r_{N-n+1}\, u_{\nu_{N-n}}(r_{N-n+1}) \quad , \quad n>1 \qquad (2.7a)$$

$$[J_1^{(N)}]^* = \int_0^\infty dr\, u^+_{\nu_N}(r) r\, u_{\nu_{N-1}}(r) \quad . \qquad (2.7b)$$

## 2.2 Numerical Integration

For $N>N_0$ neither the integrals $M^{(N)}$ nor $M^{\pm(N)}$ are formally convergent. We can, however, use Eqs. (2.3a) to express $M^{+(N)}$ as a sum of N! integrals of the type $I_N(\infty)$, where

$$I_n(r) = \int_0^r dr'\, f_n(r') I_{n-1}(r') \quad , \quad N \geq n > 1 \quad , \qquad (2.8a)$$

$$I_1(r) = \int_0^r dr'\, f_1(r') \quad . \qquad (2.8b)$$

Here $f_n(r)/r$ is the product of an irregular outgoing wave function (either $u^+_{\nu_N}(r)$ or one of the $\hat{I}^+_{\nu}(r)$) and a regular wave function (either $u_g(r)$ or one of the stationary $\hat{R}_\nu(r)$). For present purposes we need only know that the $f_n(r)$ have no poles in the finite complex r-plane and that while $f_N(r)$ and $I_{N-1}(r)$ may individually diverge for $r \sim \infty$ in the upper right quadrant, the product $f_N(r) I_{N-1}(r)$ decays exponentially in this quadrant. Hence, to evaluate $I_N(\infty)$, we can integrate along a ray inclined at an angle $\theta$ to the positive real axis, with $0<\theta<\pi/2$. Writing $r=se^{i\theta}$, with s real and positive, and defining $I_n^\theta(s) = I_n(se^{i\theta})$, we have $I_N(\infty) = I_N^\theta(\infty)$ with

$$I_n^\theta(s) = e^{i\theta} \int_0^s ds'\, f_n(s'e^{i\theta}) I_{n-1}^\theta(s') \quad . \qquad (2.9)$$

The integrand $f_N(s'e^{i\theta}) I_{N-1}^\theta(s')$ of $I_N^\theta(\infty)$ is damped exponentially and hence $I_N^\theta(s)$ converges rapidly to $I_N^\theta(\infty)$ as $s \to \infty$. We can evaluate the $I_n^\theta(s)$ rapidly by solving the coupled differential equations

$$dI_n^\theta(s)/ds = e^{i\theta} f_n(se^{i\theta})\, I_{n-1}^\theta(s) \quad , \qquad (2.10)$$

subject to the boundary conditions $I_n(0)=0$. For $N \lesssim 6$ this is a very efficient procedure for calculating $M^{+(N)}$. A similar procedure can be used to efficiently calculate the $J_n^{(N)}$. We then use Eq. (2.6) to obtain the required matrix element $M^{(N)}$. This method was recently applied [4,5] to the calculation of generalized cross-sections for 4- and 5-photon ionization of Rb and Cs at frequencies where 2 photons are sufficient to ionize the atoms.

## 2.3 Sturmian Expansion

The Sturmian functions $S^k_{n\ell}(r)$ are defined by the eigenvalue equation $[E-H_\ell + (1-\lambda^k_{n\ell})V]\ |S^k_{n\ell}) = 0$, where in coordinate space $V(r) = -Z/r+\bar{V}(r)$. The eigenvalues $\lambda^k_{n\ell}$ are fixed by the requirement that for all E not real and positive $S^k_{n\ell}$ is damped by the factor $e^{ikr}$ for $r \sim \infty$. The eigenfunctions obey the orthogonality relation $(S^k_{n'\ell}|V|S^k_{n\ell}) = \beta^k_{n\ell}\delta_{n'n}$, where $(a|b) = \int_0^\infty dr\ a(r)b(r)$. If E is real and negative, and V(r) is suitably chosen, the $S^k_{n\ell}(r)$ form a complete set in the space $L^{\ell+1}_V$ of all piecewise continuous functions F(r) for which $(F|V|F)$ is finite and $F(r)/r^{\ell+1}$ is bounded for $r \sim 0$.

Consider the matrix element $(b|g_\nu|a)$ with $\nu=(E,\ell)$; we drop the "+" superscript from the Green's operator $g^+_\nu$ since we allow E to range over complex values: $E=|E|e^{i(\pi-2\phi)}$, with $-\pi/2<\phi\le\pi/2$. This matrix element can be expanded in the form [3,6,7]

$$(b|g_\nu|a) = \sum_{n=\ell+1}^{\infty} \frac{(b|S^k_{n\ell})(S^k_{n\ell}|a)}{\beta^k_{n\ell}(\lambda^k_{n\ell}-1)} \quad . \tag{2.11}$$

We argued previously [3,8] that this expansion converges for all E such that $a_\phi(r) \equiv a(re^{i\phi})$ and $b_\phi(r) \equiv b(re^{i\phi})$ belong to $L^\ell_{1/V}$. Now consider the two-photon matrix element $M^{(2)}$, which corresponds to taking $|a) = r|u_g)$ and $|b) = r|u_{\nu_2})$. We have $E<0$ for $N=N_0$ while $E>0$ for $N>N_0$. Note that $a_\phi(r)$ is exponentially damped, and belongs to $L^\ell_{1/V}$ for $E<0$ (i.e. $\phi=0$); this remains true as E moves into the upper half complex E-plane $(0\le\phi<\pi/2)$, but ceases to be true as E approaches the positive real axis. However $|a)$ can be approximated to arbitrary accuracy by truncating the convergent expansion

$$|a) = \sum_{n=\ell+1}^{\infty} (\beta^\kappa_{n\ell})^{-1}\ (S^\kappa_{n\ell}|a)V|S^\kappa_{n\ell}) \tag{2.12}$$

after a sufficient number of terms; provided we choose $\kappa$ so that $\mathrm{Im}(\kappa e^{i\phi})>0$ this yields an accurate approximation to $a_\phi(r)$ which is damped for $\phi=\pi/2$, i.e. for $E>0$. Now b(r) is a stationary wave, and so $b_\phi(r)$ diverges exponentially for $r \sim \infty$, and does not belong to $L^\ell_{1/V}$, if $\phi\neq0$. Consequently, the Sturmian expansion of $M^{(2)}$ converges only if $E<0$. However, the outgoing wave component, $b^+_\phi$, of $b_\phi(r)$ decreases exponentially for $r \sim \infty$ if $\phi\neq\pi/2$, and merely oscillates if $\phi=\pi/2$. Unfortunately, $b^+_\phi(r)$ is irregular at r=0. Nevertheless, the Sturmian expansion of $M^{+(2)}$ converges for both $E<0$ and $E>0$ provided that $|a)$ is replaced by the (truncated) expansion of Eq. (2.12) when $E>0$. The convergence is, however, rather slow [3] when $E>0$. The convergence can be significantly accelerated by regularizing $b^+(r)$ at the origin [9]. Presumably, the convergence could be further accelerated by using the Pade method

[10]. These remarks can be extended to the case N>2 and to the matrix elements $J_n^{(N)}$, and thereby the Sturmian expansion can be used to calculate $M^{(N)}$ via Eq. (2.6). The principal drawback of using Sturmian expansions when E>0 is that terms tend to be large and to cancel, leading to a rather rapid buildup of roundoff error.

Note that $g_\nu$ has a sequence of bound-state poles with an accumulation point at E=0. Hence the point E=0 is an isolated essential singularity, and Sturmian expansions of $g_\nu$ must break down in the neighbourhood of the ionization threshold.

## 3. STRONG FIELDS

We have reviewed the problem of ionization by strong fields in detail elsewhere [11] and will concentrate here on a qualitative understanding. We mean by "strong" fields that the field cannot be treated as a perturbation.

Under the action of strong fields virtual absorption and emission significantly contribute to the probability of N-photon absorption. Furthermore, the depletion of the initial state, and the population of the continuum channels coupled to the initial state, must be taken into account. The energy of the initial state acquires a width $\Gamma$ and is shifted by $\Delta$. The theoretical description of ionization is a formidable problem at intensities such that $\Gamma \gtrsim E_0+N_0\hbar\omega+\Delta$, the minimum energy of the outgoing electron, for then the spreading of the electron wavepacket is important [11]. Since the minimum energy of the outgoing electron is less than $\hbar\omega$, a necessary condition for wavepacket spreading to be neglibible is $\Gamma \ll \hbar\omega$. Also, if we are to regard the field as monochromatic, the turn-on time must be large compared to $1/\omega$, and so again we require $\Gamma \ll \hbar\omega$, for otherwise the atom would ionize during the turn-on period. Indeed, Lambropoulos /12/ has emphasized that in some recent experiments the atoms would have ionized before the field reached its peak intensity.

Under the assumption that the field is strong but $\Gamma \ll \hbar\omega$, we now estimate the maximum number of photons absorbed by the electron. It takes $N_0$ photons to lift the electron to the continuum -- this number may, of course, vary with intensity. Once there, the kinematical effect of the interaction V between the electron and atomic core is relatively insignificant, and we may assume that the electron has only kinetic energy. The mean kinetic energy of the electron does, however, differ from its instantaneous kinetic energy because the electron oscillates back and forth in the field, or, in photon language, it absorbs and emits virtual photons. We treat the field as classical; its electric vector is $\hat{\epsilon}F_0\cos(\omega t)$, where $\hat{\epsilon}$ is the unit polarization vector, and its vector potential is $\vec{A} = -(c/\omega)\hat{\epsilon}F_0\sin(\omega t)$. The instantaneous kinetic energy of an electron whose cycle average momentum is $\vec{p}$, and whose mass and charge are e and m, respectively, is $(\vec{p}-e\vec{A}/c)^2/2m$; we subtract from this the pondermotive term $(e^2\vec{A}^2/2mc^2)$, that is, we rescale the energy so that the continuum threshold is at zero energy. Hence the mean and instantaneous kinetic energies of the electron differ by $(e/mc)\vec{p}\cdot\vec{A} \equiv n\hbar\omega$. Here we see the dynamical role played by V; for n is the number of virtual photons which, at any instant, are available to be converted by V into real photons. Squaring both sides of the previous identity, and averaging $(\vec{p}\cdot\vec{A})^2$ over all directions of $\vec{p}$, and over the cycle time, we have

$$(1/3m)(eF_0/\omega)^2(p^2/2m) = (n\hbar\omega)^2 \quad , \tag{3.1}$$

where $p = |\vec{p}|$. We now substitute for $(p^2/2m)$ the average kinetic energy that the electron has during the above-threshold absorption, and thereby obtain for n the number, $n_0$ say, of photons absorbed above threshold. Since the final kinetic energy of the electron is roughly $n_0\hbar\omega$, the average kinetic energy is roughly $\frac{1}{2}n_0\hbar\omega$ and, solving Eq. (3.1) for $n_0$, we obtain

$$n_0 \approx 4\pi e^2 I/(3m\hbar\omega^3 c) \quad , \tag{3.2}$$

where $I = cF_0^2/8\pi$ is the intensity of the field. It follows that the number of absorption peaks above threshold should increase linearly with increasing intensity, and should increase as $\omega^{-3}$ with decreasing frequency. This prediction is supported by the results of the very illuminating, but often overlooked, treatment by Berson /13/ of multiphoton ionization of a model atom -- see particularly Berson's Eq. (17). A similar relationship connecting the number of photons to the intensity and frequency was derived, and confirmed numerically, in the context of electron scattering from a potential in a radiation field [14]. For a $CO_2$ laser ($\omega$=4.4x10$^{-3}$a.u.) of intensity I=10$^{12}$W/cm$^2$ we obtain $n_0 \approx 50$ from Eq. (3.2). However, we emphasize that Eq. (3.2) provides only a rough (qualitative) estimate of $n_0$. Furthermore, it is not yet clear whether, for real atoms, the condition $\Gamma \ll \hbar\omega$ can be realized in the intensity regime where $n_0$ is large and V can be neglected in comparison with the coupling to the radiation field.

It is a pleasure to thank P. Lambropoulos for several helpful discussions. This work was supported by the National Science Foundation, Grant PHY-8512495.

1. M.J. Seaton: Mon. Not. R. Astron. Soc. 118, 504 (1958)
2. B.A. Zon, N.L. Manakov, L.P. Rapoport: Sov. Phys. - Doklady 14, 904 (1970)
3. R. Shakeshaft: Phys. Rev. A34, 244 (1986)
4. M. Edwards, X. Tang, P. Lambropoulos, R. Shakeshaft: Phys. Rev. A33, 4444 (1986)
5. ----- : Phys. Rev., to be published
6. L.C. Hostler: J. Math. Phys. 11, 2966 (1970)
7. A. Maquet: Phys. Rev. A15, 1088 (1977)
8. R. Shakeshaft: J. Phys. B18, L611 (1985)
9. ----- : Phys. Rev., to be published
10. S. Klarsfeld and A. Maquet: Phys. Lett. 78A, 40 (1980)
11. ----- : J. Opt. Soc. Am. B, to be published
12. P. Lambropoulos: Phys. Rev. Lett. 55, 2141 (1985)
13. I.J. Berson: J. Phys. B8, 3078 (1975)
14. R. Shakeshaft: Phys. Rev. A28, 667 (1983)

# Part III

# Molecules in Laser Fields

# Bound-Continuum Transitions of $Na_2$ Studied by Doppler Photofragment Spectroscopy

*G. Gerber and R. Möller*

Fakultät für Physik, Universität Freiburg, D-7800 Freiburg, Fed. Rep. of Germany

The photodissociation of alkali-metal dimers, particularly of $Na_2$, has recently been studied by several groups/1/. However, there are contradictory results concerning the population of excited fragment fine structure states in the dissociation process. We have therefore applied the technique of Doppler Photofragment Spectroscopy to study the photodissociation of $Na_2$ in a molecular beam experiment.

In the course of our experiments three different photodissociation processes were observed and identified: direct photodissociation by excitation from the $X\,^1\Sigma_g^+$ ground state to the dissociation continuum of the $B\,^1\Pi_u$ state, dissociation of quasibound levels of the B state, and induced two-photon dissociation of $Na_2$ leading to $Na^*(nl)$ fragments.

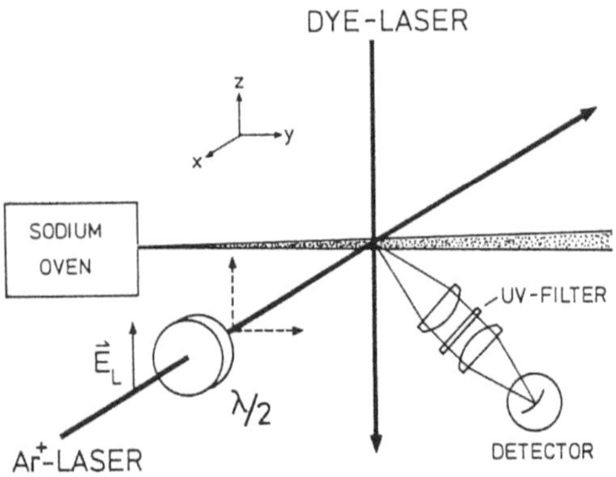


Fig.1 Experimental arrangement for Doppler photofragment spectroscopy. The $Na_2$ molecules are dissociated by an $Ar^+$-laser while a dye laser serves to analyze the dissociation products.

The basic experimental setup, partially described elsewhere /2/, consists of a sodium molecular beam crossed by two laser beams, all three being perpendicular to each other as shown in Fig.1. The sodium beam was generated by an oven operated at 800 K. The beam was collimated by a heated skimmer and by apertures resulting in a residual Doppler width of about 40 MHz. $Na_2$ molecules were dissociated by the $Ar^+$-laser, while the dye laser served to analyze the fragmentation. With several $Ar^+$-laser lines, both multimode and single mode, $Na_2$ was photodissociated into an excited $Na^*(3p)$ atom and a Na(3s) groundstate atom, recoiling in opposite directions. In our experiment dissociation was analyzed perpendicular to the molecular beam, as indicated in Fig.1, by measurement of the velocity component in direction of the analyzing laser so that the measured fragment velocity distribution coincides with the center-of-mass distribution. The excited states and the velocity components $v_z$ of the atomic fragments were derived from the Doppler shift $\Delta\nu = \nu v_z/c$ by tuning of a single frequency rhodamine-6G dye laser over the Na $3^2P_{3/2}-4^2D_{3/2,5/2}$ transitions at 568.82 nm or over the Na $3^2P_{1/2}-4^2D_{3/2}$ transition at 568.26 nm.

The fragments were monitored by the emission of the $4^2P_{1/2,3/2}-3^2S_{1/2}$ lines at 330 nm, part of the 4D-4P-3S cascade. This detection scheme is shown on the right-hand side of Fig.2, together with the schematic representation of the excitation from

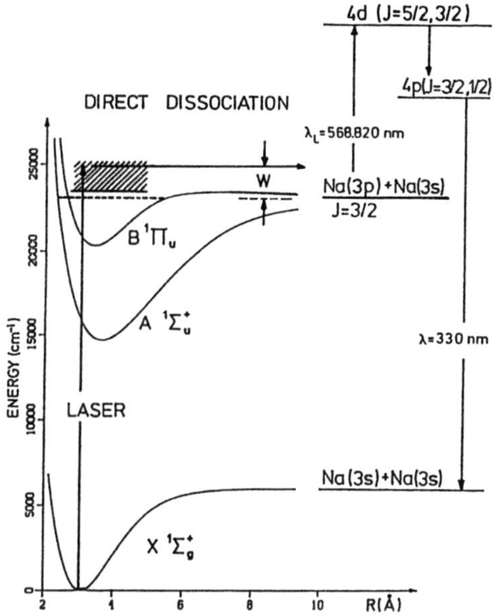


Fig.2 Schematic representation of the excitation process and the dissociation of an unbound level of the $Na_2$ $^1\Pi_u$ state. The applied scheme for the detection of the atomic fragments (including observation of the 330-nm line) is shown on the right-hand side.

the $X^1\Sigma_g^+$ state into the dissociation continuum of the $B^1\Pi_u$ state. The kinetic energy of the resulting fragments is given by the energy W of the unbound level referred to the dissociation limit.

Doppler spectra due to photodissociation with the 457.9-nm $Ar^+$-laser line are shown in Fig.3. Traces B and C are Doppler spectra obtained for parallel and perpendicular laser polarizations relative to the direction of observation. Trace A in Fig.3 shows the unshifted excitation spectrum of the monitored transitions $3^2P_{3/2}-4^2D_{3/2,5/2}$. Since the two fine-structure components are only 1.02 GHz apart, spectra B and C are the sum of two overlapping Doppler spectra.

The energy and angular distribution of the fragments produced by a dipole transition can be written in the form:

$$f(\vartheta,v) = g(v)\cdot(1 + \beta(v)\cdot P_2(\cos\vartheta)) \qquad (1)$$

where $\vartheta$ is the angle between the electric vector $\vec{E}_L$ of the polarized light and the axis of the dissociating molecule, v is the velocity of the fragment, g(v) the velo-

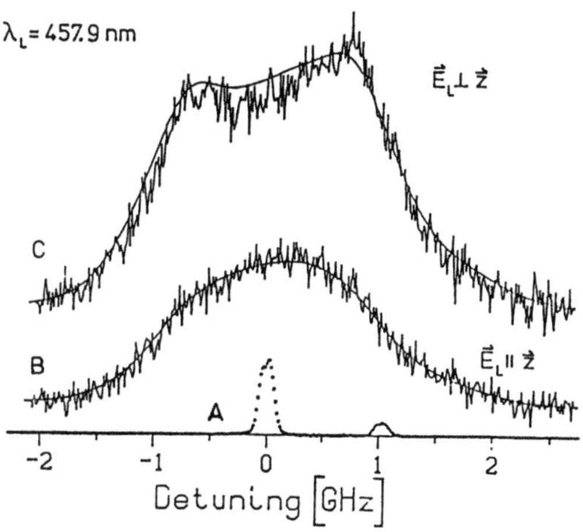


Fig.3 Doppler spectra of $Na^*3^2P_{3/2}$ fragments due to one-photon dissociation of $Na_2$. Trace A shows the finestructure components of the monitored 3P→4D transition. The solid lines are theoretical curves calculated with the distributions g(v) and ß(v) of Fig.4.

city distribution and $\beta(v)$ is the anisotropy parameter, which in general depends on v and on the type of the transition ($\Sigma$-$\Sigma$ or $\Sigma$-$\Pi$) and varies between +2 and -1 corresponding to the limiting cases of recoil parallel and perpendicular to the direction of the transition moment of the parent molecule. By projecting the distribution $f(\vartheta,v)$ on the axis of detection z the following velocity distributions are found:

$$\rho_{\parallel}(v_z) = \frac{1}{2v}\left(1+\beta(v)P_2\left(\frac{v_z}{v}\right)\right) \qquad \vec{E}_L \parallel \vec{z}$$
$$\rho_{\perp}(v_z) = \frac{1}{2v}\left(1-\frac{1}{2}\beta(v)P_2\left(\frac{v_z}{v}\right)\right) \qquad \vec{E}_L \perp \vec{z} \qquad (2)$$

with $v_z$ being the z-component of the fragment velocity.

In order to determine g(v) and $\beta(v)$ from the observed Doppler spectra $\rho_{\parallel}(v_z)$, $\rho_{\perp}(v_z)$ the inversion of these formulae can be obtained numerically by a stepwise recursion. The procedure makes use of the fact that for a given point of $\rho(v_z)$ only $v \leq v_z$ can contribute to the velocity distribution. This allows to construct $\rho(v_z)$ starting with high $v_z$. A detailed description of this procedure will be published elsewhere/3/. Figure 4 shows the velocity distribution g(v) and the function $\beta(v)$ derived from the Doppler spectra obtained with $\lambda$=457.9 nm. The velocity distribution shows a threshold of about 450m/sec$\hat{=}$E=385 $cm^{-1}$, which is approximately the height of the B-state potential barrier (379.3$cm^{-1}$) /4/.

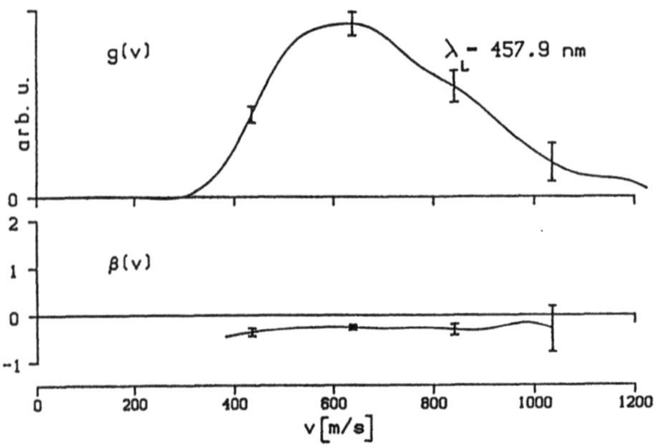


Fig.4 Velocity distribution g(v) and anisotropy parameter $\beta(v)$ obtained from the analysis of the two Doppler spectra displayed in Fig.3.

The shape of the velocity distribution g(v) with a maximum just above the threshold and a monotonic decrease for higher velocities shows that the probability for a bound-continuum transition decreases for increasing energy above the dissociation threshold. This is due to the variation of the density of continuum states $\eta(E_c)\sim 1/\sqrt{E_c}$ , where $E_c$ is the energy above the dissociation limit. The influence of the population of the initial levels v",J" of the X ground state on the shape of g(v) is rather small for a given excitation wavelength/5/.

The angular distributions can well be described by an almost constant value for the anisotropy parameter of $\beta$ = 0.4 ± 0.1, which indicates that a perpendicular electronic transition is involved in the photodissociation process. Since the ground state of $Na_2$ is a $^1\Sigma_g^+$ -state, the dissociation therefore proceeds via the $B^1\Pi_u$ state. Because of the rotation of the dissociating molecules the anisotropy parameter is reduced to values $\beta$>-1.

The influence of aligned parent molecules in the $Na_2$ beam on the determined $\beta$ parameters is negligible under our experimental conditions (800 K oven temperature and 0.2 mm nozzle diameter) since a significant molecular alignment only arises in supersonic expansion /6/.

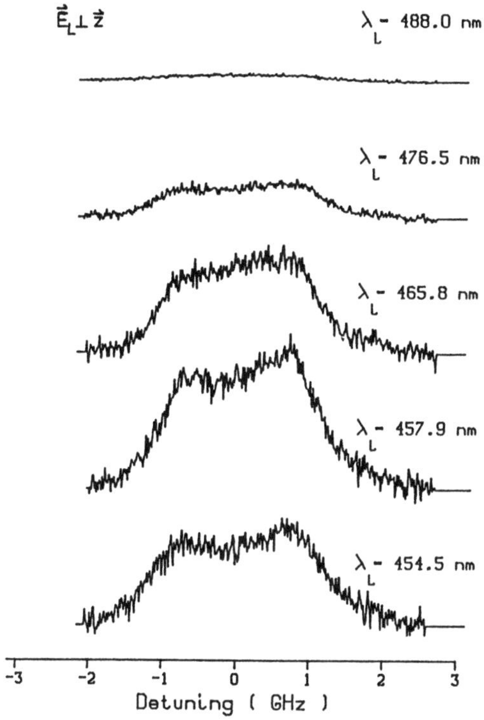


Fig.5 Doppler spectra of Na $3^2P_{3/2}$ fragments from photodissociation of $Na_2$ with different $Ar^+$-laser lines.

Doppler spectra obtained from dissociation of $Na_2$ with different $Ar^+$-laser wavelengths and perpendicular laser polarization ($\vec{E} \perp \vec{z}$) are shown in Fig.5. The spectra are taken under the same experimental conditions, therefore their intensities reflect the relative photodissociation cross-section for the different wavelengths. To understand the relative intensities of the spectra the population of the involved X-state rovibrational levels v",J" as well as the X-B bound-continuum Franck-Condon-factors have to be considered. While the intensity decreases from $\lambda$= 457.9 nm to $\lambda$= 488.0 nm by a factor of $\approx$100 which is due to the decreasing population of the available initial levels, the increase in intensity from 454.5 nm to 457.9 nm can be explained by the strong increase of the Franck-Condon-factors, which exceeds the decrease of the population.

Calculated Franck-Condon factors for X - B bound-continuum transitions (for J= 45) including the correct B-state potential barrier region /4/, shown in Fig. 6, indicate that excitation of the B-state continuum can only be performed from levels v"$\geq$10 of the ground state. The energy of the 457.9 nm line is just sufficient to reach the continuum out of v"=10 for low J. For the less energetic $Ar^+$-laser lines much higher v" levels are involved. Direct dissociation with these lines, even in a molecular beam, is still possible due to the large number of weakly populated v", J" levels. The vibrational and rotational temperatures in our experiment are $T_{vib}$=400K and $T_{rot}$=200K, respectively. The broken lines in Fig. 6 show the Franck-Condon factors for energy-conserving bound-continuum transitions excited with the two $Ar^+$-laser lines 457.9 nm and 488.0 nm. Both Franck-Condon factors are almost equal, therefore the difference in photodissociation cross-section for these two wavelengths (see Fig.5) is only due to the population of the initially available levels v",J" in the X ground state. A detailed description of the calculation of bound-continuum Franck-Condon factors has been given in /7/ . It should be noted however, that we have omitted the factor $1/\sqrt{E_c}$ in Fig. 6 in order to better represent the Franck-Condon factor variation for more energetic continuum states.

The angular distributions which are obtained from the analysis of the spectra given in Fig. 5 are characterized by values for ß(v) ranging from -1 to 0, which is expected

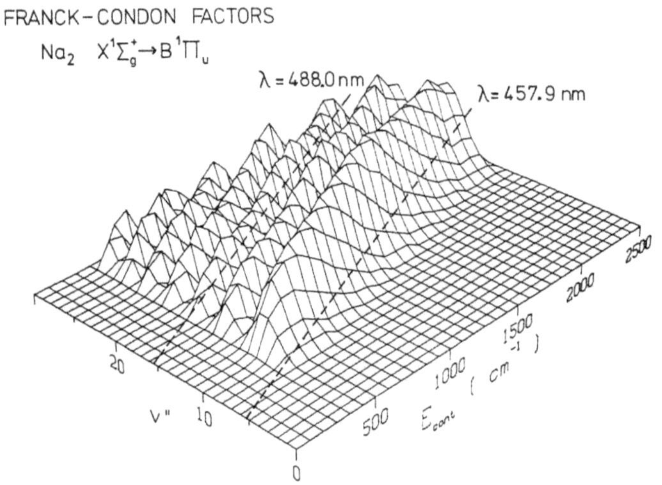


Fig. 6 Bound - continuum Franck-Condon factors for B - X transitions of $Na_2$

for the dissociation of a rotating molecule following a $\Sigma-\Pi$ transition /8/, if the rotational energy is in the same order of magnitude as the kinetic energy of the fragments released in the dissociation.

With all applied $Ar^+$- laser lines only Na 3 $^2P_{3/2}$ fragments with continuous energy distributions are observed, similar to that shown for 457.9 nm in Fig. 4. No $3^2P_{1/2}$ fragments are observed within the experimental uncertainties (5% of $3^2P_{3/2}$ signal) at comparable experimental conditions. This is in contrast to the findings of Rothe, Krause and Düren /1/ and Janson and Papermov /1/ who have reported $Na_2$ photodissociation via the B-state with the 488.0 nm and 476.5 nm lines leading to the emission of both the Na D2($^2P_{3/2}$ - $^2S_{1/2}$ ) and D1($^2P_{1/2}$ - $^2S_{1/2}$ ) lines. In an additional molecular beam experiment we have also measured the atomic line emission in order to compare our results with those of Rothe et al and Janson et al. The light emitted in the interaction region has been imaged with lenses and mirrors onto the entrance slit of a 3/4m monochromator. With this set-up the sodium D2 and D1 lines could easily be resolved, if observed.

After excitation of $Na_2$ with various $Ar^+$- laser lines we measured spectra consisting of the well-known molecular emission and of atomic line emission, but only at $\lambda$= 589.0 nm. The experimental conditions (800 K oven temperature and p·d=0.1 torr·cm) under which this 589.0 nm light was observed were the same as for the photofragment measurements. An explanation for the discrepancy between our measurements (no D1 emission) and those of Rothe et al, Janson et al and others /1/ was found when we looked how the atomic line emission depends on the temperature of the sodium molecular beam oven. Figure 7 shows this dependence for the photodissociation with $\lambda$= 488.0 nm. The intensity of the D2 - line increases up to 500°C and decreases for still higher temperature while the D1 - line emission only starts around 550°C. The strongly different curves clearly show that there must be another process responsible for the $3^2P_{1/2}$ population. Moreover, this measurement also gives a possible explanation for the wide range of the observed and reported /1/ intensity ratio I(D2)/I(D1).

Rothe, Krause and Düren /1/ suggested that a nonadiabatic interaction of the B state with electronic states correlated with the $3^2P_{1/2}$ + $3^2S_{1/2}$ dissociation limit is responsible for the observed D1 - emission. But such an interaction should be temperature independent. Another process resulting in $3^2P_{1/2}$ fragments would be dissociation via the continuum of the $A^1\Sigma_u^+$ state. This, however, can be excluded on the basis of the vanishinglysmall X→A bound-continuum Franck-Condon factors. The question therefore still remains, what is the excitation process responsible for the population of $3^2P_{1/2}$ fragment states.

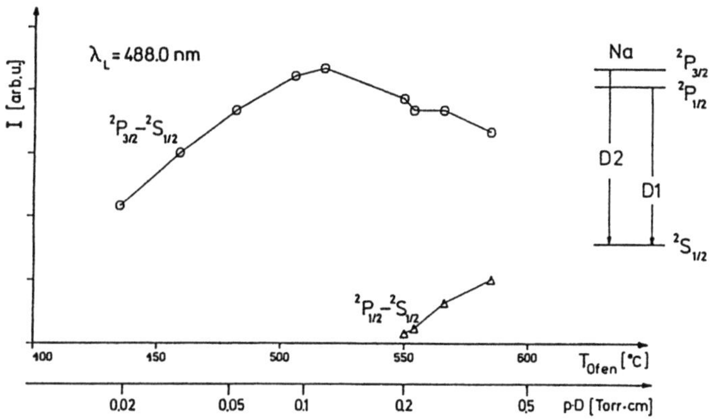


Fig. 7 Intensity of the atomic line emission $3^2P_{3/2} - 3^2S_{1/2}$ and $3^2P_{1/2} - 3^2S_{1/2}$ resulting from photodissociation of $Na_2$ with $\lambda = 488.0$ nm plotted versus sodium molecular beam oven temperature.

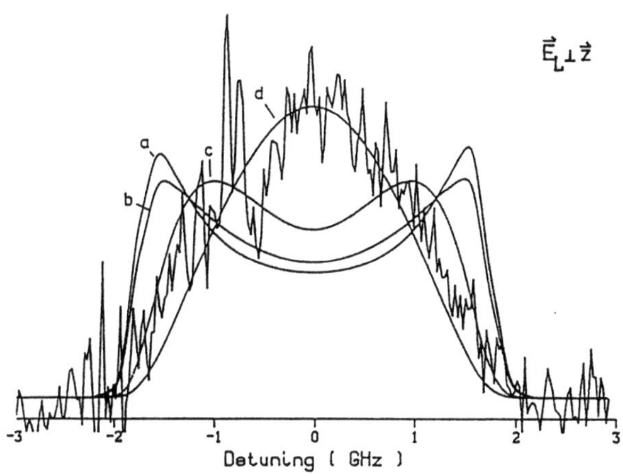


Fig. 8 Doppler spectrum of Na $3^2P_{1/2}$ fragments resulting from photodissociation of $Na_2$ with $\lambda$=488.0 nm. The curves a-d are simulations for different angular distributions.

In an additional experiment with higher oven temperature (pd=0.7 torr·cm), we measured Na $3^2P_{1/2}$ fragments after excitation with the 488.0- and the 476.5 nm lines. The $^2P_{1/2}$ fragments were detected by inducing the $3^2P_{1/2}-4^2D_{3/2}$ transition at 568.26 nm and subsequently recording the 4P-3S fluorescence at 330 nm. The Doppler spectrum of $Na^*3^2P_{1/2}$ fragments obtained with the 488.0 nm line and perpendicular laser polarization, shown in Fig. 8, clearly demonstrates the different angular distribution and higher energy of the $^2P_{1/2}$ fragments. Moreover, the much weaker $^2P_{1/2}$ signal increases with higher pd values relative to the decreasing $^2P_{3/2}$ signal. If a wavelength of 457.9 nm is used for the photodissociation under the same experimental conditions only $Na^*3^2P_{3/2}$ atoms, but no $Na^*3^2P_{1/2}$ atoms, are observed. This excludes an effect by fine-structure changing collisions. The analysis of the 488.0 nm fragmentation process leads to the conclusion that the $Na^*\ 3^2P_{1/2}$ fragments result from a two-photon excitation of a high-lying $Na_2^*$ state dissociating into two $Na^*$ 3P atoms.

In a two-photon process, the first linear polarized photon creates an anisotropic molecular sample which is then dissociated by absorption of a second linear polarized photon. The anisotropy of the process is reflected in the angular distribution $f(\vartheta)$ of the photofragments /9/

$$f(\vartheta) = (1/4\pi)\cdot\left[\ 1 + \beta_D \cdot P_2(\cos\vartheta) + \gamma_D \cdot P_4(\cos\vartheta)\ \right] \qquad (3)$$

Since the Franck-Condon factors for the $Na_2$ $X^1\Sigma_g^+ - A\ ^1\Sigma_u^+$ transitions with 488.0 nm are vanishingly small ($<10^{-8}$), we need only consider $X\ ^1\Sigma_g^+ - B^1\Pi_u$ transitions. From symmetry considerations it follows that only $^1\Sigma_g^+$, $^1\Sigma_g^-$, $^1\Pi_g$ and $^1\Delta_g$ states can be reached by two-photon excitations out of a $^1\Sigma_g^+$ state. From the different possible transitions $\Sigma-\Pi-\Sigma$, $\Sigma-\Pi-\Pi$ and $\Sigma-\Pi-\Delta$ we can exclude the $\Sigma-\Pi-\Sigma$ and $\Sigma-\Pi-\Delta$ processes, since in these cases the two perpendicular transitions result in an angular distribution $f(\vartheta) \sim \sin^4\vartheta$ (classical Limit), which is definitely not observed. The curve a represents this classical limit, whereas curve b is calculated with $\beta_D$ and $\gamma_D$ values taken from Singer and Freed /9/. Assuming a $\Sigma-\Pi-\Pi$ process it was found that the observed Doppler spectrum (Fig. 8) can be described by a distribution with a kinetic energy of $W=1900\pm300\ cm^{-1}$ and parameters $\beta_D=2.37$ and $\gamma_D=0.49$ taken from /9/. Curve c in Fig. 8 represents the classical limit for a $\Sigma-\Pi-\Pi$ transition. However, because of the low statistics of the measurement it was not possible to obtain these parameters from a fit to the observed distribution.

The two-photon excitation with the 488.0 nm line proceeds via the $B\ ^1\Pi_u$ state, the dominant first transition being (v"=3, J"=43)-(v'=6, J'=43). This particular excitation process is shown in Fig. 9. Since the excitation energy is just 1870 $cm^{-1}$ above the Na 3P + Na 3P dissociation limit, the dissociation into two Na 3P atoms is strongly suggested by the measured photofragment energy of W = 1900 $cm^{-1}$. The symmetry of the upper state must be $^1\Pi_g$ as is inferred from the observed angular distribution (Fig. 8). It should be mentioned that $Na_2^+$ formation has been reported /10/ for two-step excitation of $Na_2$ with the 488.0 and 476.5 nm lines, a process which might be in competition with the dissociation process discussed here.

In summary, our experiments show that the direct dissociation of $Na_2$ with wavelengths between 450 nm and 515 nm only populates the Na $3^2P_{3/2}$ and the Na $3^2S_{1/2}$ states. The corresponding angular and energy distributions confirm that the dissociation proceeds via the dissociation continuum of the B-state. No evidence for non-adiabatic coupling of the dissociating B-state with other electronic states correlating to the $3^2P_{1/2}$

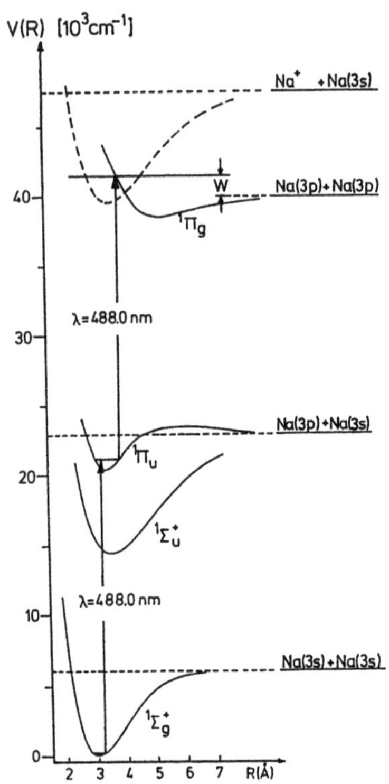


Fig. 9 Two-Photon excitation of a $^1\Pi_g$ state dissociating into Na(3p) + Na(p) (the $^1\Pi_g$-potential curve is estimated)

+ $3^2S_{1/2}$ dissociation limit is found, as suggested by several authors. We also found no evidence for a dissociation via the A-state continuum. A two-photon absorption process leading to a dissociation and to a population of the $3^2P_{1/2}$ and the $3^2P_{3/2}$ level can clearly be distinguished by the different angular and energy distributions as compared to the one-photon dissociation resulting only in $3^2P_{3/2}$ excited fragments. The reported dissociation of the quasi-bound level v′= 31, J′= 42 of the $B^1\Pi_u$ state of $Na_2$ /11/ represents a very rare example of a dissociation process where for a neutral diatomic molecule all the quantities governing the process are well determined. The quantum numbers v′, J′and the lifetime $\tau$ were obtained, and the energy and angular distributions as well as the internal states of the atomic fragments were analyzed.

This research was supported by the Deutsche Forschungsgemeinschaft.

## References

1. R.H.Callender, J.I.Gersten, R.W. Leigh, J.L. Yang Phys.Rev. A14, 1672 (1976)
V.B. Grushevskii, S.M. Papernov, M.L. Yanson Opt.Spektrosk. 44,809 (1978)
E.W. Rothe, U. Krause, R. Düren J.Chem.Phys. 72, 5145 (1980)
J.L. Gole, G.J. Green, S.A. Pace, D.R. Preuss J.Chem.Phys. 76, 2247 (1982)
M.L.Janson, S.M. Papernov J.Phys. B15, 4175 (1982)
J.Keller, J. Weiner Phys.Rev. A29, 2943 (1984)
2. G. Gerber, R. Möller, Chem.Phys.Lett, 113, 546 (1985)
3. G. Gerber, R. Möller to be published
4. H.J. Vedder, G.K. Chawla, R.W. Field Chem.Phys.Lett. 111, 303 (1984)
5. R. Möller, PhD Thesis, Freiburg 1986
6. A.G. Visser, J.P. Bekooy, L.K. Van der Meij, C.deVreigd, J. Korving Chem.Phys.20,391
7. G. Gerber, R. Möller, H. Schneider, J.Chem.Phys. 81, 1538 (1984)
8. C. Jonah, J.Chem.Phys. 55, 1915 (1971)
9. S.J. Singer, K.F. Freed, Y.B. Band J.Chem.Phys. 8i 3064 (1984)
10. V.S. Kushawa, J.J. Leventhal, J.Chem.Phys. 75, 5966 (1981)
K. Bergmann, E. Gottwald, Chem.Phys.Lett. 78, 515 (1981)
E.W. Rothe, F. Ranjbar, D. Sinha, G.P. Peck Chem.Phys.Lett. 78, 16 (1981)
11. G. Gerber, R. Möller, Phys.Rev.Lett. 55, 814 (1985)

# Mechanism of Laser Induced Decomposition of Gaseous Polyatomic Molecules

*E. Borsella*[1], *L. Caneve*[1], *A. Giardini-Guidoni*[2], *and A. Mele*[2]

[1]ENEA, Dip. TIB, Divisione Fisica Applicata, C.R.E. Frascati, C.P. 65, I-00044 Frascati Rome, Italy
[2]Dipartimento di Chimica, Università di Rome,"La Sapienza", Rome, Italy

## 1. INTRODUCTION

IR multiple photon excitation (MPE) and dissociation (MPD) in molecules have become a field of major significance over the past years after the early discovery of ISENOR et al. /1/ of a fast dissociation process occurring in $SiF_4$ irradiated with an intense $CO_2$ laser source. The dynamics of MPE and MPD have been widely studied both theoretically and experimentally in view of the possibilities offered to selective chemistry /2/ and technical applications.

Satisfactory theoretical models have been developed, which account on one hand for the interaction of laser radiation with low-lying molecular levels, and on the other hand for incoherent absorption of radiation in the upper levels. Quantum mechanical equations describe the absorption in the low-lying molecular levels, while the MPE in the quasi continuum can be accounted for by a simple model with rate equations governing the transitions between levels and dissociation /2/.

Studies of MPE and MPD dynamics have been carried out in different experimental conditions such as cell or molecular beams using various diagnostic techniques: i) final product analysis by mass spectrometry, IR spectroscopy, and gas-chromatography; ii) on-line mass spectrometry; iii) fragment luminescence detection /3/. The observation of luminescent species may be particularly advantageous because of the high efficiency of detection. This method is also advantageous for studying time-dependent phenomena.

Much of the discussion and speculation on luminescence has been concentrated on the issues herewith summarized /3/.

Possible mechanism of molecular excitation are a) collisionless multiple photon absorption: $A + nh\nu \rightarrow A^{**} \rightarrow$ products; b) collisional energy transfer, namely an energy pooling process: $A + mh\nu \rightarrow A^*$; $A^* + A^* \rightarrow A^{**} + A \rightarrow$ Products. The symbol A represents the unexcited molecules and $A^*$ and $A^{**}$ are vibrationally excited molecules below and above the dissociation limit respectively.

In the low pressure regime, MPD can occur only through mechanism a) /1/ since no collisions can take place before dissociation; in the high-pressure regime, two molecules with intermediate energy content $A^*$ collide to give a molecule $A^{**}$ which is excited above the dissociation threshold. In the collision-free limit, luminescence is expected to build up almost instantaneously with the laser pulse, the rise time being independent of substrate pressure, while the luminescence intensity is expected to increase linearly with substrate pressure. On the other hand, collisional activation gives rise to a higher than linear dependence of the product luminescence on the substrate pressure, and rise and decay times are expected to be pressure dependent.

The origin of luminescence falls into three main categories: a) inverse electronic relaxation (IER), i.e., internal conversion of vibrational excitation into electronic excitation; b) decomposition of the parent into electronically excited fragments; c) emission through reactions of photofragments with parent or added molecules /3/.

The IR laser photochemistry of $SiH_4$ and its derivatives has recently received great attention due to the possibility of growing silicon films and producing ceramic powders both of technological interest /4/. The oxidation of $SiH_4$ has a specific relevance for the chemical vapor deposition of high grade silicon oxide.

Several experiments have shown that $SiH_4$ and $SiF_4$ irradiated by IR laser may produce visible luminescence. Neither the character of the photochemical process nor the origin of the luminescence have yet been clearly understood /3/. The aim of the present work is to report and discuss additional data on the IR photochemistry of $SiF_4$ and $SiH_4$ and mixtures of $SiH_4$ and NO.

## 2. EXPERIMENTAL

### 2.1 Molecular Beam and On-line Mass Spectrometry

The supersonic molecular beam apparatus used for IR photolysis studies has been previously described /5/. In the main chamber the pulsed molecular beam can cross two pulsed TEA $CO_2$ lasers synchronized within 50 ns. The two spatially superimposed laser beams are focused on the molecular beam. Laser-induced MPD is monitored as a decrease of the molecular beam intensity by a quadrupole mass spectrometer on the beam axis.

### 2.2 Gas Cell Experiments and Luminescence

A gas cell equipped with four crossed windows has been used for dissociation experiments in the pressure range 0.1-35 torr. Final product analysis was performed by IR spectroscopy. The fluorescence light emitted was detected at a right angle to the IR laser path and focused on the entrance slit of a spectrograph. An EGG optical multichannel analyzer (OMA III) was used to detect the entire luminescence spectrum simultaneously. The total light emitted was also studied as a function of time by inserting a fast photomultiplier and a transient recorder. Multiple photon absorption was measured by an optoacoustic cell which has been previously described /6/.

## 3. RESULTS AND DISCUSSION

### 3.1 $SiF_4$

This molecule is a spherical top whose spectroscopic parameters are well known and for which multiphoton resonances in the $v_3$ level up to the third order have been calculated /7/. The high fluence absorption spectrum around the $v_3$ vibrational mode has been measured by using only line-tunable lasers. Data reported in Fig. 1a show a large absorption band whose maximum is shifted to the red with respect to the linear absorption at 1030.9 $cm^{-1}$. The fine structure of the spectrum is hidden due to the lack of a tunable source. When another laser is preexciting the $v_3$ level at a frequency of 1029.4 $cm^{-1}$, the absorption band becomes larger with respect to the one laser spectrum and a broad feature whose maximum is around 985 $cm^{-1}$ appears. These spectra can be interpreted on the basis of an increase of the absorption cross-section due to the presence of combination bands involving other vibrational modes.

Dissociation has been measured at low rotational temperature in the molecular beam apparatus by using one or two lasers tuned at different frequencies. In Fig. 1b the two-frequency dissociation is reported as a function of the second laser frequency.

It can be seen that also dissociation, when the experiment is performed preheating the molecule on the $v_3$ mode, takes place in a large frequency region on both the blue and the red sides with respect to the $v_3$ confirming that hot and combination bands, as found in other molecules /8/, can contribute to the dissociation ladder. The band on the blue side can be attributed to MPE of the $v_1 + v_2$ modes which shows

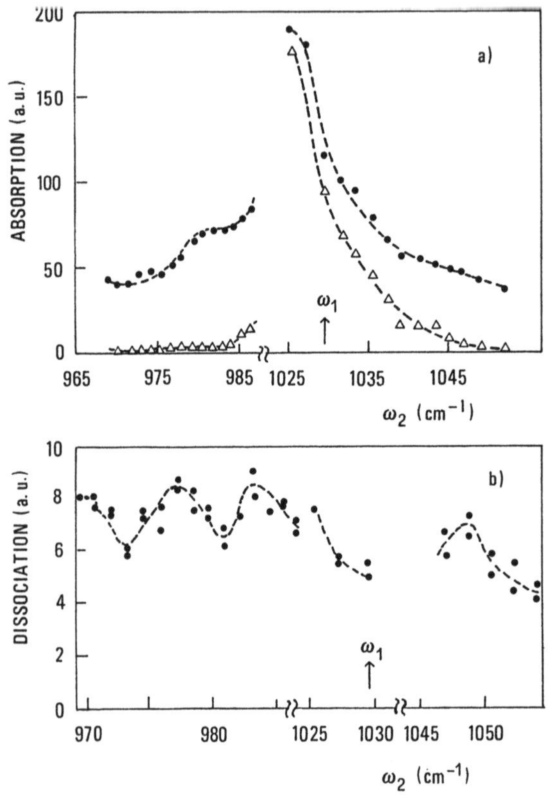


Fig. 1 (a) One (Δ) and two frequency (·) MPA spectra of $SiF_4$ at p=0.3 torr; one laser: $\phi$=1.5 J/cm$^2$; two lasers: $\omega_1$=1029.4 cm$^{-1}$, $\phi_1$=0.5 J/cm$^2$, $\phi_2$=1.0 J/cm$^2$; b) two-frequency MPD of $SiF_4$ in molecular beam: $\omega_1$=1029.4 cm$^{-1}$; $\phi_1$=1 J/cm$^2$; $\phi_2$=1.5 J/cm$^2$

a strong Coriolis interaction with the $v_3$ mode. Dissociation paths through combination bands of $v_1$,$v_4$ and $v_2$ with $v_3$ can explain the broad features at frequencies lower than the $v_3$ mode. Possible resonant paths in that region are those leading to $2v_1+2v_3+v_4$, $v_1+3v_2+2v_3+v_4$ and to $2v_2+3v_3+v_4$.

Electronic luminescence spectra of the photofragments have been obtained by MPD experiments at two different laser fluences. Figure 2 shows the luminescence spectra at 1 torr and 10 torr respectively in the experiment at lower fluence ($\phi$~10 J/cm$^2$). In the pressure regime up to 1 torr, only a broad band is observed in the spectral region from 360 to 700 nm. A similar structureless spectrum has been previously observed in the low-pressure regime /9/. As pressure increases, a sharp band localized around 450 nm emerges from the diffuse spectrum. This band could be identified as being due to emission of SiF from the $^2\Sigma^+$ to the ground $^2\Pi$ state (α band). At higher fluence, ($\phi$~25 J/cm$^2$) electronic luminescence shows the presence of several excited fragments, as shown in Fig. 3.

The nature of the excitation process leading to the dissociation of $SiF_4$ has not yet been well ascertained: both collisionless /9/ and collisional energy pooling mechanisms have been proposed /10/. The mechanism may strongly depend on the pressure conditions of the experiments. In the low-pressure range, results from luminescence detection support the unimolecular excitation character. Linear pressure dependence, prompt luminescence and rise time of the luminescence pulse independent of pressure are all evidence of this type of mechanism. The present results

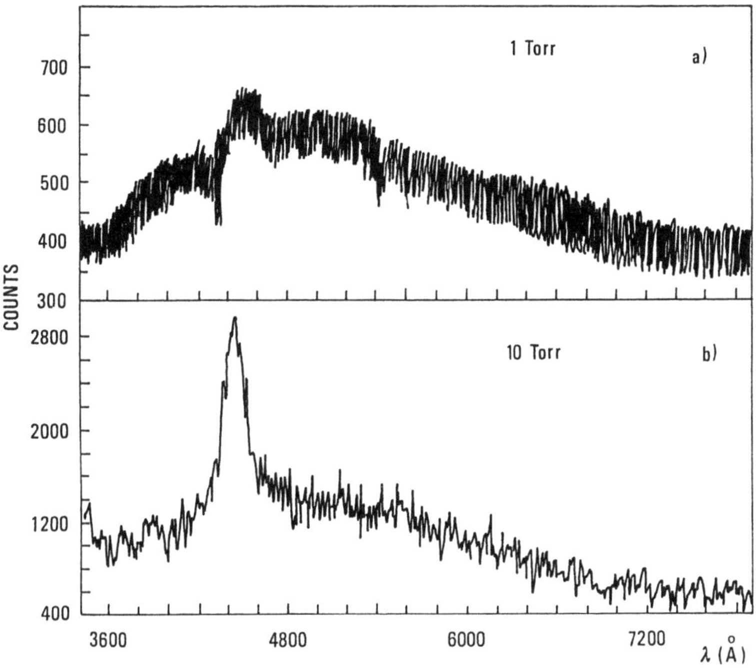


Fig. 2 OMA spectra of fragment emission after IR MPD of $SiF_4$ at low laser fluence: $\phi \simeq 10$ J/cm$^2$, $\omega$=1029.4 cm$^{-1}$

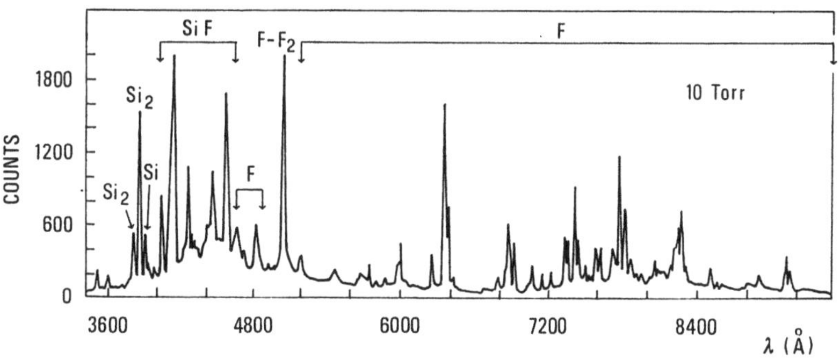


Fig. 3 OMA spectrum of fragment emission after IR MPD of $SiF_4$ at high laser fluence: $\phi \simeq 25$ J/cm$^2$, $\omega$=1029.4 cm$^{-1}$; p=10 torr

in the molecular beam would confirm the hypothesis of the unimolecular collisionless nature of the process.

The quasi-continuum structure found in the two-frequency experiments covers a large energy region where various paths can lead to dissociation. This may account both for the luminescence previously reported /9/ at 10.2 μm and for the relatively low threshold for the process (1.5 J/cm$^2$), notwithstanding with the 60 photons/ /molecule required to break the SiF bond. In the cell experiments, in the range 0.3-20 torr, the collisionless mechanism may still be valid although the energy pooling process, due to the relatively high number of collisions (~ $10^8$/sec), cannot be ruled out even at the lowest pressure where luminescence has been observed.

Further fragmentation of $SiF_4$ cannot be in accordance with the hypothesis of an IER mechanism /9/.

## 3.2 $SiH_4$

In contrast with $SiF_4$ which shows an almost continuous absorption spectrum, the MPA of $SiH_4$ is sparser due to the large spacing in the rotational manifold (Be for $SiF_4$ = 0.137 $cm^{-1}$; Be for $SiH_4$ = 2.86 $cm^{-1}$). Multiple photon resonances in $SiH_4$ have been observed at low pressure /11/.

Distinct emissions are detected when $SiH_4$ is irradiated with the 10 P(20) line of an unfocused (2 $J/cm^2 \leqq \phi \leqq 3.5$ $J/cm^2$) or focused TEA $CO_2$ laser. A faint emission around the laser beam path is observed visually in the first case; while a bright pink emission illuminating the whole cell is seen in the latter case.

Parallel optics. At low laser fluence, two different regimes can be observed in the spectral and time dependence of the total emission at $SiH_4$ pressures ranging from 1 to 30 torr.

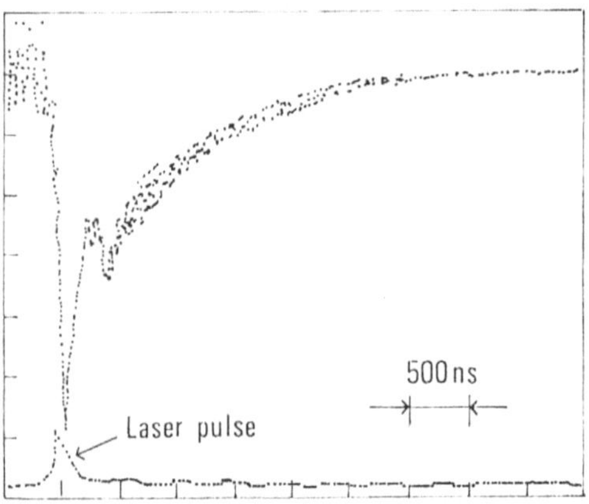


Fig. 4 Time dependence of fragment emission at p($SiH_4$)=7 torr; $\phi \cong 2.5$ $J/cm^2$

At silane pressures in the range 1-12 torr, the total luminescence shows a double exponential decay (Fig. 4), the fast component having a lifetime of 80±10 nsec and the slow component one of about 0.9 μsec. The fast component is delayed by ~ 80 nsec with respect to the maximum of the laser pulse; while the slow component is delayed by ~ 400 nsec. These delay times are not sensitive to changes in the laser fluence and gas pressure. The intensity of both components of the total emission increases almost linearly versus silane pressure. Due to the low intensity of the luminescence in these pressure conditions, spectral analysis was performed with filters ranging from 400 nm to 700 nm (with a bandwidth of ~ 50 nm). The emission falls in the range 550-700 nm and is centered at about 650 nm. Both the time components show the same trend at different wavelengths. The total luminescence yield increases drastically above 15 torr and goes to saturation at pressures greater than 30 torr. As shown in Fig. 5, drastic changes take place at p($SiH_4$)>15 torr in the time dependence, which consists of several delayed components all exhibiting the same spectrum.

Spectral analysis of the luminescence was performed with the optical multichannel analyzer and consists of two broad bands (Fig. 6). This spectral distribution was found to be independent of laser fluence in the range 0.5-1.5 $J/cm^2$.

The broad luminescence spectrum detected in the range 550-750 nm at low pressures can be attributed to decay from the $^1B_1$ electronic excited state of $SiH_2$ to the $^1A_1$ ground state. The large spread is due to unfavorable Franck-Condon factors caused

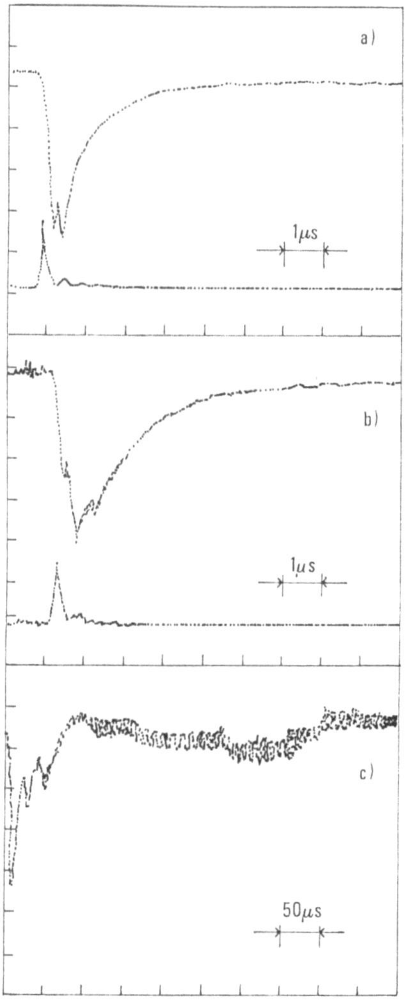


Fig. 5 Time dependence of fragment emission at different $SiH_4$ pressures; $\phi$~2.5 J/cm$^2$; a) 15 torr; b) 20 torr; c) 30 torr

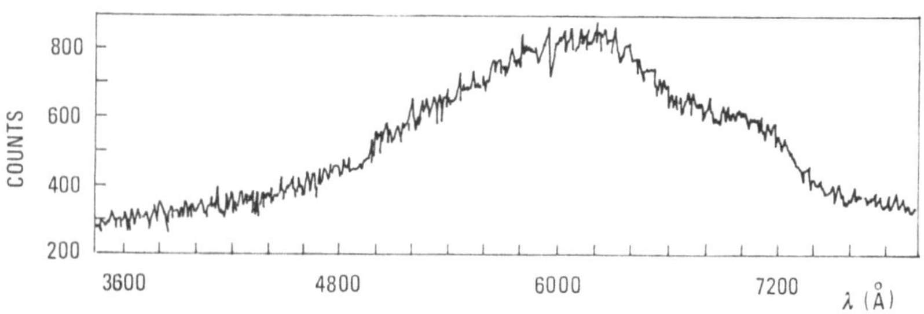


Fig. 6 OMA spectrum of fragment emission after IR MPD of $SiH_4$ at low laser fluence $\phi$~2.5 J/cm$^2$, p=30 torr

by the different equilibrium angle of the two states (92° with respect to 112°). This hypothesis is supported by the luminescence decay times which are in agreement with the results of laser-induced fluorescence of the $SiH_2$ radical excited at 580.1 nm /14/. At pressures above 15 torr, the spectrum shows a second component in the range 750-900 nm which could be tentatively assigned to emission from the $^3B_1$

electronic excited state of $SiH_2$ to the $^1A_1$ ground state. Formation and decay of this species are collisionally assisted.

Focusing Optics. A bright pink luminescence is emitted in the focusing configuration. The emission changes from a broad band to a line spectrum (Fig. 7). At p=1 torr, peaks have been attributed to emission from excited Si and $H(H_\alpha)$, while electronic excited fragments such as Si, $Si_2$, SiH, H and $H_2$ have been identified at p≧2 torr. As pressure rises, SiH emission tends to increase together with $Si_2$, H and $H_2$. A marked decrease of excited Si(at 390.5 nm) is observed. The IR analysis and the x-ray spectrum of the brownish powder deposited in the cell indicate the presence of the SiH bond and Si.

Rise and delay times of the total luminescence yield with respect to the laser pulse are dependent on the silane pressure.

The features observed in this work both in unfocused and focused optics, at different $SiH_4$ pressures, may be consistent with a dissociative channel which leads to $SiH_2$ as the possible primary fragment and also with the subsequent formation of higher silanes. In parallel optics, due to the lower energy deposited, a polymeric

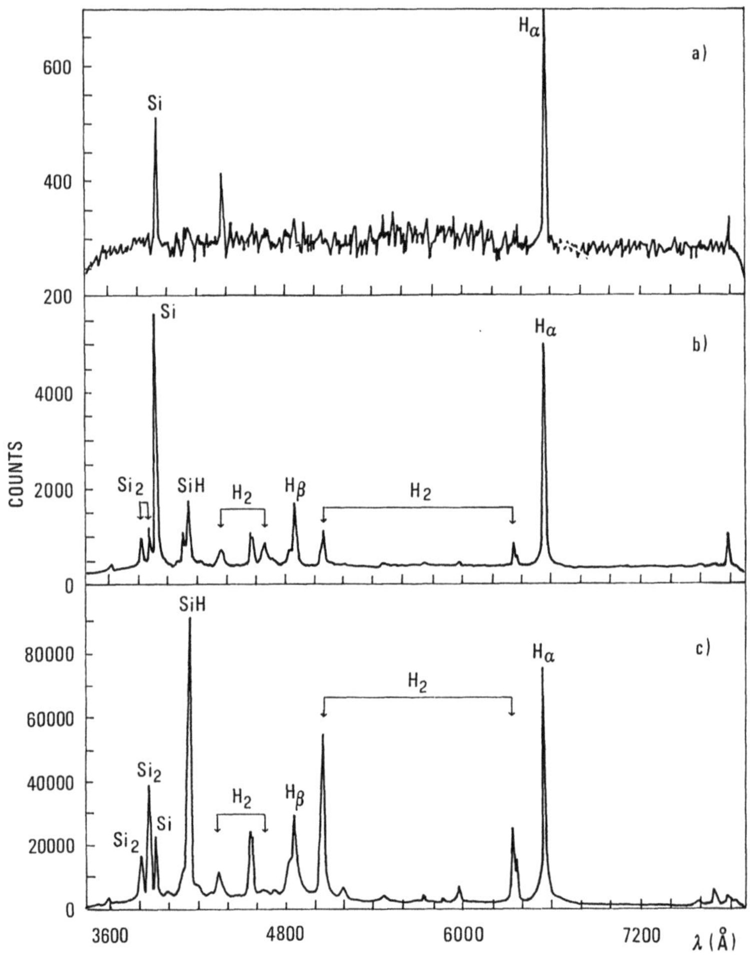


Fig. 7 OMA spectra of fragment emission after IR MPD of $SiH_4$ in focusing optics: a) p=1 torr, b) p=2 torr, c) p=12 torr

chain may grow, as occurs in conventional pyrolysis experiments. Heavy fragmentation takes place in focused optics caused both by the higher energy available and by the high exothermicity of growth reactions. Local temperature increases so fast that the high-pressure results are similar to those of shock-wave-induced decomposition of $SiH_4$ /15/ and of glow discharge plasma /16/. The analogy between the results could be attributed to the higher average vibrational temperature.

The strongly nonlinear dependence on silane pressure of the rise and delay times of the total luminescence signal, as well as of the total luminescence yield, are indicative of a complex scheme of consecutive reactions. A full understanding of the mechanism of decomposition in this laser fluence regime requires an accurate inspection of the time evolution of each fragment in order to sort out the exact sequence of reactions taking place after laser irradiation.

### 3.3 $SiH_4$ + NO

Mixtures of $SiH_4$ and NO in different NO/$SiH_4$ ratios (from 0.05 to 2) and at different total pressures (from 3 to 37 torr) have been irradiated in a cell by a TEA $CO_2$ laser on the 10P(20) line (944.2 $cm^{-1}$) in the focusing optical configuration. A single laser shot irradiation produces a strong luminescent flash which expands throughout the cell. Spectral analysis of this luminescent flash shows a broad band in the range 600-900 nm peaking at about 800 nm. IR spectroscopic analysis of gaseous mixtures has been performed before and after the irradiation of $SiH_4$. No IR active species other than $SiH_4$ and NO have been observed. The pressure change following the reaction has been measured and the trend of the reaction is reported in Table I.

Table I Pressure changes in $SiH_4$+NO reaction at different $SiH_4$/NO ratios

| Initial Pressures | | Pressure Increase | Residual Pressures | |
|---|---|---|---|---|
| $P_{SiH_4}$ torr | $P_{NO}$ torr | $\Delta P$ torr | $P_{SiH_4}$ torr | $P_{NO}$ torr |
| 17.5 | 0 | 0 | 17.5 | 0 |
| 17.5 | 1 | 0 | 17.5 | 1 |
| 17.5 | 2 | 6.9 | 8.5 | 0.7 |
| 17.5 | 4 | 8.8 | 5 | 0.7 |
| 17.5 | 6 | 7.5 | 4 | 0.7 |
| 17.5 | 12 | 5.8 | 4 | 0.7 |
| 17.5 | 15 | 5.0 | 4 | 0.7 |

A gray-white powder is formed which is more abundant than in the case of pure $SiH_4$. The IR analysis of the powder shows the presence of absorption bands of SiO and SiH, while x-ray analysis indicates the presence of both amorphous and crystalline silicon.

The data of Table I are consistent with the following reactions:

$$2\ SiH_4 + 2NO \rightarrow 2SiO + N_2 + 4H_2 \qquad (1)$$

$$14\ SiH_4 + 2NO \rightarrow 2SiO + 12Si + N_2 + 28H_2 \qquad (2)$$

which fit with the high and low NO/$SiH_4$ reactant ratios respectively. Decomposition kinetics of mixtures of $SiH_4$ and NO have already been studied. Various dissociating means such as UV photolysis, γ-ray irradiation and glow discharge were used, but a comparison between these results would not give a simple picture of the reaction mechanism. A chain reaction appears to be generally accepted /17/. This also seems to be the case for the $CO_2$ laser-induced process. Higher silane production is completely inhibited and a very efficient oxidation process by reaction of NO with $SiH_4$ takes place through a radical chain reaction with branching playing a relevant role. The chain lenght of the reaction appears to be controlled by the NO/$SiH_4$ ratio. Chain termination at an early stage takes place when the ratio is higher than 0.5. Longer chain and subsequent fragmentation leading to silicon production is predominant at a lower ratio. This seems to be a very important aspect for a laser synthesis that would produce, preferentially, pure silicon with respect to other products. A description of this intriguing process needs the identification of the chain carrier. Studies are in progress following this line.

## 4. CONCLUSIONS

A comparison of our present results with the energetics of $SiH_4$ and $SiF_4$, allow to draw some conclusions. It appears that in $SiH_4$ IR photolysis, the first dissociation channel leads to $SiH_2$ formation. This radical produces higher silanes by insertion reactions with $SiH_4$ and ultimately amorphous silicon. Chain growth is also energetically favoured.

The dissociation thresholds for $SiF_4$ are much higher than for $SiH_4$ and $SiF_3$ appears to be the energetically favoured radical. Chain growth is an endothermic process for both $SiF_3$ and $SiF_2$ radicals and polymer formation and silicon production are more difficult, as observed.

## 5. ACKNOWLEDGMENTS

The authors gratefully acknowledge the help of R. Belardinelli and M. Nardelli throughout the experiments. Technical assistance of P. Cardoni, I. Cenciarelli, G. Schina is also appreciated.

## 6. REFERENCES

1. N.R. Isenor, V.E. Merchant, R.S. Hallsworth and M.C. Richardson: Can J. Phys.
2. V.S. Lethokov: Nonlinear Laser Chemistry, Springer Series in Chemical Physics, Vol. 22 (Springer-Verlag 1983)
3. H. Reisler, C. Wittig: in Photoselective Chemistry, Ed.s by J. Jortner, R.D. Levine, S.A. Rice (Interscience Publ., 1981) p 679-711
4. W.R. Cannon, S.C. Danforth, J.H. Flint, J.S. Haggerty, R.A. Marra,: J. Am. Ceram. Soc. 65, 324 (1982)
5. E. Borsella, R. Fantoni, A. Giardini-Guidoni and C.D. Cantrell: Chem. Phys. Lett. 87 284 (1982)
6. E. Borsella, R. Fantoni, A. Giardini-Guidoni, C.R. Adams and C.D. Cantrell: Chem. Phys. Lett. 101, 86 (1983)
7. C.W. Patterson and A.S. Pine: Opt. Comm. 44, 170 (1983)
8. A. Giardini-Guidoni, E. Borsella and R. Fantoni: Ber. Bunsen-Gerellschaft Phys. Chem. 89, 286 (1985)
9. G. Yahav and Y. Haas: Chem. Phys. Lett. 83, 493 (1985)
10. V.E. Merchant: Opt. Comm. 25, 259 (1978)
11. M. Snels, R. Larciprete, R. Fantoni, E. Borsella, A. Giardini-Guidoni: Chem. Phys. Lett. 122, 5 (1985)
12. T.F. Deutsch: J. Chem. Phys. 70, 1187 (1979)
13. P.A. Longeway and F.W. Lampe: J. Am. Chem. Soc. 103, 6813 (1981)
14. G.Inoue and M. Suzuki: Chem. Phys. Lett. 105, 641 (1984)
15. J. Steinwandel and J. Hoeschele: Chem. Phys. Lett. 116, 25 (1985)
16. T. Hamasaki, H. Kurata, M. Hirose and Y. Osaka: Appl. Phys. Lett. 47, 1084 (1980)
17. J.P. Bare and F. W. Lampe: J. Phys. Chem. 81, 1437 (1977)

# Coupled Equations and the Bloch-Nordsieck Representation for Molecules

*A.D. Bandrauk*

Département de chimie, Faculté des Sciences,
Université de Sherbrooke, Sherbrooke, Quebec, Canada J1K 2Rl

## I. INTRODUCTION

We have recently shown in a series of papers [1-7] that coupled equations methods of quantum collision theory can be conveniently used to investigate processes such as direct photodissociation [1-2], Resonance Raman Scattering in weak and strong fields [3-5], and higher order nonlinear spectroscopies [6-7] for diatomics involving several well isolated excited electronic states. This was done using the dressed molecule picture [8-12] of molecule-radiation interactions, wherein photon states are explicitly included into the theoretical description. All these methods are adequate only in the case of well isolated electronic states. There is an urgent need to derive a priori the most efficient representation for general electron-nuclear-radiation field systems such as occurs in intense field laser chemistry. In recent work we have examined this problem in an effort to incorporate as much as possible the electromagnetic field into the dynamics [13-14]. One might surmise that classical approaches should work sufficiently well at the high field intensities described here, and much work has been pursued in that direction. As we have pointed out previously, this involves treating both the molecule and the field classically. For electron-radiation interactions, one would prefer a quantum formulation since electronic states are, as a result of their large excitation energies, true quantum states. Furthermore, quantum mechanics leads to a linear theory of interactions whereas classical mechanics is a highly nonlinear theory [15]. Thus using a quantum formulation of matter-field interactions [16-17], we have been able to exploit methods of early (noncovariant) QED. In particular we have shown that the Bloch-Nordsieck (BN) representation [18-20], which leads naturally to the concept of coherent states in momentum space, was very convenient as a method of introducing strong field effects into the quantum dynamics of molecular systems. Coulomb gauge (A.p) and Electric Field (EF) gauge (E.r) representations were shown to be poor zeroth order approximations for the dressed molecular eigenstates in the presence of intense fields, i.e., the adiabatic coupled equations for the later two gauges define the appropriate adiabatic state as unperturbed (zero field) molecular states [13-14].

In the present work, we will explore the application of the BN representation to intense field-molecule interactions. We shall exàmine corrections beyond the dipole approximation which could become important in intense fields. Thus, for the case of a circularly polarised incident wave of amplitude E, the electron follows the field in a circle with angular velocity $\omega$ equal to that of the wave, so that $eE = m\omega^2 r = m\omega v$. This gives $v/c = eE/m\omega c = (eE/k)/mc^2$. $eE/k$ is essentially the potential difference across one wavelength and $mc^2$ is the electron rest mass energy. Comparing with kr, we obtain $kr = (\omega/c)(eE/m\omega^2) = v/c$. This simple exercise shows us that field-induced distortions of electron orbits can render the dipole approximation inoperative at high fields, especially for low frequencies which give rise to field-induced tunnelling [21]. As an example, one obtains $v/c = 1$ for $E = 3 \times 10^6$ volts/$\lambda$, so that for a $CO_2$ laser ($\lambda = 10^{-3}$ cm), $E = 3 \times 10^9$ V/cm which is essentially the hydrogen ground stàte atomic Coulomb potential. Thus corrections to the dipole approximation must be considered in intense fields approaching the internal fields of atom, and molecules.

## II. BLOCH-NORDSIECK TRANSFORMATIONS

We start with the minimal coupling molecule-field Hamiltonian [22] $H = H_{mol} + H_R$, where,

$$H_{mol} = \sum_{\alpha=1}^{N} (2M_\alpha)^{-1} [P_\alpha - \frac{eZ_\alpha}{c} A(R_\alpha)]^2 + \sum_{i=1}^{n} (2m)^{-1} [p_i + \frac{eA(r_i)}{c}]^2$$

$$+ V_{NN} (R_\alpha - R_\beta) + V_{nN} (R_\alpha - r_i) + V_{nn} (r_i - r_j) \quad . \tag{1}$$

$(P_\alpha, R_\alpha)$ are the canonically conjugate dynamical variables for the N nuclei of charge $eZ_\alpha$ and mass $M_\alpha$; $(p_i, r_i)$ are the corresponding 6 n electronic variables and finally $V_{NN}$, $V_{nN}$, $V_{nn}$ are the nuclear-nuclear, electron-nuclear, and electron-electron Coulomb interactions, respectively. $H_R$ is the Hamiltonian for the free electromagnetic field which we prefer to write in a coordinate representation of the field operators as a harmonic oscillator for a single mode,

$$H_R = \frac{1}{2} (P_k^2 + \omega^2 Q_k^2) \tag{2}$$

where $\omega = |k|C$ is the frequency of a transverse radiation mode of polarization $\varepsilon$, satisfying the transversality condition $\varepsilon \cdot k = 0$. In terms of the photon canonically conjugate variables $P_k$ and $Q_k$, the quantum vector potential $A(r)$ is a time-independent vector operator whose single mode component can be written for a general particle coordinate r,

$$A_k(r) = 2C(\pi/V)^{\frac{1}{2}} \varepsilon[Q_k \sin(k\cdot r) + \omega^{-1} P_k \cos(k\cdot r)] \quad , \tag{3}$$

where $[Q_k, P_k] = i\hbar$. In the dipole approximation we normally have the longwavelength condition $k\cdot r \simeq 0$, so that (3) simplifies to $A_k(r) = 2C(\pi/V)^{\frac{1}{2}} \omega^{-1} P_k \varepsilon$.

We now look for unitary transformations which will eliminate certain matter-field couplings in equation (1). Since these are of the form $A\cdot p$, one tries to eliminate this term by translating the particle momenta in order to cancel the vector potential terms. This procedure leads to the Electric Field (EF) gauge [22]. It is possible to find another type of transformation which is unitary equivalent to the above EF one. This transformation leaves the particle momenta invariant but displaces only the particle coordinates in the dipole approximation [13-14]. This is essentially the BN transformation [18-20] which we generalize here beyond the dipole approximation. From the transversality condition $\varepsilon\cdot k = 0$, one readily proves the commutation relations $[k\cdot r_i, \varepsilon\cdot p_i] = [k\cdot R_\alpha, \varepsilon\cdot P_\alpha] = 0$. We can therefore focus attention on a transformation operator $U = \exp(iS)$ where $S = (\varepsilon\cdot p)[g(k\cdot r)Q_k - f(k\cdot r)P_k]$. Then $(\varepsilon\cdot p)$ commutes with g and f. As a consequence of this one obtains that terms beyond the second commutator in the expansion of $O' = U^{+}OU$ vanish exactly for all the dynamical variable operators $O = p_i, P_\alpha, R_\alpha, Q_k, P_k$, i.e.,

$$O' = O + [O,S] + \frac{1}{2} [[O,S],S] \quad , \tag{4}$$

becomes exact only for those operators.

From the definition (2) of the vector potential and the transformation equation (4), one finds that the generation function S which cancels all $A\cdot p$ terms is given by

$$S = -(\Theta/k) \{\sum_i^n (p_i\cdot\varepsilon) [-Q_k \cos(k\cdot r_i) + \omega^{-1} P_k \sin(k\cdot r_i)]$$
$$- \sum_\alpha^N \frac{mZ_\alpha}{M_\alpha} (P_\alpha\cdot\varepsilon) [- Q_k \cos(k\cdot R_\alpha) + \omega^{-1} P_k \sin(k\cdot R_\alpha)]\} \quad , \tag{5}$$

where $\Theta = (2e/m\omega)(\pi/V)^{\frac{1}{2}}$ is a coupling parameter which relates directly particle displacements to the field amplitude $Q_k$ or the electric field E. Using S of (5) in (4) one obtains the following relations between the new displaced (primed) and original (unprimed) dynamical variables.

$$P'_\alpha = P_\alpha + \frac{m\Theta Z_\alpha}{M_\alpha}(P_\alpha\cdot\varepsilon)k\,([Q_k \sin(k\cdot R_\alpha) + \omega^{-1} P_k \cos(k\cdot R_\alpha)] +$$

$$(\Theta/4\omega)\,\{\sum_i (p_i\cdot\varepsilon)\cos[k\cdot(R_\alpha - r_i)] - \sum_{\alpha'} (\frac{mZ_{\alpha'}}{M_{\alpha'}})(P_{\alpha'}\cdot\varepsilon)\cos[k\cdot(R_{\alpha'} - R_\alpha)]\}) \quad ,$$

$$R'_{\alpha'} = R_\alpha - \frac{\varepsilon m\Theta Z_\alpha}{M_\alpha}([-Q_k \cos(k\cdot R_\alpha) + \omega^{-1} P_k \sin(k\cdot R_\alpha)] + \frac{\Theta}{2\omega}\{\sum_i (p_i\cdot\varepsilon)$$

$$\sin[k\,(R_\alpha - r_i)] - \sum_{\alpha'} (\frac{mZ_{\alpha'}}{M_{\alpha'}})(P_{\alpha'}\cdot\varepsilon)\sin[k\cdot(R_\alpha - R_{\alpha'})]\}) \quad ,$$

$$p'_i = p_i - \Theta(p_i\cdot\varepsilon)\,k\,([Q_k \sin(k\cdot r_i) + \omega^{-1} P_k \cos(k\cdot r_i)] + (\Theta/4\omega)\,\{\sum_{i'}$$

$$(p_{i'}\cdot\varepsilon)\cos[k\cdot(r_i - r_{i'})] - \sum_\alpha (\frac{mZ_\alpha}{M_\alpha})(P_\alpha\cdot\varepsilon)\cos\cdot[k\,(r_i - R_\alpha)]\}) \quad ,$$

$$r'_i = r_i + \varepsilon\Theta\,([-Q_k \cos(k\cdot r_i) + \omega^{-1} P_k \sin(k\cdot r)] + (\Theta/2\omega)\,\{\sum_{i'} (p_{i'}\cdot\varepsilon)$$

$$\sin[k\cdot(r_i - r_{i'})] - \sum_\alpha (\frac{mZ_\alpha}{M_\alpha})(P_\alpha\cdot\varepsilon)\sin[k\cdot(r_i - R_\alpha)]\}) \quad , \qquad (6)$$

$$Q'_k = Q_k + (\Theta/\omega)\,[\sum_i (p_i\cdot\varepsilon)\sin(k\cdot r_i) - \sum_\alpha (\frac{mZ_\alpha}{M_\alpha})(P_\alpha\cdot\varepsilon)\sin(k\cdot R_\alpha)] \quad ,$$

$$P'_k = P_k + \Theta\,[\sum_i (p_i\cdot\varepsilon)\cos(k\cdot r_i) - \sum_\alpha (\frac{mZ_\alpha}{M_\alpha})(P_\alpha\cdot\varepsilon)\cos(k\cdot R_\alpha)] \quad .$$

We observe that in the BN representation all particle momenta are displaced to first order in the parameter $\Theta$ along the direction of the propagation of the electromagnetic wave, and may be therefore viewed as particle recoil. These terms which are proportional to $k\Theta p \sim ev/c$ will give rise to multipole contributions to the Hamiltonian. The particle positions are all displaced along the direction of the field polarization $\varepsilon$. Both particle momenta and positions also undergo displacements proportional to order $e^2v^2/c^2$ which are in dependent of field strength. These field-independent terms will give rise to mass renormalization and vacuum many-body forces [23-24].

It should be noted that invoking the dipole approximation on the generating function S gives $S_D = (\Theta Q_k/\hbar)\,[\sum_i (p_i\cdot\varepsilon) - \sum_\alpha (\frac{mZ_\alpha}{M_\alpha})(p_\alpha\cdot\varepsilon)]$. Clearly this commutes with the particle momenta, i.e. using $S_D$ one would obtain $p'_i = p_i$, $P'_\alpha = P_\alpha$. This differs from the results applying the dipole approximation after the transformation is performed. Thus applying the dipole approximation to equation (6), one finds that recoil terms of order $k\Theta p \propto ev/c$ remain (e.g.: $p'_i = p_i - k\Theta\,(p_i\cdot\varepsilon)\,\omega^{-1} P_k$). The operations of carrying out the unitary transformations and the dipole approximation are therefore not commutative in the quantum approach. A similar noncommutativity of gauge transformations and dipole approximations was first noticed in the semiclassical EF gauge by Reiss [25].

Finally it should be remarked that the new photon momenta and coordinates $P'_k$ and $Q'_k$ are both modified as a result of coupling to the particles. In particular the new coordinate $Q'_k$ are displaced by terms dependent on particle momenta. Since coherent photon states correspond to photon coordinates $Q_k$ displaced by particle dipoles in the EF gauge [13-14] [26], we see from equation (6) the emergence of generalised coherent states in the BN representation which privileges momentum exchanges between particles and photons.

## III. EFFECTIVE POTENTIALS

Equation (6) furnishes us the field-modified momenta and coordinates. These expressions can be further simplified by using the electric field amplitude operator in terms of the coordinate representation of the photons, i.e.,

$$E(r) = -\frac{1}{c}\dot{A}(r) = (-i/\hbar c)\,[H_R, A(r)]$$

$$= \frac{m\omega\Theta\varepsilon}{e}\,[\omega\, Q_k \cos(k\cdot r) - P_k \sin(k\cdot r)] \quad . \tag{7}$$

Neglecting field-independent terms which arise from vacuum fluctuation corrections [14], we obtain the following expressions relating the field-perturbed r' and zero field coordinates r,

$$r_i' = r_i - (e/m\omega^2)\,E(r_i), \quad R_\alpha' = R_\alpha - (eZ_\alpha/m_\alpha\omega^2)\,E(R_\alpha) \quad . \tag{8}$$

This is consistent with the classical particle displacements expected in an electromagnetic field which gives rise to ponderomotive forces [27]. We note that in our development, the quantities E(r) are field operators, enabling us to go from the vacuum and thus highly quantum regime to the high-field semiclassical regime.

The effect on the Coulomb potentials is seen in the new potentials

$$\begin{aligned} V_{nn}'(r_i-r_j) &= V[r_i-r_j - (e/m\omega^2)\,(E(r_i) - E(r_j))] \quad , \\ V_{nN}'(r_i-R_\alpha) &= V[r_i-R_\alpha - (e/m\omega^2)\,(E(r_i) + \frac{mZ_\alpha}{M_\alpha}E(R_\alpha))] \quad . \end{aligned} \tag{9}$$

In the dipole approximation $E(r) = \dot{E}(o) = \frac{m\omega^2\Theta}{e}Q_k$ and one obtains

$$V_{nn}'(r_i-r_j) = V_{nn}(r_i-r_j); \; V_{nN}'(r_i-R_\alpha) = V[r_i-R_\alpha - \varepsilon\Theta Q_k\,(\frac{1+mZ_\alpha}{M_\alpha})] \quad , \tag{10}$$

i.e., only for opposite charged particles do the fields add coherently. Corrections to like charged particle interactions ($V_{nn}$, $V_{NN}$) appear only when recoil terms are retained, i.e.,

$$\begin{aligned} V_{nn}'(r_i-r_j) &= V[(r_i-r_j)\,(1 + \varepsilon k\Theta\omega^{-1}\,P_k)] \quad , \\ V_{nN}'(r_i-R_\alpha) &= V[(r_i-R_\alpha) - \varepsilon\Theta Q_k\,(1 + mZ_\alpha/M_\alpha) \\ &\quad + \varepsilon\Theta\omega^{-1}\,P_k\,(k\cdot r_i + \frac{mZ_\alpha}{M_\alpha}k\cdot R_\alpha)] \quad . \end{aligned} \tag{11}$$

Since $k/\omega = 1/c$, we see that recoil terms can dominate at high fields for particles of same charge.

The <u>dipolar</u> electron-nuclear Coulomb interaction $V_{nN}'$ was used in early work to discuss radiative corrections such as the Lamb shift [18-20] by expanding $V_{nn}'$ in a Taylor series in $Q_k$. We show here that a convergent series expansion is possible to obtain the effective vacuum Coulomb potential. Thus in our photon coordinate representation, the general integral we need to evaluate is

$$I_{mm'} = \langle m|V(r-\alpha Q)|m'\rangle = \int_{-\infty}^{\infty} U_{m'}(Q)\,\frac{1}{|r-\alpha Q|}\,U_m(Q)\,dQ \quad , \tag{12}$$

where $U_m(Q)$ are harmonic oscillator functions and $\alpha = \varepsilon\Theta\,(1 + \frac{mZ_\alpha}{M_\alpha})$. In particular the vacuum Coulomb potential is given by the simple integral (for zero photons)

$$I_{00} = \frac{c}{r}\int_{-\infty}^{\infty} \frac{e^{-x^2}}{(1 + ax + bx^2)^{\frac{1}{2}}}\,dx \quad , \tag{13}$$

whereas the general field-modified Coulomb potential matrix element is given by

$$I_{mm'} = \frac{c'}{r} \int_{-\infty}^{\infty} \frac{e^{-x^2} H_m(x) H_{m'}(x)}{(1 + ax + bx^2)^{\frac{1}{2}}} dx \tag{14}$$

The vacuum-corrected Coulomb potential (Lamb shift) $I_{00}$ was tried on MAXSYMA unsuccessfully. Only a series expansion was found useful in evaluating the integral (13):

$$I_{00} = \frac{c}{r} \exp(\tfrac{1}{2}b) \sum_{k=0}^{\infty} \binom{-\frac{1}{2}}{2k} a^{2k} \left(\frac{1}{b}\right)^{3/2k} \Gamma(k+\tfrac{1}{2})\, W_{-3/2k,-1/2k}(1/b) \quad , \tag{15}$$

where W are the Wittaker functions [28]. The series was obtained by expanding in powers of $(ax/(1+bx^2))$ which is always convergent [29]. We recall that $a = 2\alpha\cos\phi/r$, $b = \alpha^2/r^2$, where $\phi$ is the angle for the unit vector product $\hat{E}\cdot\hat{r}$ (i.e., the angle between the field and the electron position vector). We note that the case of orthogonal field configuration, $\phi = \pi/2$ or $a = 0$ gives a closed expression for the Lamb shift (vacuum) Coulomb potential,

$$I_{00}\ (a = 0) = \frac{Cb^{-\frac{1}{2}}}{r} \exp(\alpha/2b)\, K_0\,(\alpha/2b) \quad , \tag{16}$$

where $K_0$ is a modified Bessel function.

The dressed Coulomb potential for <u>any</u> number of photons n is given by the infinite series

$$I_{nn} = \frac{c}{r} \exp(1/2\ b) \sum_{k=0}^{\infty} \binom{-\frac{1}{2}}{2k} a^{2k} \left(\frac{1}{b}\right)^{\frac{k}{2}+n/2} \sum_{2S\leq n} \binom{n}{2S}^2 \left[\frac{(2S)!}{S!}\right]^2 2^{-4S}$$

$$b^S\, \Gamma(k - 2S + n + 1/2)\, W^{(1/b)}_{-3/2\,k + S - n/2,\ -k/2 - S + n/2} \quad , \tag{17}$$

which reduces to (16) for $n = 0$.

For large n, corresponding to the semiclassical limit of intense fields, one can go through the Fourier Transform of Coulomb potentials to obtain

$$I_{nn} = \frac{1}{2\pi^2} \int \frac{d^3k}{k^2} \exp i\, k\cdot r \langle n|\exp ik\cdot\alpha Q|n\rangle \quad . \tag{18}$$

The matrix element in the integral (18) is a Laguerre polynomial [13a] of order n, which for large n reduces to a Bessel function of order zero [30]. The integral (18) then reduces to

$$I_{nn} = \frac{c}{r} \int_{-1}^{+1} \frac{dx}{[(1-x^2)\ (1+ax+bx^2)]^{\frac{1}{2}}} \tag{19}$$

where $a = \frac{2\alpha\cos\phi}{r}$, $b = (\alpha/r)^2$, $\alpha = \Theta(\frac{\hbar}{2\omega})^{\frac{1}{2}} \sqrt{4n+2}$. The integral is expressible in terms of hyperelliptic functions F [29].

In conclusion, we have shown here that the Bloch-Nordsieck representation enables one to express multiphoton transitions in terms of field-modified Coulomb potentials for all field strengths. In zero field, one obtains the vacuum corrections to Coulomb potentials in a molecule. These are normally called Lamb shifts. For large number of photons n, one expects these to go over into the quantum version of the space translation approach [13a], [31]. These matrix elements need to be extended to include the first order recoil corrections (11) obtained beyond the dipole approximation. Finally the matrix elements (14), (17) will serve as effective photon transition matrix elements and potential renormalizations for the full molecular coupled equations developed in our previous work on the subject [13].

REFERENCES


1. A.D. Bandrauk, M.L. Sink, J. Chem. Phys. 74, 1110 (1981).
2. A.D. Bandrauk, G. Turcotte, J. Phys. Chem. 87, 5098 (1983).
3. A.D. Bandrauk, G. Turcotte, R. Lefebvre, J. Chem. Phys. 76, 225 (1982).
4. A.D. Bandrauk, G. Turcotte, J. Chem. Phys. 77, 3867 (1982).
5. G. Turcotte, A.D. Bandrauk, Chem. Phys. Lett. 94, 175 (1983).
6. A.D. Bandrauk, G. Turcotte, J. Phys. Chem. 89, 3039, 4473 (1985).
7. A.D. Bandrauk, N. Gelinas, Chem. Phys. Lett. 129, 362 (1986).
8. N.M. Kroll, K.M. Watson, Phys. Rev. A13, 1018 (1976).
9. A.I. Voronin, A.A. Samokhin, Sov. Phys. JETP 43, 4 (1976).
10. T.F. George, I.H. Zimmerman, J.M. Yuan, J.R. Laing, P.L. Devries, Acc. Chem. Res. 10, 449 (1977).
11. J.M. Yuan, T.F. George, J. Chem. Phys. 68, 3040 (1978).
12. T.F. George, J. Phys. Chem. 86, 10 (1982).
13. T.T. Nguyen-Dang, A.D. Bandrauk; a) J. Chem. Phys. 79, 3256 (1983); b) 80, 4926; c) 83, 2840 (1985).
14. A.D. Bandrauk, O.F. Kalman, T.T. Nguyen-Dang, J. Chem. Phys. 84, 6761 (1986).
15. G.E.O. Giacaglia, Perturbation Methods in Non-Linear Systems (Springer, N.Y. 1972), vol. 8.
16. R.P. Feynman, Quantum Electrodynamics (Benjamin Publishers, N.Y. 1962).
17. T.D. Lee, Particle Physics and Introduction to Field Theory, (Harwood Academic, N.Y. 1981), chapter 6.
18. F. Bloch, A. Nordsieck, Phys. Rev. 52, 54 (1937).
19. W. Pauli, M. Fierz, Nuovo Cimento 16, 167 (1938).
20. T.A. Welton, Phys. Rev. 74, 1157 (1948).
21. L.V. Keldysh, Sov. Phys. JETP 20, 1307 (1965).
22. E.A. Power, T. Thirunamachandran, Phys. Rev. A28, 2649 (1983).
23. H. Primakoff, T. Holstein, Phys. Rev. 55, 1218 (1939).
24. G. Chanmugan, S. Schweber, Phys. Rev. A1, 1369 (1939).
25. H.R. Reiss, Phys. Rev. A19, 1140 (1979); 22, 770 (1980).
26. E. Goldin, Waves and Photons (Wiley, N.Y. 1982).
27. M.H. Mittleman, Introduction to the Theory of Laser-Atom Interactions (Plenum Press, N.Y. 1982), chap. 6.
28. H. Bucholz, The Confluent Hypergeometric Function (Springer Verlag, Berlin, 1969), page 107.
29. A.D. Bandrauk, O.F. Kalman, to be published.
30. I. Byalynicki-Birula, Z. Bialynicka-Birula, Phys. Rev. A8, 3146 (1973).
31. W.C. Henneberger, Phys. Rev. Lett. 21, 838 (1968).
32. M. Gavrila, J.Z. Kaminski, Phys. Rev. Lett. 52, 613 (1984).

# The Franck-Condon Principle

*J. Tellinghuisen*

Department of Chemistry, Vanderbilt University,
Nashville,TN 37235, USA

## 1. Introduction

The Franck-Condon Principle is a thread running through all of molecular spectroscopy and much of atomic and molecular physics. Many dynamical problems require the evaluation of nuclear motion wavefunctions and overlap integrals; and product distributions are often governed by the squares of such integrals, i.e. by Franck-Condon factors. In this paper I address the question: What is the range and nature of the Franck-Condon distributions occurring in such problems? I will focus my attention on electronic transitions in diatomic molecules, but the results are applicable to other processes. Note that the term "distribution" is appropriate, because at the outset we know from closure that the Franck-Condon factors from a given initial level ($v_i$) to all possible levels of a second electronic state must sum (for discrete final states, $v_f$) and integrate (for the continuum, $\epsilon_f$) to unity.

The Franck-Condon Principle and the Born-Oppenheimer approximation go hand-in-hand: Both start with the premise that the nuclei in a molecule are so massive compared with the electrons that the motions of the nuclei and electrons may be treated separately. In electronic transitions the Franck-Condon Principle thus states that nuclear positions and momenta must be conserved during the transition. Until recently spectroscopists tended to interpret the Principle in an oversimplified fashion, in terms of vertical transitions from the classical turning points of the initial level [1]. Then about 1970 two interesting examples of highly structured diffuse spectra were treated -- in $H_2$ by DALGARNO, HERZBERG, and STEPHENS [2]; and in $I_2$ by MULLIKEN [3]. These examples pointed out the importance of conserving nuclear momentum as well as position, and they have led to an appreciation of the role of the *difference potential* in the Franck-Condon Principle [4-6].

I will consider the Franck-Condon Principle from the classical, quantal, and semiclassical standpoints. In particular, the last of these leads to the categorization of Franck-Condon distributions as displaying *reflection* or *interference* structure, depending on whether the initial wavefunction samples a monotonic or "polytonic" region of the difference potential. Reflection structure is the case where the spectrum "reflects" the radial probability distribution in the initial state, i.e. it conserves the peak and node count. There is no such simple relationship known for the case of interference structure. I will illustrate both types of structure with examples, and I will demonstrate how the stationary phase approximation at the heart of the primitive semiclassical treatment is manifested in quantal calculations of the overlap integrals.

## 2. Reflection and Interference Structure

Consider the case illustrated in Fig. 1. For the indicated initial level of the excited state, the dotted curve $X(R)$ represents the locus of points which classically conserve nuclear position and momentum in the transition to the lower state. This curve is given by

$$X(R) = U''(R) + E' - U'(R), \tag{1}$$

and is generally known as the "Mulliken difference potential." Since the

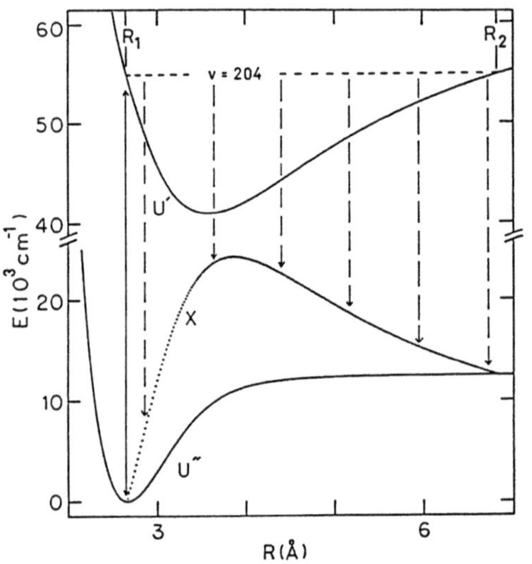


Fig. 1. Potential diagram for $D \to X$ fluorescence spectrum in $I_2$

transitions from $E'$ must terminate on $X(R)$, we can immediately write a simple expression for the *classical transition frequency* (with all quantities in $cm^{-1}$),

$$\nu(R) = E' - E''(R) = E' - X(R) = U'(R) - U''(R) \equiv V(R). \tag{2}$$

For obvious reasons $V(R)$ is known simply as the difference potential. Note that with this classical interpretation of the Franck-Condon Principle, we already have an answer for one part of the question posed at the outset: The *range* of the FC distribution is limited to values of $V(R)$ falling within the sampling region, $R_1 \le R \le R_2$. The only role of the excited state energy $E'$ is in the determination of this sampling range. However, changes in the sampling range can lead to profound changes in the *nature* of the FC distribution. For example, as we will see below, emission from the high $v'$ level illustrated in Fig. 1 yields a highly structured interference-type spectrum. On the other hand, emission from low $v'$ gives clear reflection structure, since low levels sample only a monotonic region of $V(R)$. It is interesting to note that the $I_2$ spectrum associated with the potential diagram of Fig. 1 has been known experimentally since 1913. The near constancy of the long-wavelength region of the spectrum with respect to changes in $E'$ was particularly baffling to early workers. Now it is clear that this effect is attributable to the minimum in $V(R)$ -- a near-trivial consequence of the Franck-Condon Principle!

Of course we realize that for quantitative assessment of Franck-Condon factors, we must go beyond this simple classical picture and employ quantal wavefunctions for the vibrational motion. The quantities to be evaluated are

$$\mathrm{FCF} = \left(\int_0^\infty \psi_{v'}(R)\psi_{v''}(R)\mathrm{d}R\right)^2 \tag{3}$$

and for bound-free transitions, the Franck-Condon density,

$$\mathrm{FCD} = \left(\int_0^\infty \psi_{v'}(R)\psi_{\epsilon''}(R)\mathrm{d}R\right)^2. \tag{4}$$

The numerical methods commonly used in such calculations have been developed and honed for over two decades, and for most purposes can be considered exact [6]. However, these computer codes tend to take the user from problem to answer with very little insight into underlying principles. For a better appreciation of the latter, I have found the semiclassical treatment of the FC overlap integral invaluable.

In the "primitive" semiclassical treatment of the integrals in (3) and (4), the wavefunctions are represented as (omitting normalization constants)

$$\psi_v = k_v^{-1/2} \cos \phi_v \quad \text{and} \quad \psi_\epsilon = k_\epsilon^{-1/2} \cos \phi_\epsilon , \tag{5}$$

where $k$ is the local wavenumber and $\phi$ is the phase [6,7]. The functional part of the Franck-Condon integral in (4) is

$$\langle \epsilon | v \rangle \approx (1/2) \int (k_\epsilon k_v)^{-1/2} (\cos \phi_+ + \cos \phi_-) \, dR , \tag{6}$$

where $\phi_\pm = \phi_\epsilon \pm \phi_v$. To evaluate this integral we invoke the stationary phase approximation [7], which recognizes that the integral of the product of two sinusoidal functions can only accumulate significantly in regions where the two functions have the same periodicity, or equivalently where the phases $\phi_\pm$ in (6) are changing slowly with $R$. The condition for a point of stationary phase is $\partial\phi_\pm/\partial R = 0$. No solution exists for $\phi_+$. For $\phi_-$ the condition becomes $k_v(R^*) = k_\epsilon(R^*)$, which is precisely the Franck-Condon Principle. If we consider all possible final state energies $\epsilon$, the points of stationary phase $R^*$ trace out the curve $X(R)$ defined in (1) and depicted in Fig. 1. Thus the primitive semiclassical treatment of the FC integral leads directly to the classical FC Principle.

When the upper and lower potentials are configured as shown in the inset to Fig. 2, the curve $X(R)$ is *monotonic* in the region $R_1 \leq R \leq R_2$. The semiclassical treatment predicts oscillations in the FC integral, as the upper- and lower-state wavefunctions move in and out of phase with each other. Furthermore, the FC integral changes sign $v_i$ times, leading to $v_i$+1 peaks in the FC distribution. Thus the spectrum "reflects" the radial probability distribution of the initial level $v_i$, hence the designation *reflection* structure for this case. Quantal calculations (*e.g.*, Fig. 2) show that this prediction of the primitive semiclassical treatment is substantially correct.

Now let us return to the case of Fig. 1, illustrated for another transition in Fig. 3. Here the Mulliken difference potential displays a *maximum*, and for all continuum energies below this maximum there are *two* points of stationary phase. The maximum in $X(R)$ becomes a minimum in $V(R)$, as was noted earlier, and the FC Principle predicts that there will be a long-wavelength cutoff of the emission. Specifically, there will be no significant emission to the red of this cutoff wavelength, *no matter what the value of* $v_i$. In classical spectroscopy this potential configuration leads to a "head of heads" in a discrete band spectrum; in

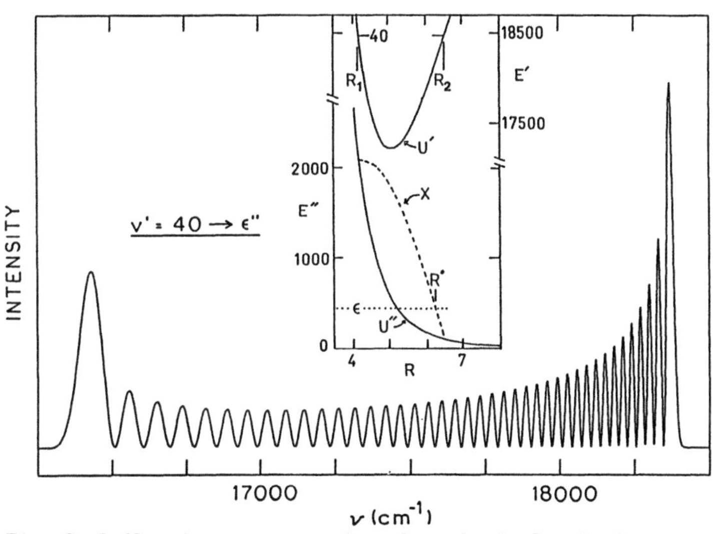


Fig. 2. Reflection structure in a hypothetical emission spectrum of $Cs_2$

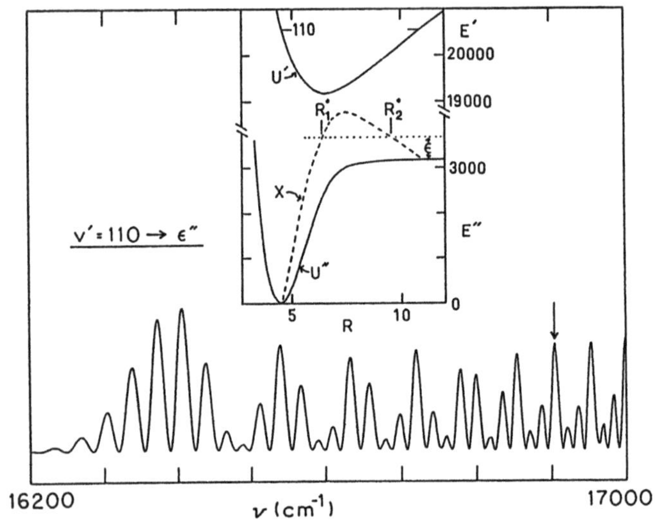


Fig. 3. Interference structure in the $E \to X$ spectrum of $Cs_2$

atomic line-broadening parlance the extremum in $V(R)$ becomes a "satellite" [4]. It is in fact this case which was treated in [2] and [3]. Since there are now two points of stationary phase for each allowed $\epsilon''$, there are two regions of $R$ where the FC overlap integral can accumulate. These two contributions can add either constructively or destructively. The resulting "interference" leads to a very distinctive, characteristic structure in the FC distribution, which I have accordingly designated *interference* structure. For situations involving two stationary phase points, as in Fig. 3, this interference is manifested as a low-frequency modulation of the high-frequency oscillatory structure. It is interesting to note that, in his paper on the quantal version of Franck's principle, CONDON [8] discussed a type of transition which he called "internal diffraction," which is equivalent to that of Fig. 3. However, Condon was apparently unaware of the earlier work on the $I_2$ spectrum, and the full properties of this type of transition remained unappreciated for four decades.

Although the primitive semiclassical treatment correctly predicts the range and nature of the Franck-Condon distributions, one may rightly wonder whether the principle of stationary phase is valid in any real sense. To investigate this question I have examined the way in which the overlap integrals accumulate as functions of $R$ -- an approach used also by GERBER *et al.* [9] in conjunction with a study of diffuse spectra in $Sr_2$. The results of such calculations (e.g. Fig. 4, and others to be shown, which are already available in the literature [6,10]) show that these integrals indeed build up in relatively local regions of $R$ near the stationary phase points. As might be expected, the stationary phase principle holds better for large mass than it does for small, in keeping with the more nearly classical behavior for large mass. However, it should be noted that it is in the *approach* to the classical limit that we observe the richest quantum structure, and it is mainly heavy diatomics that have provided the best illustrations of reflection and interference structure in FC distributions.

Before closing this section, there are several additional points I wish to make. First the classification of FC distributions into reflection and interference structure holds for discrete transitions as well as for bound-free. Thus, for example, the discrete spectrum adjoining the continuum in Fig. 3 displays reflection structure; however, in this case the reflection pattern in incomplete, since only a portion of the total distribution is associated with the discrete spectrum. In some discrete spectra the FC structure may be "finer" than the discrete structure, in which case the characteristic reflection or interference pattern may not be

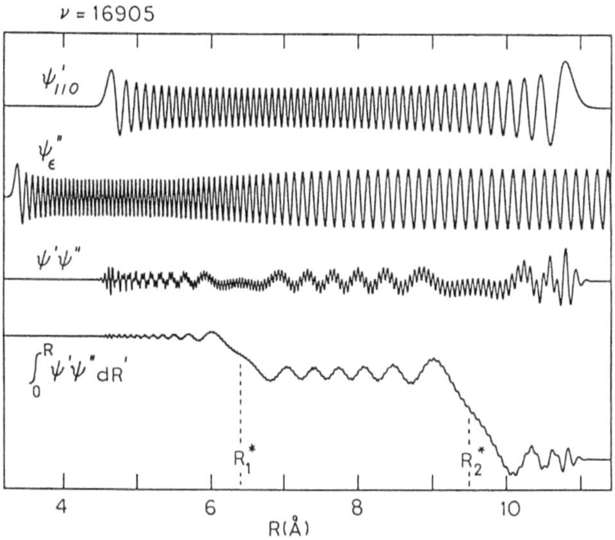


Fig. 4. Constructive interference in the overlap integral at 16905 $cm^{-1}$ in Fig. 3

discernible, and the classification becomes moot. Second, while I have dealt with only the two-point problem in my treatment of interference structure, it is clear that more than two points of stationary phase might occur for a given energy under some circumstances. Quantal calculations show that the structure can be intriguingly complex in such cases [6,10]. Finally, the occurrence of reflection *structure* should not be taken as justification for using the simple reflection *method* [1] to calculate the spectrum. Numerical tests suggest that as a rule the reflection method is reliable only for $v_i = 0$.

## 3. Applications

In the foregoing I have dealt with the Franck-Condon Principle in its most fundamental form. In most cases of practical interest, the calculation of product distributions involves the evaluation of integrals slightly more complicated than those of (3) and (4). For example in radiative transitions the intensities are proportional to quantities of form $\langle v_f|\mu_e(R)|v_i\rangle^2$, where $\mu_e(R)$ is the transition moment function. If the $R$-dependence of $\mu_e$ can be neglected, the intensities are proportional to the FCFs, and we have the case often called the Condon approximation. More commonly the $R$-dependence *cannot* be neglected, especially for large $v_i$, where the wavefunction spans a large region of $R$. As long as $\mu_e$ varies much more slowly with $R$ than does $\psi_v$, the treatment and conclusions of the preceding section remain valid; and $\mu_e^2(R)$ plays the role of an intensity scaling function superimposed upon the Franck-Condon distribution.

The transition moment function in fact remains one of the most elusive observables in molecular spectroscopy. In efforts to relate $\mu_e(R)$ to observed intensities, it is customary to go one step beyond the Condon approximation and invoke the $R$-centroid approximation [11],

$$\langle v_f|\mu_e(R)|v_i\rangle \approx \mu_e(\bar{R})\langle v_f|v_i\rangle \ , \tag{7}$$

where $\bar{R}$ is the $R$-centroid,

$$\bar{R} = \langle v_f|R|v_i\rangle / \langle v_f|v_i\rangle \ . \tag{8}$$

It is easy to show that this approximation becomes rigorous when $\mu_e$ is linear in $R$. Furthermore, in cases where a single stationary phase point contributes to the overlap integral (reflection structure), that point *is* the $R$-centroid -- $\bar{R} \approx R^*$ -- to a very good approximation [6,11-13]; and (7) is then quite reliable. When there are multiple transition points (7) is not trustworthy; however, by segmenting the range of integration it is possible to define a modification of (7) and (8) which agrees much better with exact calculations [13].

I have completely neglected the role of rotation in the Franck-Condon Principle, and for a good reason: The range and structure of Franck-Condon distributions are determined entirely by the vibrational motions and wavefunctions. Of course rotation must be taken into account in accurate calculation of numerical FCFs and FCDs. One typically finds that the FC distribution simply shifts along the spectral axis as the $J$-dependent centrifugal term is included in the effective potentials used to calculate the wavefunctions. For example the reflection spectrum for $J' = J'' = 0$ in Fig. 2 will spread out somewhat toward higher $\nu$ with increasing $J$. It should be evident that in regions of sharp structure, the overlap integrals for specific $\epsilon_f$ or $v_f$ can depend sensitively on $J$, even though the overall FC distribution is *qualitatively* independent of $J$.

Often in radiation problems one needs to evaluate the total radiative decay rate from some initial excited level $v'$ to all final levels of a second electronic state. For diatomics this calculation may be expressed

$$A_T = (64\pi^4/3h)\Big(\sum_{v''} \nu^3_{v'v''}|\langle v''|\mu_e(R)|v'\rangle|^2 + \int \nu^3_{v'\epsilon''}|\langle \epsilon''|\mu_e(R)|v'\rangle|^2 d\nu\Big) , \tag{9}$$

where $\nu$ is the wavenumber of the $v'{\rightarrow}v''$ or $v'{\rightarrow}\epsilon''$ transition. Although this computation is in principle straightforward, it can be tedious in cases like that of Fig. 3, where a highly structured continuum adjoins a discrete spectrum. Recently I have shown [14] that the sum and integral in (9) can be collapsed to an initial-state expectation value,

$$A_T \approx (64\pi^4/3h)\langle v'|\nu^3\mu_e^2(R)|v'\rangle , \tag{10}$$

which is evaluated with the aid of the Franck-Condon Principle, by substituting $V(R)$ for $\nu$. Although this approximate sum rule was actually derived long ago by LAX [15], its remarkable reliability (better than 1%) was not recognized. Similar sum rules hold for other Franck-Condon-dominated matrix elements; and further, expectation values like that in (10) can be evaluated classically and extremely efficiently through powerful quadratures [16].

There are now many known examples of both reflection and interference structure in diatomic spectra -- in both diffuse (bound-free or free-free) and discrete (bound-bound) transitions. A number of these will be included in the oral portion of this paper, and I refer the reader to my recent review [6] for details of these and other cases in the literature. Here I would like to make a few comments on the Franck-Condon Principle in "radiative collisions." It sometimes happens that developments in one field parallel those in another, with widespread ignorance of this condition of the part of the players in both fields. Such is, I believe, the case in the development of the Franck-Condon Principle by molecular spectroscopists on the one hand and atomic physicists studying spectral line broadening on the other. Spectroscopists, who deal explicitly and daily with molecular potentials, have been slow to appreciate the significance of the difference potential. In line broadening theory the difference potential enters as early as 1932 [17], and plays a prominent role in quasistatic classical methods [18]; however, with the notable exception of Gallagher's group, the line-broadening community have tended to consider the difference potential as just a mathematical formality and have been reluctant to accept the fact that atoms in collision (at least in "slow" collision) *are* molecules. This provincialism extends also to the study of radiative collisions, which began about a decade ago as a faddish preoccupation with processes such as

$$A^* + B + hc\nu \rightarrow A + B^* \tag{11}$$

The study by FALCONE *et al.* [19], with A = Sr and B = Ca, is often cited as the first case of a radiative collision in experimental "captivity." However entirely equivalent processes (at least in the weak-field limit) have been known to spectroscopists for decades as free-free transitions. The special feature of process (11) is that $\mu_e(R)$ vanishes at large $R$; but this condition holds also in many well-characterized molecular transitions, both diffuse and discrete -- for example in the rare gas oxides, which have even been made to *lase* on such transitions. It is true that radiative collision people have tended to emphasize the fate of the heavy bodies, while molecular spectroscopists generally focus their attention on the photons. However, there is usually a direct mapping between the spectrum and the collision dynamics. Other distinctions are really just matters of viewpoint. For example, the collision dynamicist speaks of (11) as occurring off resonance by virtue of the presence of the radiation field, while the spectroscopist thinks of the collision as tuning the transition *into* resonance, by finding $R$ such that $V(R) = \nu$ for the AB *molecular* transition. In a very real sense we might even say that radiative collisions were already understood by the time of FINKELNBURG's 1938 monograph [20], since for example his Fig. 45 depicts potential configurations covering free-free and free-bound transitions as well as discrete and bound-free systems.

## 4. Conclusion

The primitive semiclassical treatment of Franck-Condon overlap integrals serves to unify the classical and quantal versions of the Franck-Condon Principle. The classical version states that transitions terminate at the point or points $(R^*, E^*)$ which conserve nuclear position and momentum. Quantum mechanics says that particles cannot be localized in this manner, and that one must calculate overlap integrals employing wavefunctions which extend over the classical region and into the nonclassical region. The semiclassical stationary phase approximation tells us that these integrals nevertheless accumulate mainly in the vicinity of the classical transition points. The examples I have presented show that this picture is fundamentally correct.

It is interesting to note that in the same paper in which Condon coined "internal diffraction," he also introduced the reflection method, whereby spectra can be approximated by simply reflecting the initial probability distribution onto the final potential curve. Although the reflection method is now known to be of limited reliability, it is clear that by 1928 Condon already had an understanding of both reflection and interference structure in Franck-Condon distributions. In the intervening years we have learned a great deal about the mechanical application of the Principle, through computational methods made possible by the advent of the computer. However, I think it is fair to say that the only significant new *qualitative* feature we have learned in this time is the all-important role played by the difference potential.

## 5. References


1. G. Herzberg: *Spectra of Diatomic Molecules*, 2nd edition (Van Nostrand, Toronto 1950)
2. A. Dalgarno, G. Herzberg, T. L. Stephens: Astrophys. J. *162*, L49 (1970)
3. R. S. Mulliken: J. Chem. Phys. *55*, 288, 309 (1971)
4. A. Gallagher: In *Atomic Physics*, ed. by G. zu Putlitz, E. W. Weber, A. Winnacker, Vol. 4 (Plenum, New York 1975) p. 559
5. J. Tellinghuisen, G. Pichler, W. L. Snow, M. E. Hillard, R. J. Exton: Chem. Phys. *50*, 313 (1980)
6. J. Tellinghuisen: Adv. Chem. Phys. *60*, 299 (1985)
7. L. D. Landau, E. M. Lifshitz: *Quantum Mechanics -- Non-Relativistic Theory*, 2nd.ed. (Pergamon, Oxford 1965)
8. E. U. Condon: Phys. Rev. *32*, 858 (1928)
9. G. Gerber, R. Möller, H. Schneider: J. Chem. Phys. *81*, 1538 (1984)

10. J. Tellinghuisen: J. Mol. Spectrosc. *103*, 455 (1984)
11. P. A. Fraser: Can. J. Phys. *32*, 515 (1954)
12. A. L. Smith: J. Mol. Spectrosc. *28*, 269 (1968)
13. C. Noda, R. N. Zare: J. Mol. Spectrosc. *95*, 254 (1982)
14. J. Tellinghuisen: Chem. Phys. Lett. *105*, 241 (1984)
15. M. Lax: J. Chem. Phys. *20*, 1752 (1952)
16. J. Tellinghuisen, P. S. Julienne: J. Chem. Phys. *81*, 5779 (1984)
17. H. Margenau: Phys. Rev. *40*, 387 (1932)
18. R. E. M. Hedges, D. L. Drummond, A. Gallagher: Phys. Rev. *A6*, 1519 (1972)
19. R. W. Falcone, W. R. Green, J. C. White, J. F. Young, S. E. Harris: Phys. Rev *A15*, 1333 (1977)
20. W. Finkelnburg: *Kontinuierliche Spektren* (Springer, Berlin 1938)

Part IV

# Atomic Collisions in Laser Fields

# Reactive Collision Dynamics of the Collision System Mg + $H_2$ Using Far Wing Laser Scattering

*A.M. Lyyra, P.D. Kleiber, K.M. Sando, and W.C. Stwalley*

Departments of Physics and Chemistry and the Iowa Laser Facility,
University of Iowa, Iowa City, IA 52242, USA

## 1. Abstract

The far wing absorption of laser light to the collision complex $MgH_2$ leads to nonreactive formation of excited state magnesium, Mg*, and reactive formation of MgH [1]. Far wing absorption of laser light is monitored by observation of the final product (Mg* or MgH) as a function of pump laser detuning from the magnesium resonance line. We monitor production of MgH (v" = 0) rotational levels J" = 6 and J" = 23, close to low J and high J maxima of a bimodal rotational distribution, respectively. For this we use laser-induced fluorescence in a two laser pump and probe experiment. We also monitor the competing nonreactive channel Mg* by observation of resonance fluorescence Mg ($3^1P_1^0$ - $3^1S_0$). The onset of the reactive channel in the Mg + $H_2$ system is highlighted through a comparison with far wing profiles from the nonreactive collision system Mg + He and the reactive system Mg + $H_2$ [1]. We have developed a simple excited state dynamics model for the theoretical interpretation of our data. This model uses standard quasistatic theory to estimate the absorption probability as a function of detuning of the laser frequency from resonance. The potential energy curve data [2] is limited to fixed geometries, $C_{2v}$ and $C_{\infty v}$. Based on this, only rough comparisons are possible between the dynamics model and experiment.

## 2. Introduction

It is well known that the far wing absorption profiles of pressure-broadened transitions are sensitive to the intermolecular potential curves of the collision system [3]. In the quasistatic approximation, absorption of radiation occurs at a fixed internuclear separation $R_0$ between the potential curves of the collision partners so that the potential curves are shifted to resonance with the radiation field. The quasistatic absorption profile is related to the probability of finding an atom-perturber pair near $R_0$

$$k(\Delta) \sim 4\pi R_0^2 (e^{-V_g(R_0/kT)})/(dV(R)/dR)_{R_0} \quad (1)$$

where $V(R) = V_e(R) - V_g(R)$ is the difference potential and the exponential term is the ground state Boltzmann factor.

The far wing line broadening techniques have been extended to atom-molecule collisions. It is no longer possible to invert the line profile to obtain the potential energy surfaces [4] since there are many points on the multidimensional surfaces, which contribute to the signal at a certain detuning. However, the wing profile data can be used to test models of the potential surfaces and it is possible to obtain qualitative information about the dynamics of the collision. These techniques have been applied to reactive collisions [5, 6]. Recently we presented initial results of far wing laser absorption into the $MgH_2$ collision complex [1]:

$$\begin{aligned} Mg\,(3^1S_0) + H_2 + \hbar\omega \to (MgH_2)^*_{complex} &\to Mg^*\,(^1P_1^0) + H_2 \\ &\to MgH\,(v'', J'') + H \end{aligned} \quad (2)$$

The ground state reaction is 25000 $cm^{-1}$ endoergic and is unlikely to occur in the absence of laser pumping (Figure 1). The kinetics of the excited state reaction

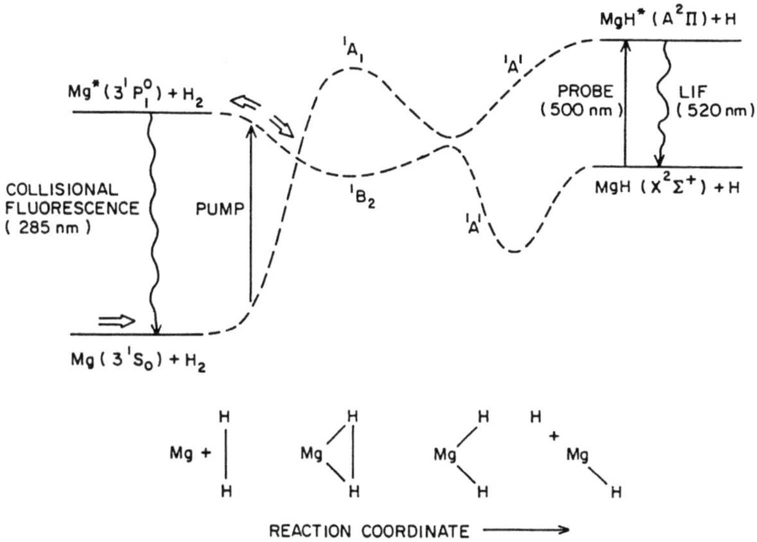


Figure 1. Representation of reactive collision mechanism involving side-on attack of $H_2$ by Mg and consecutive laser absorption to the complex. Final products are monitored via laser-induced fluorescence (MgH) and collisional fluorescence (Mg*).

$$Mg^* (3^1P_1^0) + H_2 \rightarrow MgH + H \tag{3}$$

are already well known [7]. This reaction is entrance channel-controlled occurring with gas kinetic cross-section. Breckenridge et al. [8] observe a bimodal, nonstatistical product rotational distribution suggesting that high rotation component is due to side-on attack of Mg* to the $H_2$ bond (insertion reaction) and low rotation component is due to end-on attack (abstraction reaction). In the following we discuss the implications of our results (both low J and high J) for the understanding of the dynamics of this excited state reaction. These preferred geometries simplify our theoretical modeling by limiting the orientational averaging.

## 3. Experimental

The experimental set-up, based on a standard pump-probe technique, is described in detail in [9]. A frequency-doubled and tripled 30 Hz Nd:YAG laser pumps simultaneously two pulsed tunable dye lasers, one of which is frequency doubled to a wavelength region of the magnesium resonance line at 285 nm. The probe dye laser is operated in the A-X band region of MgH near 506 nm. The beam intensities are limited so that the observed signal is linear in pump laser power, typically 1 MW/cm$^2$ for the pump and 10 MW/cm$^2$ for the probe for Mg densities $\leq 1 \cdot 10^{13}$/cm$^3$. The probe pulse is optically delayed by 7 ns from the pump laser pulse. This delay is short enough to minimize secondary collision processes leading to MgH formation via allowed excited state reaction. The final state distribution was studied using the probe laser to produce laser-induced fluorescence of the MgH $A^2\Pi$ - $X^2\Sigma^+$ (0, 0) band. Detuning the pump laser from Mg resonance transition, the far wings were measured for either the low (J" = 6) and high (J" = 23) rotation component of the bimodal product rotational distribution. The competing nonreactive channel of excited state Mg* ($3^1P_1^0$) was detected at the resonance wavelength of 2852 A using the same detection system. The photomultiplier signal was then fed to a gated integrator system. The pump-laser intensity was monitored with an energy detector and the far wing profiles were normalized to constant laser power.

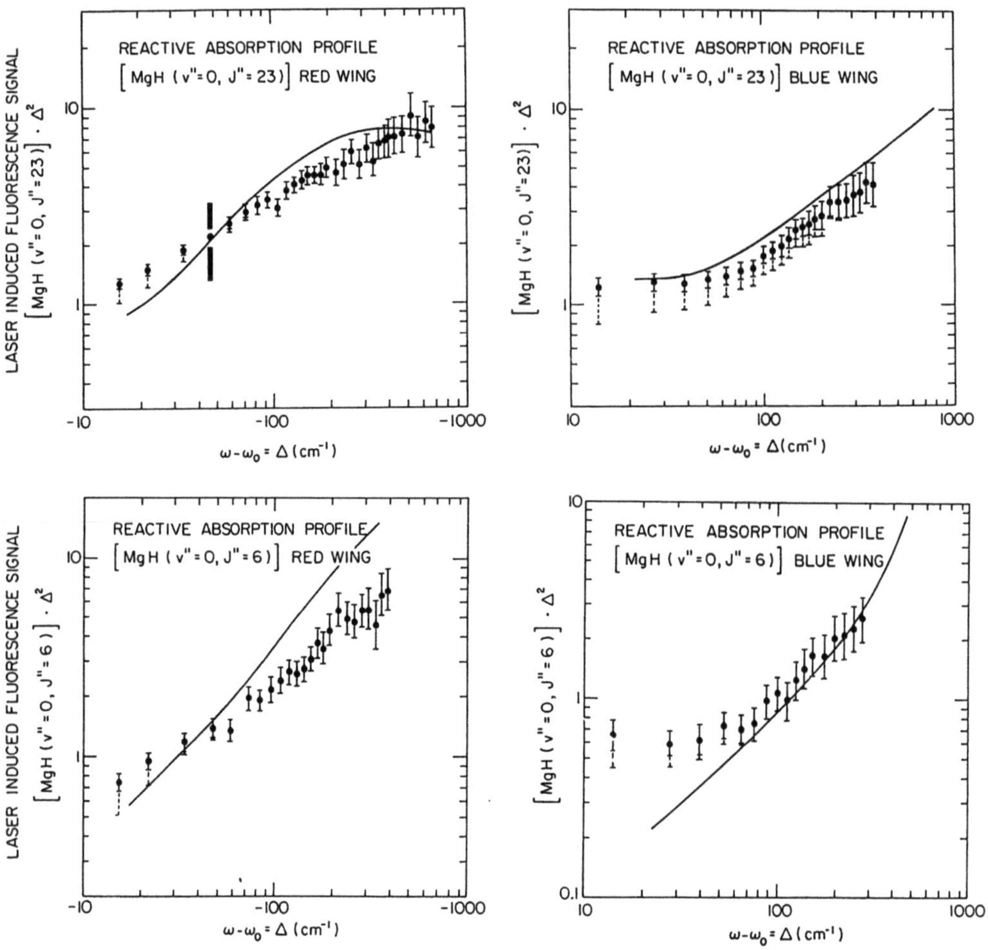


Figure 2a. Experimental far wing profiles for the reactive Mg + $H_2$ collision channel for v" = 0, J" = 6, 23. Laser-induced fluorescence signal has been plotted versus detuning $\omega - \omega_0 = \Delta$. Profiles have been multiplied by $\Delta^2$.

## 4. Results

Figures 2a and 2b display absorption profiles of the reactive and nonreactive channel, respectively. The absorption profiles have been multiplied (customarily) by $\Delta^2$ to enhance the structure. Also shown are the results of a simple semiquantitative theoretical model (solid lines, Figure 2a). The theoretical curves of Figure 2b are based on a single perturber quantum mechanical calculation of the absorption profiles. Due to weak signal levels in the reactive channel, extensive signal averaging was necessary and data runs extend over a period of several weeks. It is difficult to keep experimental conditions fixed under these circumstances and we have compensated for this by normalizing different runs to a constant signal at a red detuning $\Delta = 47\ cm^{-1}$. The solid error bar in Figure 2a illustrates the magnitude of this long-term drift in the signal.

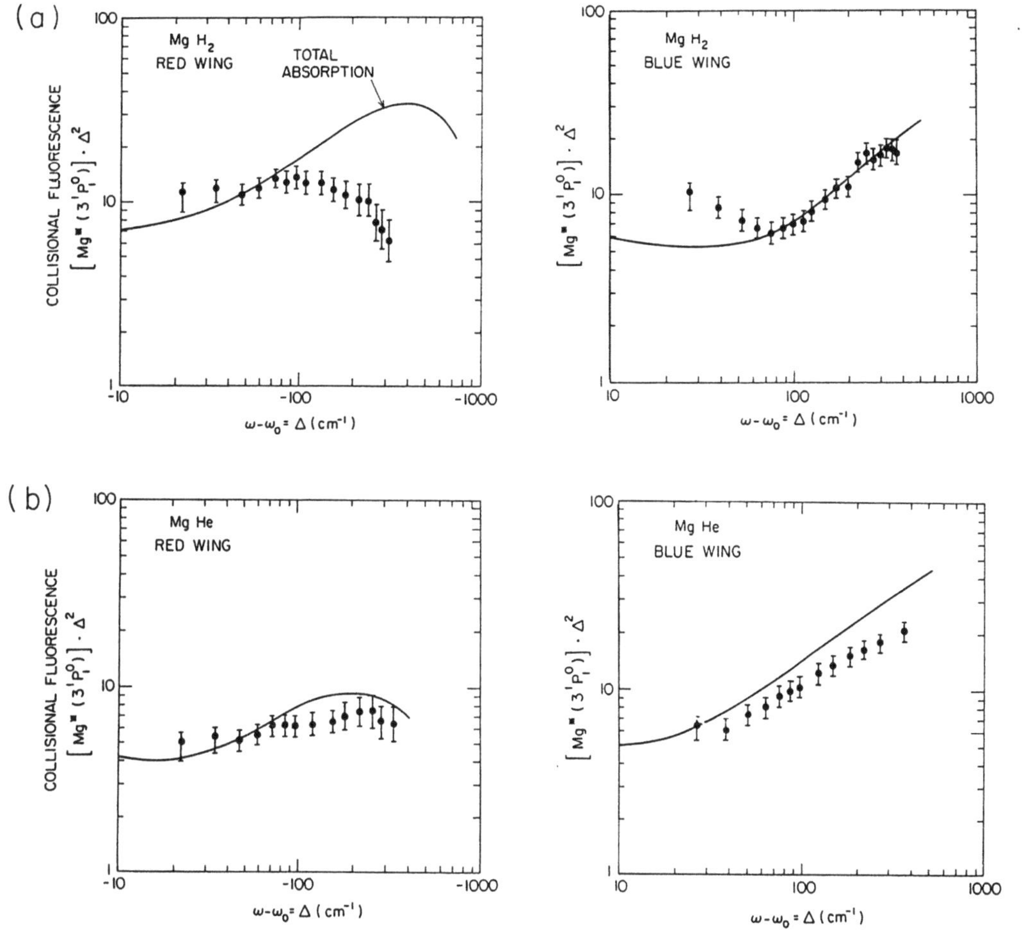


Figure 2b. Nonreactive Mg* channel. Comparison of the far wing profiles of Mg + $H_2$ and Mg + He.

## 5. Theoretical Modeling of Results

In the following we try to explain the experimental observations in terms of collision processes on the potential energy surfaces of this system. The initial step of electronic excitation of the Mg atom and hydrogen molecule involved in the collision is identical to that for absorption in the wings of a pressure-broadened atomic line.

This initial step can be modelled using quasistatic theory based on the classical Franck-Condon principle. The dynamics on the excited state surface then determines the product final state distribution.

In order to predict the absorption spectrum and the dynamical branching ratios accurately, complete potential energy surfaces would be required. The ab initio SCF-CI potentials for Mg + $H_2$ by Chaquin et al. [2] are only available for limited geometries ($C_{2v}$ and $C_{\infty v}$). This limits our approximations for the excited state dynamics following the initial absorption step. The theoretical data is also limited at long internuclear distance and we fit a $C_6/R^6$ type of long-range potential to

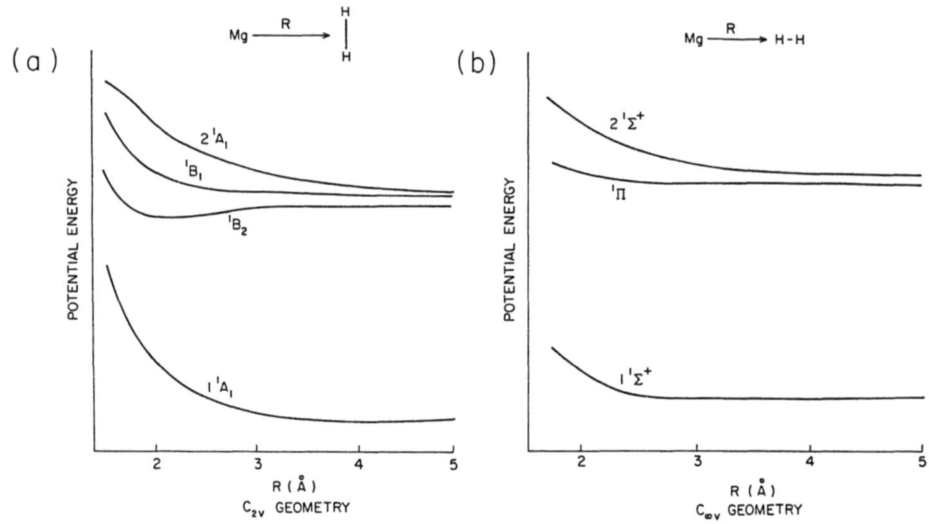


Figure 3. Theoretical potential energy curves for fixed H-H bond length (equilibrium). Adapted from [2].

these surfaces. The $C_{2v}$ geometry curves (Figure 3a) approach this limiting case. However, the $C_{\infty v}$ geometry curves do not approach the simple van der Waals form in the long-range part. For these we use a Buckingham potential fit

$$V(R) = A\, e^{-BR} + C\, R^{-6} \qquad (4)$$

to the data of [2]. This may introduce some additional uncertainties for the linear geometry. Due to limited potential surface data we have to make several limiting assumptions about the dynamics involved. First we assume fixed H-H distance. This is reasonable according to theoretical estimates [2], since the minimum energy approach of Mg to $H_2$ on the ground state surface maintains the H-H distance near the equilibrium value. The second assumption follows from the dynamical arguments by Breckenridge et al. [8]. Here we correlate the higher rotational part of the bimodal MgH rotational distribution with an insertion reaction i.e. Mg approaching $H_2$ perpendicular to the $H_2$ bond ($C_{2v}$). Similarly we correlate the lower rotational part of the rotational MgH final state distribution with Mg approaching $H_2$ end-on ($C_{\infty v}$). This just simply means that the potential energy surfaces in Figure 3(a) have been used to model the J" = 23 data and the surfaces in Figure 3(b) for J" = 6 data. We do not expect more than a rough semiquantitative comparison with the experimental data from this simplified model.

Mg* can be produced from any approach geometry but an average of the two geometries $C_{2v}$, $C_{\infty v}$ (weighted by the degeneracies) should be justified if the anisotropy is not too significant. Figure 2b also presents Mg* far wings from collisional broadening by He gas. The theoretical absorption profiles are based on model potential energy curves by Malvern [10] fit to a Buckingham potential as described above. The Mg* profiles have been calculated using single perturber quantum mechanical calculations. The theoretical description of this method is given in [11]. The Mg + He and Mg + $H_2$ data can be used for a rough comparison of the Mg* profiles. At larger internuclear distances both profiles are dominated by long-range interactions and weak collisions and should be roughly similar. For larger detunings we expect differences due to the onset of the reactive channel in Mg + $H_2$ collision.

In the following we describe the assumptions we make concerning the excited state dynamics and the reactive versus nonreactive branching ratio. We assume that all free-bound and free-quasibound absorption to $^1B_1$ and $^1B_2$ in $C_{2v}$ geometry and to $^1\Pi$ in $C_{\infty v}$ geometry (Figure 3) leads to reaction. We also assume that free-free absorption to levels above the rotational barrier leads to reaction on incoming trajectories, while those on outgoing trajectories do not react. Blue wing absorption to $2^1A_1$ ($C_{2v}$) and $2^1\Sigma^+$ ($C_{\infty v}$) is in both cases free-free absorption.

The standard quasistatic theory of line broadening has been used to calculate the line profiles and results are given in Figure 2a together with experimental results. The red wing theoretical profiles have been arbitrarily normalized to experimental data at $\Delta = 47\ cm^{-1}$ and the same normalization is used for the blue wing comparison to maintain red/blue asymmetry.

## 6. Discussion

The overall agreement between calculated and experimental far wing profiles (Figure 2) seems reasonable. At small detunings the relatively poor agreement could be either due to the uncertainties associated with the long-range form of the calculated potential energy curves, or perhaps long-range breakdown of approximations made concerning the dynamics. The experimental results agree at least qualitatively with calculated profiles. The fact that the absorption profiles are sensitive to potential energy curve shape is evident from comparison of the theoretical profiles of Figure 2a for J" = 6 and J" = 23. For these we use the $C_{\infty v}$ and $C_{2v}$ potential energy curves respectively as described above.

The reactive profiles show a pronounced red wing extending to detunings that correspond to absorption into attractive excited state potential surfaces. The nonreactive red wing collisional redistribution profile of Figure 2b on the other hand shows a fairly sharp drop-off at laser detunings greater than 100 $cm^{-1}$. We suggest that this fall-off could be correlated with nonreactive vs reactive channel competition. The comparison of $MgH_2$ and MgHe profiles of Figure 2b reinforce this view. There is close agreement between the $MgH_2$ and MgHe collisional redistribution profiles to detunings of 100 $cm^{-1}$ in the red wing. Beyond this point the $MgH_2$ profile falls off rapidly while MgHe profile exhibits the appearance of a weak satellite near 200 $cm^{-1}$ detuning. The properly normalized sum of the reactive (MgH) and nonreactive (Mg*) profiles should be compared to the <u>total</u> theoretical <u>absorption</u> profile shown in Figure 2b ($MgH_2$). Since we have data for so few MgH final product states it is not possible to construct the total experimental channel. Furthermore the branching ratios are not known accurately enough so that one could estimate the relative contributions of reactive and nonreactive components. For these reasons we have not presented a comparison between the total theoretical absorption profile and the sum of reactive and nonreactive experimental line profiles.

Figure 4 shows the red/blue wing asymmetry for the reactive profiles of Figure 2a. There the arbitrary normalization factors have cancelled out for the comparison between theoretical and experimental results. This absolute fit of the data to the theory is surprisingly good considering that our dynamics model assumes a unit reaction probability for absorption on incoming trajectories into the repulsive $2^1A_1$ state. The large reaction probability for trajectories on the repulsive curve is surprising since the attractive $1^1B_2$ level should be the primary channel leading to excited state reaction. It is less surprising considering the fact that Breckenridge <u>et al</u>. [8] have reported essentially gas kinetic cross-section for the excited state reaction. Currently the experimentally observed blue wing reactive profiles lack theoretical explanation.

Figure 5 shows the low J/high J experimental and theoretical ratios of the red wing profiles leading to formation of MgH within the low J and high J components of the bimodal final product rotational distribution.

The theoretical curve follows from correlating the low J component with $C_{\infty v}$ collisions geometry and the high J component with $C_{2v}$ collision geometry. The dashed line is the expected result if the entrance channel geometry is completely irrelevant. Our results fall between these two limiting models. Given the inadequate nature of our theoretical model, it is impossible to make a definite choice between these extreme cases. However, the results suggest some correlation between entrance channel geometry and final product rotational distribution.

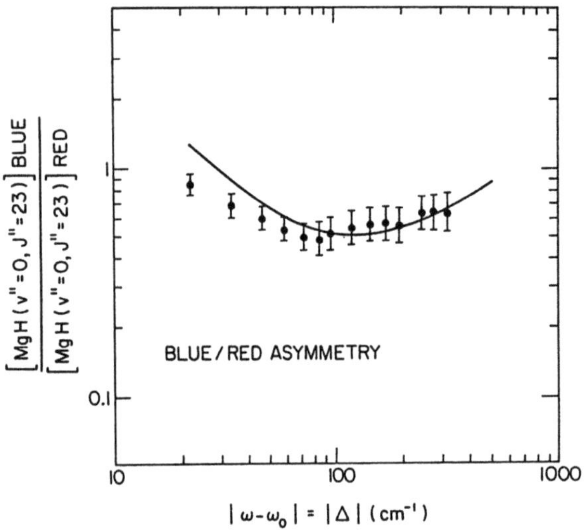


Figure 4. Blue wing/red wing asymmetry for MgH (v" = 0, J" = 23) reactive absorption profile.

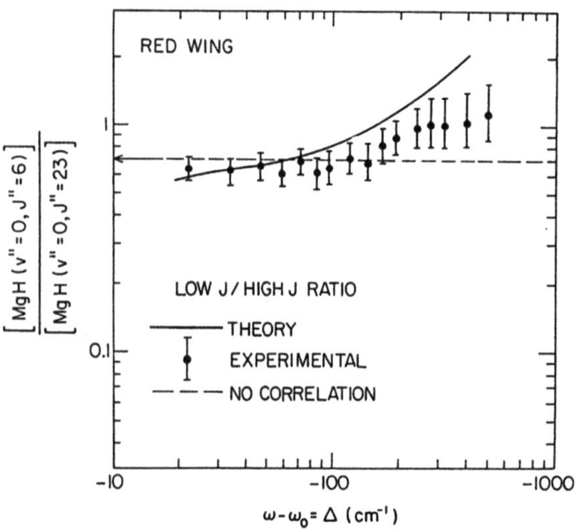


Figure 5. Branching ratio as a function of detuning for low rotation and high rotation reaction product. Solid line assumes correlation between entrance channel geometry and dashed line is for the case of no correlation between entrance channel geometry and final product rotational state.

## 7. Conclusions

The far wing absorption profiles of the $MgH_2$ collision system have been observed to lead to two distinct final product channels. The nonreactive channel leads to electronically excited magnesium Mg* and the reactive channel leads to formation of MgH in a bimodal rotational final state distribution. We have compared the nonreactive Mg* profiles with Mg* profiles from collisional broadening by helium, the latter system being entirely nonreactive. From this comparison we have concluded that our results exhibit reactive versus nonreactive channel competition effects in the far red wings of $MgH_2$. Strong far blue wing absorption also leads to reactive channel suggesting significant reaction probability for trajectories on repulsive surfaces.

We have developed a simple dynamics model in order to have a semiquantitative theoretical interpretation of our results. The available potential energy curve data [2] is limited to fixed geometries and we have been forced to fit the long range parts of these potential to a simplified van der Waals form. Therefore only rough comparisons between theory and experiment can be expected. We are able to conclude, however, that some correlation exists between entrance channel geometry and final product rotational distribution.

## 8. Acknowledgements

We gratefully acknowledge the support of the National Science Foundation.

## 9. References

1. P. D. Kleiber, A. M. Lyyra, K. M. Sando, S. P. Heneghan and W. C. Stwalley, Phys. Rev. Lett. 54, 2003 (1985); A. M. Lyyra, K. M. Sando and P. D. Kleiber, Phys. Rev. A XX, xxxx (1986).
2. P. Chaquin, A. Sevin and H. Yu, J. Phys. Chem. 89, 2813 (1985).
3. A. Jablonski, Phys. Rev. 68, 78 (1945).
4. A. Gallagher, in Spectral Lines Shapes, Volume II, edited by K. Burnett (de Gruyter, Berlin, 1982), p. 755.
5. P. Arrowsmith, S. H. P. Bly, P. E. Charters and J. C. Polanyi, J. Chem. Phys. 79, 283 (1983).
6. T. F. Maguire, P. R. Brooks and R. F. Curl Jr., Phys. Rev. Lett. 50, 1918 (1983); P. Hering, P. R. Brooks, R. F. Curl Jr., R. S. Judson and R. S. Lowe, Phys. Rev. Lett. 44, 687 (1980).
7. W. H. Breckenridge and H. Umemoto, J. Chem. Phys. 75, 698 (1981).
8. W. H. Breckenridge and H. Umemoto, J. Chem. Phys. 80, 4168 (1984).
9. P. D. Kleiber, A. M. Lyyra, K. M. Sando, V. Zafiropulos and W. C. Stwalley, submitted to J. Chem. Phys.
10. A. R. Malvern, J. Phys. B 11, 831 (1978).
11. P. S. Herman and K. M. Sando, J. Chem. Phys. 68, 1153 (1978).

# Laser Pulse-Induced Molecular Collisions: Effect of the Temporal Width of the Pulse

*H.-W. Lee*[1] *and T.F. George*[2]

[1]Department of Physics, Oakland University, Rochester, MI 48063, USA
[2]Departments of Chemistry and Physics & Astronomy, 239 Fronczak Hall, State University of New York at Buffalo, Buffalo, NY 14260, USA

A theoretical investigation of the effect of the pulse duration on laser-induced collision processes is carried out within the framework of the Landau-Zener model. Collision cross-sections for laser-induced electronic transitions in atom-atom collisions are calculated for two different pulse shapes, square and gaussian. For sufficiently high laser intensities ($\geq 10^{10}$ W/cm$^2$), there is a pronounced change in the cross-section as the pulse duration is shortened from 10 to 1 ps for both types of pulse shapes.

## 1. INTRODUCTION

It has been shown both theoretically and experimentally that the presence of strong laser radiation can greatly alter the outcome of molecular collision processes. We have suggested in the past that details of laser radiation, such as the temporal variation of the intensity and the linewidth, need to be carefully examined before the effect of radiation is accurately determined [1,2]. In particular, we showed that, if laser radiation consists of pulses whose duration is comparable to or shorter than the collision time, transition probabilities may be significantly higher than for the case of long laser pulses [3]. An experimental study of this short-pulse effect has recently been reported [4].

The short-pulse effect can be best illustrated with the Landau-Zener model, as we described in our earlier paper [3]. If one considers a dipole transition taking place in the presence of laser radiation of intensity I, the Landau-Zener transition probability is given by

$$P_1 = 2\exp(-\beta I)[1 - \exp(-\beta I)] \quad , \tag{1}$$

where $\beta$ is a constant depending on the details of the system. If, however, the transition occurs in the presence of a short pulse so that the interaction exists on the way in or out only, the probability becomes

$$P_2 = 1 - \exp(-\beta I) \quad . \tag{2}$$

If the laser intensity is sufficiently high ($\beta I >> I$), $P_2$ is significantly higher than $P_1$.

In this paper we report the result of further investigation of the effect of the pulse duration on laser-induced collision processes. Within the Landau-Zener formalism, we calculate the collision cross-section (or the total number of atoms that make the transition) as the duration of the pulse is varied. Calculations have been carried out at different intensities for two different pulse shapes, square and gaussian.

## 2. THEORY

We assume that the molecular system is prepared in its lowest electronic state $u_1$ and consider a transition from $u_1$ to an upper state $u_2$ with an absorption of a photon from laser radiation. The laser frequency is assumed to be in resonance with the two states $u_1$ and $u_2$ at the internuclear distance $q = q_c$. Assuming a dipole transition, the Landau-Zener probability can be written as

$$P = \exp\{-\gamma\varepsilon_+^2[1 - (\frac{b}{q_c})^2]^{1/2}\} (1 - \exp\{-\gamma\varepsilon_-^2[1 - (\frac{b}{q_c})^2]^{1/2}\})$$

$$+ \exp[-\gamma\varepsilon_-^2[1 - (\frac{b}{q_c})^2]^{1/2}\} (1 - \exp\{-\gamma\varepsilon_+^2[1 - (\frac{b}{q_c})^2]^{1/2}\}) \quad , \qquad (3)$$

where

$$\gamma = \frac{\pi\mu^2}{2\hbar v\alpha} \quad ,$$

$\varepsilon_\pm$ is the electric field strength at the crossing point $q_c$ on the way in and out, respectively, $\mu$ is the transition moment, $\alpha$ is the slope at $q_c$ of the energy difference of the two states, b denotes the impact parameter, and for simplicity a straight-line, constant-velocity (= v) trajectory is assumed. We also have assumed that the pulse is linearly polarized along the direction of v and the transition moment is parallel to the internuclear axis as in a $\Sigma$ - $\Sigma$ transition.

We first consider the case of a square pulse with intensity $I_0$ (and corresponding electric field strength $\varepsilon_0$) and duration T. If the pulse duration T is longer than the collision time $\tau_b$, where $\tau_b$ is defined as the time between two crossing points,

$$\tau_b = 2(q_c^2 - b^2)^{1/2}/v \quad , \qquad (4)$$

the total number of excited atoms per target atom with b between b and b + db is

$$dN = (T - \tau_b)v\rho 2\exp\{-\gamma\varepsilon_0^2[1 - (\frac{b}{q_c})^2]^{1/2}\}(1 - \exp\{-\gamma\varepsilon_0^2[1 - (\frac{b}{q_c})^2]^{1/2}\})$$

$$\times 2\pi b db + 2\tau_b v\rho(1 - \exp\{-\gamma\varepsilon_0^2[1 - (\frac{b}{q_c})^2]^{1/2}\}) 2\pi b db \quad , \qquad (5)$$

where $\rho$ is the number density of projectile atoms. The first term on the right-hand side arises from those molecules illuminated both on the way in and out ($\varepsilon_+ = \varepsilon_- = \varepsilon_0$), while the second term accounts for those illuminated either on the way in or out only ($\varepsilon_\pm = \varepsilon_0$, $\varepsilon_\mp = 0$). If the pulse duration T is shorter than the collision time $\tau_b$, then we have $\varepsilon_\pm = \varepsilon_0$, $\varepsilon_\mp = 0$, and thus

$$dN = 2Tv\rho(1 - \exp\{-\gamma\varepsilon_0^2[1 - (\frac{b}{q_c})^2]^{1/2}\})2\pi b db \quad . \qquad (6)$$

When Eq. (5) or (6) is integrated over b from b = 0 to $b = q_c$, we obtain the total number of excited atoms per target atom, N. In order to study the effectiveness of a pulse in inducing the transition, we divide this number N by the total energy of the pulse. Thus we obtain

$$N_e \equiv \frac{N}{TI_0} \quad . \qquad (7)$$

This number $N_e$ represents the total number of excited atoms per target atom per unit pulse energy, and can be considered as a measure of the effectiveness of the laser pulse in inducing the transition.

We now consider a more realistic case, a gaussian pulse of peak intensity $I_0$ and the width (FWHM) T,

$$I(t) = I_0 \exp[-(\frac{t}{T/2})^2 \ln 2] \quad , \tag{8}$$

or

$$\varepsilon^2(t) = \varepsilon_0^2 \exp[-(\frac{t}{T/2})^2 \ln 2] \quad . \tag{9}$$

In this case, different pairs of colliding atoms sample different intensities of the pulse, and therefore the averaging over the pulse intensity needs to be performed. We thus have

$$dN = 2\pi b db v \rho \int_{\infty}^{\infty} dt \exp\{\{-\gamma\varepsilon^2(t)[1 - (\frac{b}{q_c})^2]^{1/2}\}(1-\exp\{-\gamma\varepsilon^2(t + \tau_b) [1 - (\frac{b}{q_c})^2]^{1/2}\})$$

$$+ \exp\{-\gamma\varepsilon^2(t + \tau_b)[1 - (\frac{b}{q_c})^2]^{1/2}\}(1-\exp\{\gamma\varepsilon^2(t) [1 - (\frac{b}{q_c})^2]^{1/2}\}) \tag{10}$$

The total number of excited atoms per target atom, N, is obtained by integrating Eq. (10) over b. Finally, the total number of excited atoms per target atom per unit pulse energy is given by

$$N_e = \frac{N}{(TI_0\sqrt{\pi}/2\sqrt{\ln 2})} \quad . \tag{11}$$

## 3. CALCULATIONAL RESULTS AND DISCUSSION

In this section we report the results of calculations for $N_e$ for both square and gaussian pulses. The parameters used for our calculation are: $q_c = 5$ Å, $v = 4$ Å/ps, $\alpha = 0.002\ a_0$ and $\mu = 3$ a.u. With the above $q_c$ and v, the collision time $\tau_b$ is 2.5 ps at $b = 0$ and 2.12 psec at $b = 5$ a.u. For both square and gaussian pulses we have computed $N_e$ for different pulse intensities ($I_0 = 10^8, 10^9, 10^{10}, 10^{11}$ W/cm$^2$) and for different pulse durations (T = 1000, 100, 10, 1, 0.1 ps). The last two values of T (1 and 0.1 psec) correspond to the case of a short pulse ($t < \tau_b$ for most of b values).

Table 1 shows the results of our calculation. At $10^8 - 10^9$W/cm$^2$, there is little difference between long and short pulses. This is expected because at these low intensities, the factor $\beta I$ which appears in Eqs. (1) and (2) is small and $P_1 \approx 2P_2$. The difference, however, begins to show up at $10^{10}$W/cm$^2$, as a noticeable jump is seen as one goes from 10 to 1 ps. The difference is more pronounced at $10^{11}$ W/cm$^2$. The significant difference at $10^{11}$ W/cm$^2$ between 1000, 100, 10 ps pulses for the case of a square pulse is due to the sudden decrease in intensity at each edge of the pulse, characteristic only of a square pulse. We believe the data with a gaussian pulse are much more realistic.

In Table 2 we show the effectiveness of a short pulse vs. a long pulse, defined as

$$R_{eff} \equiv \frac{N_e(T = 0.1 \text{ ps})}{N_e(T = 1000 \text{ ps})} \quad . \tag{12}$$

The ratio is significantly high at high intensities, confirming that the

Table 1. Total number of excited atoms per target atom per unit pulse energy (in arbitrary units)

| | T(ps) | 1000 | 100 | 10 | 1 | 0.1 |
|---|---|---|---|---|---|---|
| $I_0 = 10^8 W/cm^2$ | square pulse | 19.8 | 19.8 | 19.9 | 20.2 | 20.2 |
| | gaussian pulse | 20.0 | 20.0 | 20.0 | 20.3 | 20.3 |
| $I_0 = 10^9 W/cm^2$ | square pulse | 15.1 | 15.1 | 15.7 | 18.5 | 18.5 |
| | gaussian pulse | 16.5 | 16.5 | 16.6 | 19.0 | 19.0 |
| $I_0 = 10^{10} W/cm^2$ | square pulse | 1.45 | 1.57 | 2.81 | 8.75 | 8.84 |
| | gaussian pulse | 3.99 | 3.99 | 4.24 | 10.8 | 11.3 |
| $I_0 = 10^{11} W/cm^2$ | square pulse | 0.00188 | 0.0186 | 0.186 | 1.05 | 1.07 |
| | gaussian pulse | 0.217 | 0.217 | 0.266 | 1.70 | 2.21 |

Table 2. Effectiveness of a short pulse compared with a long pulse in inducing the transition

| $I_0(W/cm^2)$ | $10^8$ | $10^9$ | $10^{10}$ | $10^{11}$ |
|---|---|---|---|---|
| square pulse | 1.02 | 1.23 | 6.10 | 571 |
| gaussian pulse | 1.01 | 1.15 | 2.84 | 10.2 |

transition probability is much higher with a short pulse at high intensities. One, however, notes that the dramatically high ratio that exists for the case of a square pulse, which was also reported in our earlier paper [3], is considerably softened if a gaussian pulse is considered.

Finally, we note that our study is carried out within the Landau-Zener formalism. We expect that, as we consider higher pulse intensities and/or shorter pulse durations, the Landau-Zener model becomes less accurate. At extremely high intensities, Rabi oscillations will occur to bring down the probability, no matter how short the laser pulse is. If the pulse is extremely short, the linewidth becomes large, which in turn increases the size of the interaction region, invalidating the assumption of the model that the transition is limited to a narrow region.

ACKNOWLEDGMENTS

This research was supported in part by the National Science Foundation under Grant Nos. CHE-8512406 and CHE-8519053 and the Air Force Office of Scientific Research (AFSC) under Contract F49620-86-C-0009. The United States Government is authorized to reproduce and distribute reprints for government purposes notwithstanding any copyright notation hereon.

REFERENCES

1. H. W. Lee and T. F. George: J. Chem. Phys. 69, 2596 (1978).
2. H. W. Lee, P. L. DeVries, I. H. Zimmerman and T. F. George: Mol. Phys. 36, 1693 (1978).
3. H. W. Lee and T. F. George: J. Phys. Chem. 83, 928 (1979).
4. T. Sizer, II and M. G. Raymer: Phys. Rev. Lett. 56, 123 (1986).

# Charge Transfer in Collisions and in Complexes Involving Excited Atoms

*W. Kamke, E.E.B. Campbell*, R. Witte*, B. Kamke, H. Schmidt, Z. Wang, H.U. Kiefl, and I.V. Hertel**

Institut für Molekülphysik, Freie Universität Berlin,
Arnimallee 14, D-1000 Berlin 33
*and: Fakultät für Physik, Universität Freiburg,
Herrmann-Herder-Str. 3, D-7800 Freiburg, Fed. Rep. of Germany

## 1 Introduction

In the understanding of chemical reactions the transfer of energy or charge between atoms or molecules plays an important role. In general, atoms or molecules must be in excited states for a transfer reaction to take place. Also the orientation and alignment of the electronic charge clouds will influence the probability for such processes. Initial electronic excitation is most easily achieved by photon absorption, using lasers or synchrotron radiation sources. This leads us to a second approach: Photophysical reactions, i.e. processes initiated by photon absorption, are fundamental in every living substance. The detailed dynamics of such reactions are far from being understood. Basic questions that have to be asked are : Where does the energy go? How is it transferred? Are photons reemitted? Are free charges generated? Are electrons or holes or protons exchanged? Which are the reactants? Which steps are involved? What is the timescale? etc. The path to follow in order to eventually reach a "detailed understanding" is to study a few representative cases on simple systems like isolated molecules or atoms. In experiments, we can fairly well define internal states and the relative kinetic energy of the reactants and products by combining modern techniques. To illustrate the state of the art we would like to discuss three representative processes : Associative Ionisation, Resonant Charge Exchange and Intramolecular Penning Ionisation. However, the discussion on associative ionisation is deferred in this contribution and the reader is referred to the paper by J.Weiner /1/ on this topic.

## 2 Resonant Charge Exchange

As an example of resonant charge exchange we focus on the system using laser aligned $Na^*(3p)$ atoms :

$$Na^+ + Na^*(3p,\lambda) \to Na_2^+(\Lambda) \to Na^*(3p,\lambda') + Na^+$$

which may be considered as one of the simplest resonant charge transfer processes incorporating nonadiabatic transitions between the $\Sigma$ and $\Pi$ molecular states (designated by $\Lambda$). The quantum numbers $\lambda$ and $\lambda'$ refer to the 3p substates at large internuclear separation. The final states, $\lambda'$, remain unobserved in the experiment and are averaged over. However, the initial distribution of $\lambda$ substates can be controlled by exciting the atomic target prior to the collision with polarised laser light. We are thus able to determine the complete set of alignment and orientation parameters for the collision process.

It is sometimes convenient to describe the polarisation dependence of the above reaction in terms of the inverse process : Assume we were to start on the right hand side of the above equation with an isotropic distribution of $\lambda'$ substates. The charge transfer process would then leave the $Na^*$ atom after the collision in an aligned and oriented state with a nonisotropic distribution of $\lambda$ states. This collision-induced charge distribution is illustrated in fig.1. Our measurements

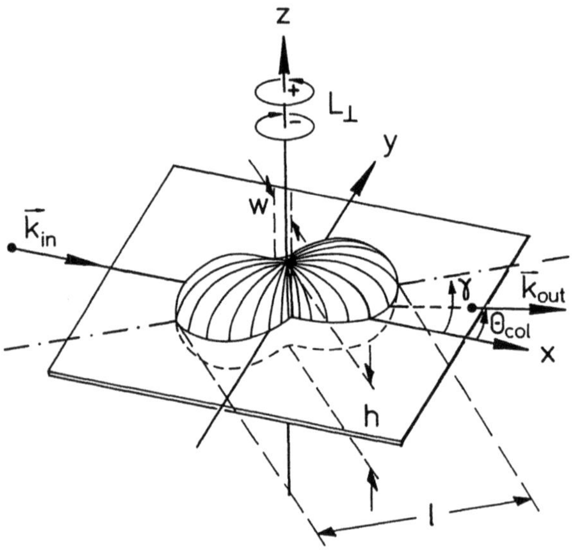


Fig.1: Collision - induced charge distribution in the "natural" frame, giving the definitions of the alignment angle $\gamma$ and the angular momentum transfer $L_\perp^+$. The scattering plane is defined by the directions of the incoming and outgoing particles. l, w and h refer to the length width and height of the charge cloud measured at the intersections of the main axes.

allow us to determine, as a function of scattering angle $\Theta_{col}$, the density matrix elements related to the length, width and height of the charge distribution, its alignment angle $\gamma$ and the transfer of angular momentum, $L_\perp^+$.

The experimental details and description of the detailed analysis of the experimental results have been given elsewhere /2/.

The alignment parameters ($\gamma$, $P_l^+$, $\rho_{00}$), determined by linear polarisation studies, are plotted in fig.2 for $E_{c.m.}$= 100 eV. At small scattering angles, T = $E_{c.m.}\Theta_{c.m.}$ < 20 eV°, the alignment angle $\gamma$ is very small, indicating that a p$\sigma$ atomic orbital prepared at large internuclear separation of the interacting particles is most efficient for the charge transfer process. This is a substantial effect since $P_l^+ = ((1-w)/(1+w))$ is 0.4 at T = 15 eV°, (fig.2b), implying that p$\sigma$ is more than twice as effective for charge transfer than p$\pi^+$. As the scattering angle increases, both orbitals become more and more equivalent until at T = 40 eV°, $P_l^+$ almost vanishes and $\gamma$ is no longer well defined. The anisotropy increases again as the scattering angle is increased further, however this time $\gamma$ is nearly orthogonal to the situation at small scattering angles. For T > 60 eV° we find $\gamma$ = 80 eV and $P_l^+$ = 0.8 which means that a nearly p$\pi^+$ preparation of the atomic orbital at large internuclear distances is about nine times more effective for the process than the p$\sigma$ orbital.

Careful consideration of the observed differential cross-sections and calculated classical deflection functions for scattering from the $Na_2^+$ molecular states (see /2/) indicates that scattering from the $1\Pi_u$ molecular potential dominates at small scattering angles (T = 20 eV°) whereas the $2\Sigma_g$ potential dominates at larger angles (T > 40 eV°). Thus an asymptotically prepared $\sigma$ ($\pi^+$) atomic orbital becomes a $\Pi$ ($\Sigma$) molecular state at small internuclear distances (<20 a.u.). This conclusion is borne out by considering the relative height of the electronic charge cloud, $\rho_{00} = (h/(1+w+h))$, shown in fig.2c. This is a measure of the relative probability of the $\pi^-$ contribution to the charge transfer. Since reflection symmetry has to be conserved in the collision , no transitions to $\Sigma$ states are possible for this initial preparation of the atomic orbital, and we thus have an image of pure $\Pi$

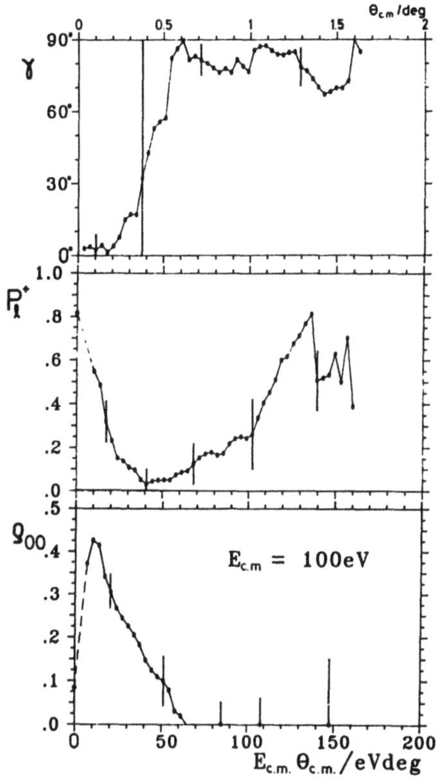


Fig.2: Alignment parameters γ, $P_1^+$ and $\rho_{00}$ for $E_{c.m.}$=100eV.

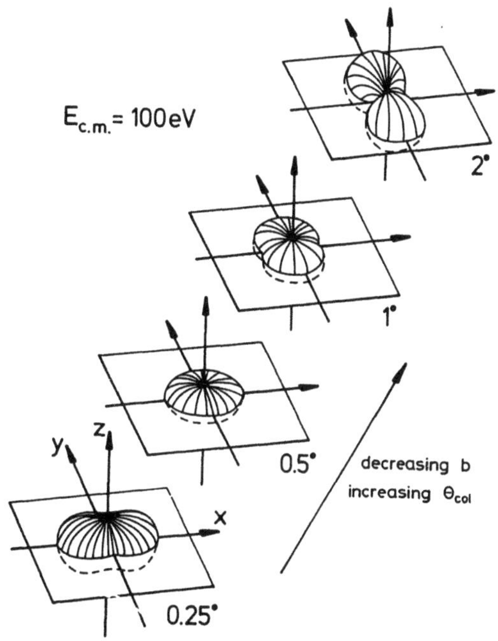


Fig.3: Charge distribution after resonant charge exchange from an isotropically populated $Na^*(3p)$ state as derived from the alignment parameters for different $\Theta_{col}$.

scattering. This is largest for the small scattering angles and reaches almost zero for T > 60 eV°.

Figure 3 summarises the experimental results obtained from linearly polarised light showing the preferred shape of the electronic charge cloud for charge transfer at different values of $\Theta_{col}$.

## 3 Intramolecular Penning Ionisation

We now focus our attention onto another quite different type of charge exchange that occurs in a bound complex rather than during a free collision. In a supersonic beam apparatus the aromatic benzonitrile (Bn,$C_6H_5CN$) was seeded into argon resulting in extensive formation of homogeneous as well as heterogeneous clusters. The molecular beam is intersected by a monochromatic VUV-photon beam from the dedicated synchrotron source BESSY at Berlin. The wavelength was tuned from 55 nm to typically 140 nm (wavelength resolution 2-4 Å) and the ions thus produced by single step photoionisation were mass selected in a quadrupole mass analyser before being detected.

Figure 4 shows photoionisation efficiency (PIE) spectra of Bn for various Ar stagnation pressures. Increasing the Ar contents in the beam apparently leads to an overall increase in the ion yield caused by extensive formation of resonance

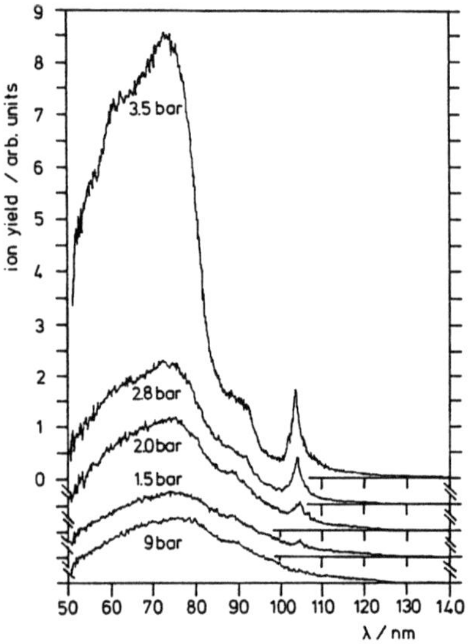


Fig.4: PIE spectra for Bn seeded into Ar at different stagnation pressures.

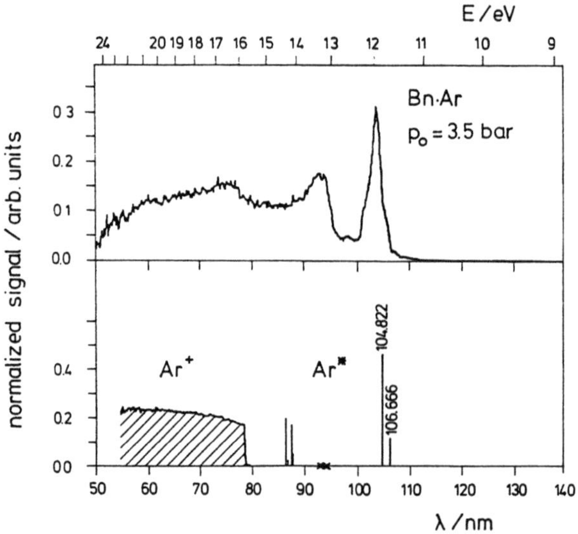


Fig.5: PIE spectrum of Bn.Ar compared with the (schematic) absorption spectrum of Ar.

structures in the region around 104 nm and a strong signal enhancement at about 80 nm. Benzonitrile PIE-curves using Kr instead of Ar exhibit the resonance structures at 116 nm and a large step around 90 nm /3/ proving that these features must be caused by resonance excitation of the rare gas (RG) component ($3p^6(^1S)$->$4s(^3P)$) as well as of Rydberg levels before reaching the RG ionisation threshold followed by ionisation of the Bn-molecule.

Since the resonance lines in the $Bn^+$-spectra are shifted and broadened with respect to the transitions in a free RG, the ionisation cannot be explained by associative ionisation in a free collision but must be due to the processes occurring within a mixed cluster followed by fragmentation of this cluster, i.e. a process of the type

$$Bn_x.RG_y^* \rightarrow Bn_x^+.RG_y \rightarrow Bn_x^+ + RG_y$$

To check this assumption the PIE-spectra of the smallest mixed clusters were measured. Figure 5 shows the Bn.Ar spectrum as an example. The spectrum is dominated by the resonance at 104 nm and by a second "hump" at transitions that are "forbidden" in free Ar. There is only a very weak enhancement observed due to excitation of the RG Rydberg levels showing that such high excess energies lead to complete vaporisation of the clusters ( large signal enhancement in the monomer spectrum, $Bn^+$). Because of its resemblance to Penning ionisation we call the ionisation mechanism "Intramolecular Penning Ionisation" (IPI). Figure 6 gives a qualitative sketch of the relevant two-body interaction between the molecule MOL,and the rare gas, A. Photoabsorption from the ground to the excited state, MOL.A$^*$, of the complex lying in the ionisation continuum of $MOL^+$.A will, in general, lead to some vibrational excitation and to ejection of an electron within less than one vibrational period leaving the complex either in a bound or in a repulsive state leading to extensive fragmentation of the complex.

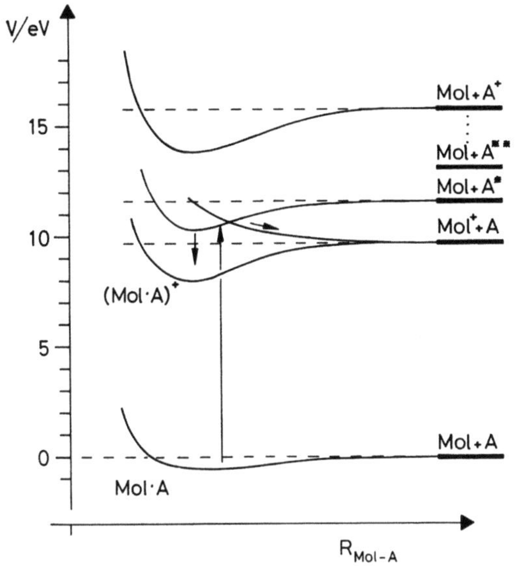


Fig.6: Two-body interaction for an (aromatic) molecule and a rare gas atom (qualitative).

To see whether the IPI process is unique to benzonitrile we used other aromatics also /4/ and show the results in fig. 7 for benzene (Bz), toluene (Tol) and pyridine (Pyr) for fairly high stagnation pressures of Ar. Obviously IPI structures dominate all the spectra in a very similar manner. In fact except for pyridine which has a somewhat different ring structure, all the monomer spectra are very much alike. We believe that at these high Ar pressures they are products of large $Mol.Ar_n$ clusters and thus represent basically the $Ar_n$ absorption spectra.

We now draw our attention to the lineshapes of the IPI-lines. Whereas our measurements on a first glance yield broad and, with respect to the atomic

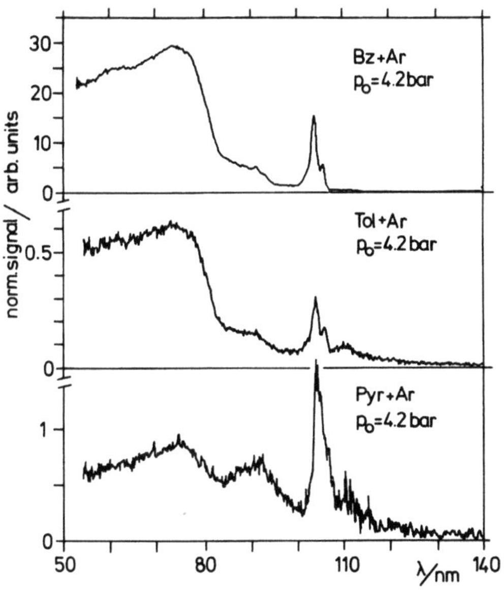


Fig.7: PIES of benzene (Bz), toluene (Tol) and pyridine (Pyr) seeded into Ar.

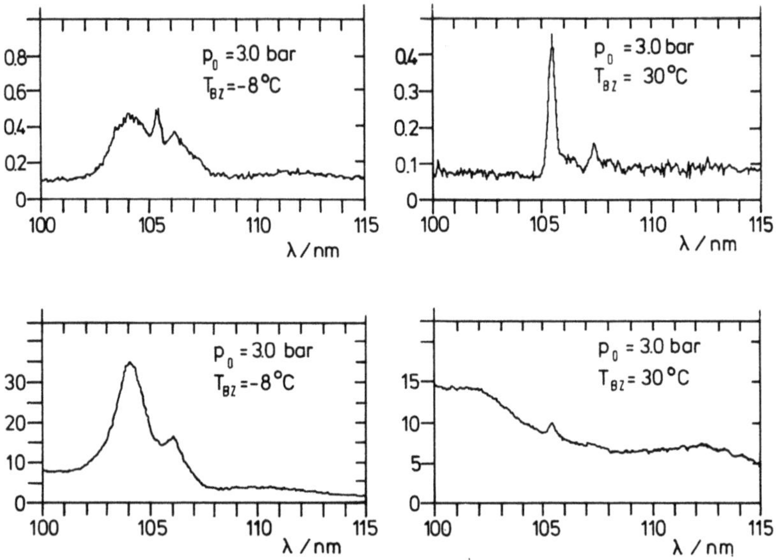


Fig.8: IPI resonances observed in the $Bz.Ar^+$ spectrum (upper graphs) and in the $Bz^+$ spectrum (lower graphs) for two different seed ratios.

resonance, blue shifted lines, Brutschy and coworkers /5/ observed narrow and slightly red shifted lines in the Bz.Ar spectrum. First hints that these discrepancies depend on the neutral cluster distribution in the beam led us to an extensive study of the dependence of the lineshapes on the seed ratio benzene/Ar. A few results are depicted in fig.8: When relatively little Ar is in the beam (small Ar pressures or high temperatures of the benzene reservoir) preferably homogeneous clusters $Bz_n$ and only the smallest mixed clusters Bz.Ar are produced. In this case the excitation of a single Ar bound to Bz leads to a narrow and red shifted line , the red shift being in accord with Penning ionisation data in $Bz,Ar^*$ collisions /5/. The "fragmentation channel" i.e. the Bz-spectrum shows only a very weak IPI peak. With higher Ar pressures and low Bz temperatures clusters of the type $Bz.Ar_n$,

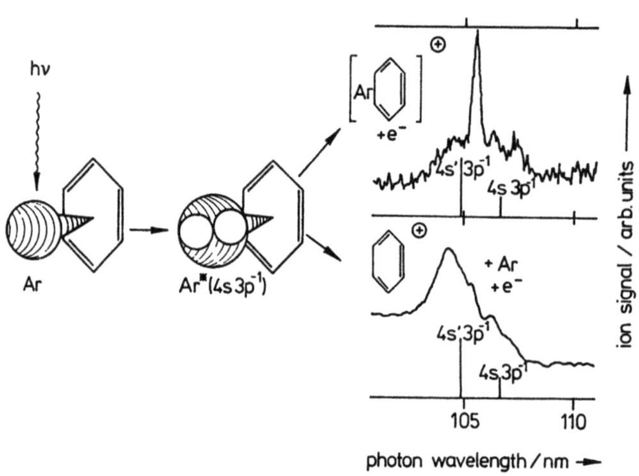


Fig.9: Illustration of the IPI process as an example of a charge transfer process.

n>>2 are preferentially produced. Apparently fragmentation of these clusters upon IPI leads to broad blue shifted humps in the Bz.Ar as well as in the Bz spectrum. The magnitude of the broad structure in the Bz ion yield indicates complete vaporisation of the larger clusters. A detailed discussion of the origin of the blue shift will be given in a forthcoming paper.

Finally, fig.9 illustrates how the IPI process can be understood as a charge transfer process : Excitation of one electron in the Ar atom in principle leaves a positively charged $Ar^+$ atom plus the electron in an outer shell. An electron from the aromatic will then hop into the hole produced (charge exchange) thus leaving the aromatic positively charged. The outer electron will be ejected.

## 4 References


1. J. Weiner , this volume
2. R. Witte, E.E.B. Campbell, C. Richter, H. Schmidt,I.V. Hertel: Z.Phys.D., to be published
3. W. Kamke, B. Kamke, H.U. Kiefl, I.V. Hertel: Chem.Phys.Lett. 122,356 (1985)
4. W. Kamke, B. Kamke, H.U. Kiefl, Z. Wang, I.V. Hertel: Chem.Phys.Lett. 128,512 (1986)
5. E. Ruehl, P. Bisling, B. Brutschy, K. Beckmann, O. Leisin, H. Morgner: Chem.Phys.Lett. 128,399 (1986)

# Laser-Induced Charge Transfer in Ion-Atom Collisions

*L.F. Errea, L. Méndez, and A. Riera*

Departamento de Química Física y Química Cuántica,
Universidad Autónoma de Madrid, Cantoblanco, E-28049 Madrid, Spain

We present laser-induced charge transfer cross-sections for $O^{8+}$ + H(1s) collisions in the range of impact energies in which a quasimolecular description is appropiate. We consider in detail the particular case of radiative transitons taking place near a large-R pseudocrossing of two adiabatic molecular energy curves. For this particular case we point out the importance of (usually neglected) diagonal ac Stark terms

## 1. Introduction

In recent years, the potential ability of intense optical fields to strongly modify charge exchange ion-atom cross-sections has stimulated considerable interest (see e.g. WEINER /1/). In particular, attention has been drawn /2/-/9/ to the possibility of using powerful laser beams to determine the impurity content in tokamak plasmas by enhancing charge transfer cross-sections for reactions between atomic hydrogen and the multiply charged ions that make up the impurities in the plasma core. An interesting feature of this diagnostic method is that it does not disturb the plasma, in contrast with other methods (for example particle beams). On the other hand, the wavelengths required to produce some of these reactions (like $C^{6+}$ + H(1s) + $h\omega$ /9/ or $O^{8+}$ + H(1s) + $h\omega$ /10/) lie in the infrared region, while the radiation produced after the collision lies in the soft X-ray region. this property may be useful in the development of short wavelength lasers.

In this paper we shall start (section 2) by setting up the basic equations of our method that have been explained in detail in previous work /4/,/11/. In section 3, we shall consider photon-induced charge transfer calculations for $O^{8+}$ + H(1s) collisions. The practical interest of these reactions comes from the fact of $O^{8+}$ ions are dominant impurities in magnetically confined fusion plasmas /12/. We shall finally study (section 4) the specific case of radiative transitions induced by a strong field whose frequency is resonant to the splitting at a sharp pseudocrossing of two molecular energies. Radiative transitions taking place near the pseudocrossing of $6h\sigma$ and $7i\sigma$ $OH^{8+}$ adiabatic energy curves will be considered as an illustration of this last point.

## 2. Theory

We consider charge transfer reactions taking place in presence of a very intense laser radiation. Because of the large number of photons in the laser mode, we can describe it as a classical electromagnetic field. Also for not too low nuclear velocities, a semiclassical description can be used to treat the collision process. In this approximation, the nuclei are assumed to follow a classical trajectory $\mathbf{R}(t)$, while the electronic motion is treated quantum-mechanically, and is described by a wavefunction $\psi(\mathbf{r},t)$, that is solution of the semiclassical equation

$$(H_{el} + V)\ \psi(\mathbf{r},t) = i\ \delta\psi/\delta t|_{\mathbf{r}} \tag{1}$$

where V is the radiation-matter interaction term. For nuclear velocities lower than the velocity of the active electrons it is useful to employ a molecular expansion in which the wavefunction $\psi(\mathbf{r},t)$ is expressed as a linear combination of the eigenfunctions $\chi_k$ of the Born-Oppenheimer hamiltonian Hel

$$\psi(\mathbf{r},t) = \sum_k a_k(t)\, \chi_k \exp(-i \int E_k\, dt). \tag{2}$$

In practice, computer programs usually employed to evaluate cross-sections in the field-free case can be easily modified to include the radiation-matter interaction term; however, some additional approximations are often introduced in actual calculations in order to reduce the computational effort. In particular, in the method developed by COPELAND and TANG /13/ and ourselves /4/ the following approximations are introduced: Firstly, the usual dipole and rotating wave approximations are used. In addition, for not too large radiation intensities, Stark shift and broadening of molecular energies is neglected. The two further approximations of this method are to neglect dynamical couplings and to employ first order perturbation theory to calculate transition probabilities. We have found /4/, /11/ that these two last approximations can be employed in the range of laser intensities, I, and nuclear velocities, v, such that $Iv^{-1} < 10^6$ W s $cm^{-3}$ (see also HO et al /6/). Use of perturbation theory considerably reduces the computational effort in averaging cross-section over field orientations. In particular, for rectilinear trajectories $\mathbf{R}(t) = \mathbf{b} + \mathbf{v}\, t$, the orientation-averaged cross-sections $\bar{\sigma}$ are given by the expressions /4/:

$$\bar{\sigma} = (2\pi/3)\epsilon_o^2 \int_0^\infty bdb\, (|A|^2 + |B|^2) \tag{3}$$

for $\Sigma$-$\Sigma$ transitions, and

$$\bar{\sigma} = (2\pi/3)\epsilon_o^2 \int_0^\infty bdb\, (|A|^2 + |B|^2 + |C|^2) \tag{4}$$

for $\Sigma$-$\Pi$ transitions, with

$$A,B,C = \int_{-\infty}^{\infty} dt\, f_{A,B,C}(t)\, \mu \exp\left[-i \int_0^t (E_1 - E_2 \pm \omega)\, dt'\right] \tag{5}$$

where $\epsilon_o$ is the field strength, $\mu$ the transition dipole moment and $f_A = b/R$, $f_B = vt/R$, $f_C = 1$. A generalization of eqs. (3), (4), (5) for general trajectories is given in ref. /11/

In order to evaluate the integrals A,B,C some additional approximations may sometimes be employed. In particular, the stationary phase approximation (SPA) assumes that the more important contributions to the integrals (5) come from the points of stationary phase, where $\omega$ is equal to the energy difference $\pm(E_1 - E_2)$. One then obtains, for $b<R_o$

$$|A_{SPA}|^2 = (8\pi b^2\mu^2)/(|a|R_o^2 v_R)\cos^2\left[\int_0^{t_o} (E_1 - E_2 \pm \omega)\, dt + \pi/4\right] \tag{6}$$

with $E_2(R_o) - E_1(R_o) = \pm\,\omega$; $t_o^2 = (R_o^2 - b^2)\, v^{-2}$; $a = [d\,(E_1 - E_2)/dR]_{R=R_o}$; $v_R = dR/dt$

and similar expressions for the B and C integrals. For $b>R_o$ there are no points of stationary phase and the contribution of order $v^{-1}$ vanish. We notice that the expression (6) diverges for $b = R_o$. Ignoring this divergence and replacing the oscillatory factors by their average value ½, we can integrate the transition probabilities to obtain the perturbative Landau-Zener (LZ) formula

$$\bar{\sigma}_{LZ} = 8\,\pi^2\,\epsilon_o^2\,\mu^2(R_o)\,R_o^2/3\,|a|\,v \tag{7}$$

for Σ-Σ transitions and twice for Σ-Π transitions. The SPA, and also the LZ formula, breaks down when the transition dipole moment $\mu$ presents a peak at a given value ofR. In this case, transitions take place not only at the points of stationary phase, but also in the peak region. Besides, when energies are quasidegenerate, transitions occur over a range of internuclear distances and the SPA and LZ expressions break down. However, even in these cases, the LZ formula can be employed to obtain a rough estimate of cross-sections /9/,/10/.

## 3. Calculations

As an illustration of calculations of laser-induced charge transfer cross-sections, we have chosen the specific case of $O^{8+}$ + H(1s) + $\hbar\omega$ collisions for the range of relative nuclear velocities $2.0\times10^{6} \leqslant v \leqslant 2.0\times10^{7}$ cm $s^{-1}$ ;other examples can be found in refs. /4/-/9/. To evaluate the cross-sections, we employed eqs. (4) and (5) without additional approximations.

The correlation diagram of $OH^{8+}$ quasimolecule (see Fig. 1) shows that the entrance channel for the collision process corresponds to the 7iσ molecular orbital. The electronic energy of this orbital presents a sharp avoided crossing with that of the 6hσ state at R ≃ 16.75 a.u.. For the impact energies considered this avoided crossing is traversed diabatically , so that for R < 16 a.u., we can take the collision system as initially described by the 6hσ molecular orbital. In Fig. 2 we present the transition dipole moments between the 6hσ molecular orbital and other molecular states; we can observe the sharp peak in $\langle 6h\sigma|\mu|7i\sigma\rangle$ that lies in the pseudocrossing region and that will be considered in the next section.

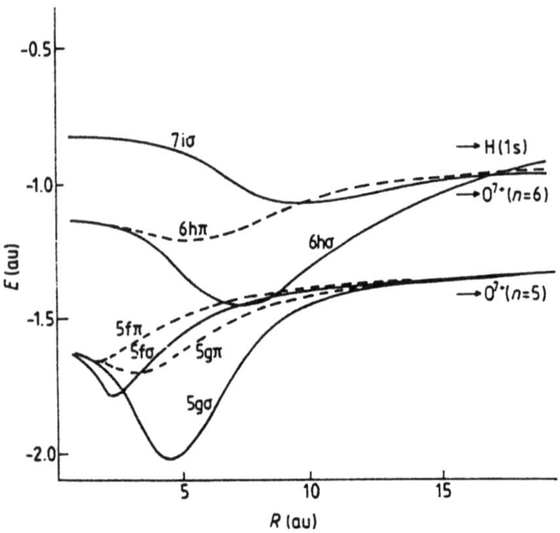


Fig. 1. Correlation diagram of the $OH^{8+}$ quasimolecule for R < 20.0 a.u.

Our calculated cross-sections are plotted in Fig. 3 as functions of the laser wavelength for a focused intensity I = 0.1 TW $cm^{-2}$. The four processes considered have different characteristics. 6hσ-7iσ transitions take place only at the points of stationary phase. 6hσ-5gσ and 6hσ-5gπ transitions take place in a domain of internuclear distances within the range $7.0 \leqslant R \leqslant 10.5$ a.u., and not only at the points of stationary phase. 6hσ-5fσ transitions occur within the range of internuclear distances $7 \leqslant R \leqslant 8.5$ a.u., where the energy curves are quasidegenerate, the SPA breaks down completely, and the cross-section is nearly independent of the laser wavelength for $\lambda > 10^{5}$Å /9/,/10/.

We compare in Fig. 4 the laser-induced cross-sections for I = 0.1 TW $cm^{-2}$ with the unassisted cross-sections (SHIPSEY et al /14/, FRITSCH and LIN /15/) for the

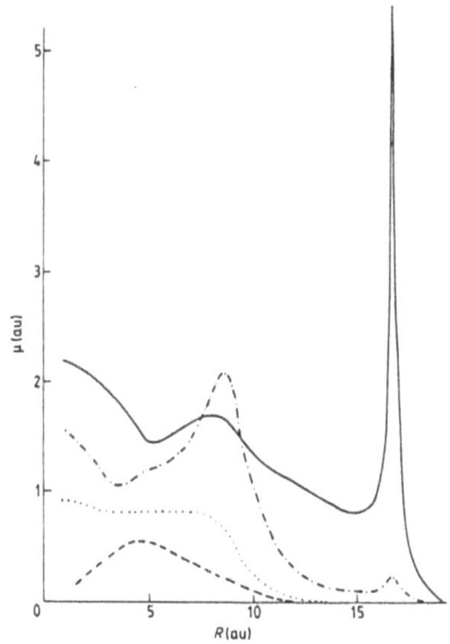


Fig. 2. Values of the transition dipole moments of $OH^{8+}$ between the 6hσ molecular orbital and the other molecular states: ———, 7iσ; -·-·- , 5gσ; ...., 5gπ, - - - , 5fσ

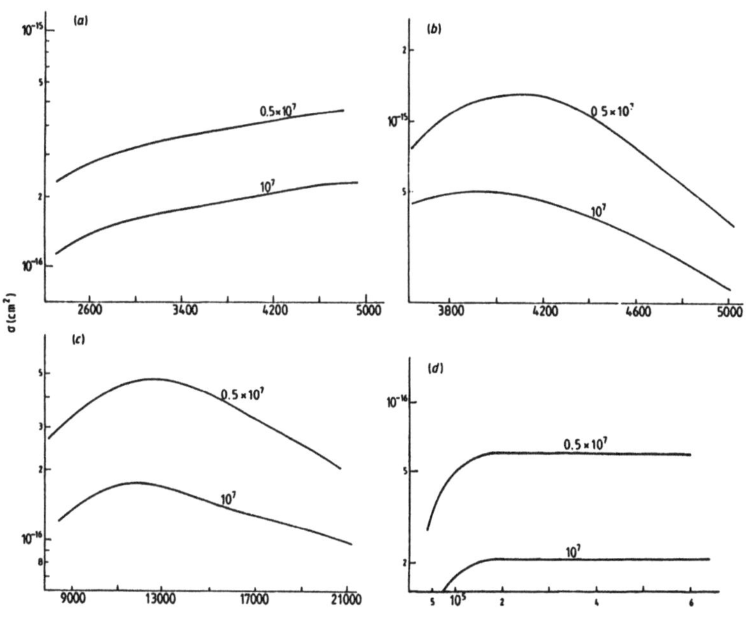


Fig. 3. Orientation-averaged cross-sections as functions of the laser wavelength for two nuclear velocities given on the curves in cm $s^{-1}$. (a) 6hσ–7iσ transitions; (b) 6hσ–5gσ transitions; (c) 6hσ–5gπ transitions; (d) 6hσ–5fσ transitions

reactions:

$$O^{8+} + H(1s) \rightarrow O^{7+}(n=5) + H^{+} \qquad (8)$$
$$O^{8+} + H(1s) \rightarrow O^{7+}(n=6) + H^{+} \qquad (9)$$

In this figure we can see that collision-induced cross-sections are larger than those of the laser-assisted processes. At first sight, this fact could be taken as a signal that the neglect of the dynamical couplings in our treatment is not justified; however, this large value of the collision-induced cross-section is due /14/ to transitions that occur at very large internuclear distances (at the 7i$\sigma$-6h$\sigma$ pseudocrossing) with a transition probability close to one. Hence, the population of the 6h$\sigma$ molecular orbital is practically equal to one for R < 16.0 a.u. and, for radiative transitions taking place at relatively small internuclear distances, neglect of the dynamical couplings is fully justified/10/. For charge transfer to $O^{7+}$(n=5), laser and collision-induced transitions take place in the same range of internuclear distances; hence for this process the accuracy of our results is only ensured for values of nuclear velocities and radiation intensities such that laser assisted cross-sections are greater than the collision-induced ones.

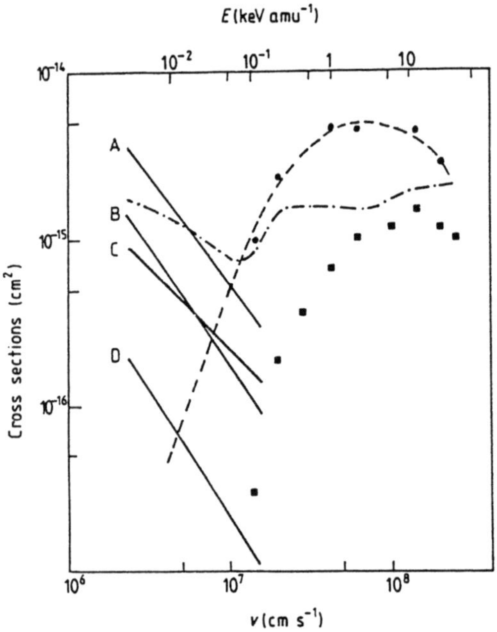


Fig. 4. Comparison of laser-induced (——•) and collision-induced cross-sections corresponding to reactions (8) (-•-•- ,SHIPSEY et al /14/;● , FRITSCH and LIN /15/) and (9) (- • - • -, SHIPSEY et al /14/; ■ ,FRITSCH and LIN /15/ and for the laser wavelengths; λ=4560Å for 6h$\sigma$-7i$\sigma$ transitions (C); λ= 4145Å for 6h$\sigma$-5g$\sigma$ transitions (A); λ=11400Å for 6h$\sigma$-5g$\pi$ transitions (B); λ= 152000Å for 6h$\sigma$-5f$\sigma$ transitions (D)

## 4. Laser-Induced Transitions Near Molecular Energy Pseudocrossings

We have seen that the 6h$\sigma$-7i$\sigma$ transition dipole moment ,$\mu_{12}$, presents a very sharp peak where the corresponding energies pseudocross. Its origin can be understood by a straightforward application /16/to the treatment of MACIAS and RIERA /17/; this treatment also shows that the diagonal dipole matrix elements are large and vary rapidly around the pseudocrossing position.

At first sight, one would expect radiative transitions with high probability at that position. To study this problem, we have calculated charge transfer cross

Table 1 Calculated values of the orientation-dependent cross-sections for $7i\sigma$-$6h\sigma$ transitions for two laser intensities and a nuclear velocity $v = 10^7$ cm $s^{-1}$. (a) with ac Stark terms included; (b) with ac Stark terms neglected

| (A) $I = 10^{12}$ W $cm^{-2}$ | (a) | (b) |
|---|---|---|
| $(\theta, \phi)$ | $\sigma$ ($cm^2$ $10^{-16}$) | $\sigma$ ($cm^2$ $10^{-16}$) |
| $(\pi/2, \pi/2)$ | 6.0 | 6.0 |
| $(\pi/2, 0)$ | 25.9 | 45.6 |
| $(\pi/2, \pi)$ | 1.51 | 29.13 |
| $(0, \phi)$ | 12.52 | 9.26 |
| $(\pi, \phi)$ | 12.57 | 9.26 |
| $\bar{\sigma}$ | 9.89 | 14.06 |
| (B) $I = 10^{11}$ W $cm^{-2}$ | | |
| $(\theta, \phi)$ | $\sigma$ ($cm^2$ $10^{-16}$) | $\sigma$ ($cm^2$ $10^{-16}$) |
| $(\pi/2, \pi/2)$ | 6.0 | 6.0 |
| $(\pi/2, 0)$ | 8.5 | 9.1 |
| $(\pi/2, \pi)$ | 4.0 | 5.4 |
| $(0, \phi)$ | 6.0 | 6.4 |
| $(\pi, \phi)$ | 6.0 | 6.4 |
| $\bar{\sigma}$ | 5.1 | 6.4 |

sections for $O^{8+}$ + H(1s) collisions in the presence of a strong (I= 0.1 and I= 1 TW $cm^{-2}$) laser field whose frequency ($\omega$ = 0.005 a.u.) is quasiresonant with the energy defect at the pseudocrossing. For those intensities, use of perturbation theory is not justified, and therefore not used in our calculation; similarly, diagonal ac Stark terms and dynamical couplings were retained in the interaction matrix. Our calculated values of the cross-sections are shown in Table 1 as functions of the polar angles $\theta$ and $\phi$ that define the direction of the radiation field in a laboratory fixed reference frame.

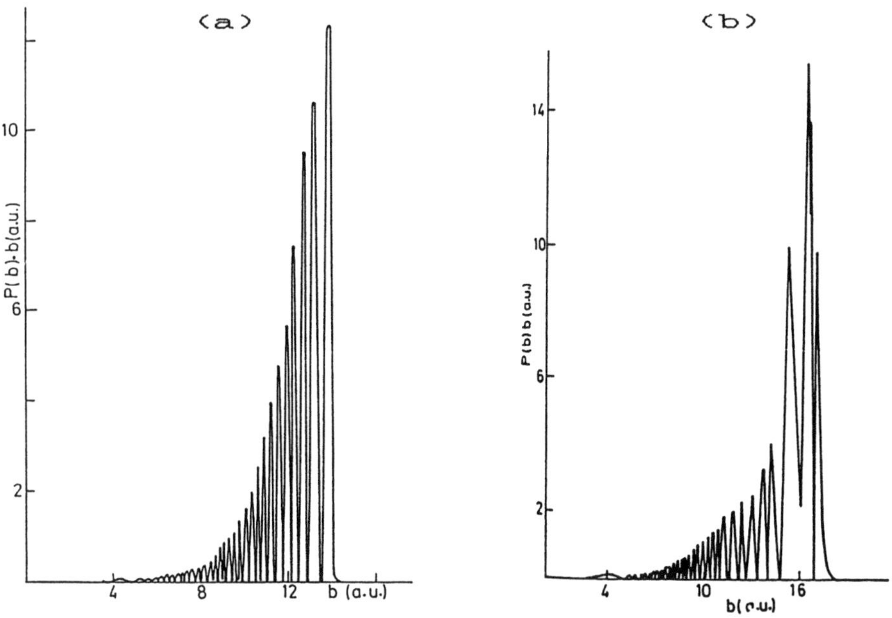


Fig. 5. Charge Exchange transition probability P(b) times the impact parameter b for $I = 10^{12}$ W $cm^{-2}$, $v = 10^7$ cm $s^{-1}$, $\theta = \pi/2$, $\phi = 0$ (a) ac Stark terms included; (b) ac Stark terms neglected

For very intense laser fields ($I=10^{12}$ W $cm^{-2}$) the results obtained neglecting the ac Stark terms notably differ from the exact ones. Besides, the cross-sections turn out to be strongly anisotropic and averaging over all field orientations does not cancel the effect of these terms. For lower intensities the effect of the Stark diagonal terms is less important and then to neglect them is justified. An analysis of the photon-induced processes is provided by the corresponding transition probabilities (See Fig.5). For $\theta = \pi/2$, $\phi = 0$ and $I = 10^{11}$ W $cm^{-2}$, found that radiative transitions take place at internuclear distances lower than $R_0 \simeq 16.75$ a.u.which can be interpreted by considering the Stark shifted energies, $h_i$ , (see fig. 6a) which present two crossings, at $R = R_0$ and $R = R_1$. At $R_0$ no radiative transitions occur due to the effect of the rapidly varying Stark terms, while radial coupling induces non-adiabatic transitions at $R_0$ with a probability $\geq 0.95$ (the crossing is traversed diabatically), and photon-induced processes take place near the second crossing, $R_1$, whose position depends on the field orientation. A different situation is presented in Fig. 6b for the shifted energies with $\theta = \pi/2$, $\phi = \pi$. In this case $R_1 > 25.0$ a.u.and inspection of Fig. 2 shows that the transition dipole moment is very small at $R \simeq R_1$; Consequently, transition probabilities and cross-section are very small too.

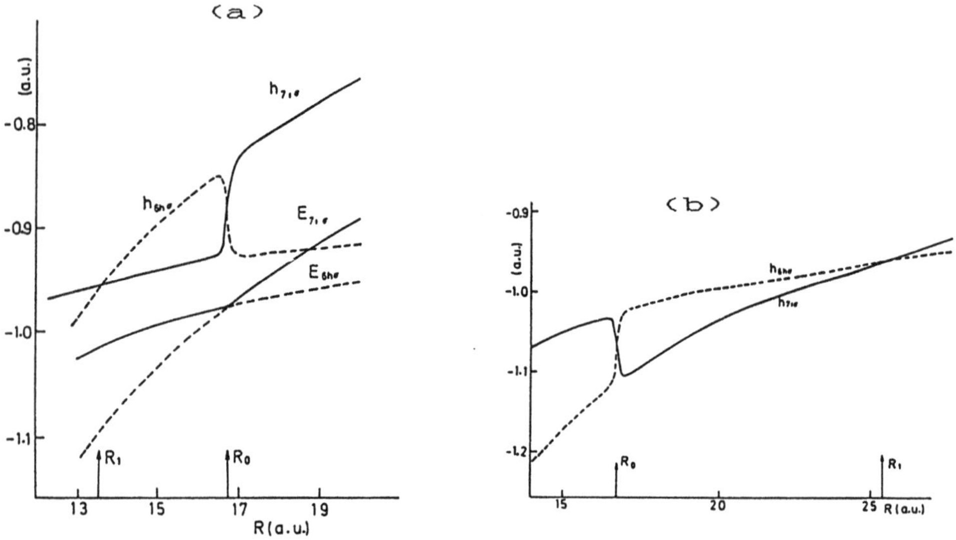


Fig. 6. Molecular energies ($E_i$) and ac Stark shifted energies ($h_i$) of the $6h\sigma$ (- - -) and $7i\sigma$ (——) molecular states of $OH^{8+}$ for b = 12.85 a.u.; $\omega$ = 0.005 a.u.; $I = 10^{12}$ W $cm^{-2}$ as functions of the internuclear distance. (a) ($\theta = \pi/2$, $\phi = 0$); (b) ($\theta = \pi/2$, $\phi = \pi$)

## References


1. J. Weiner: Proc. Summer School on Chemical Photophysics. Les Houches (1979)
2. S. I. Yakovlenko: Sov. J. Quantum Electr., 8, 151 (1978)
3. J. F. Seely: J. Chem. Phys., 75, 3321 (1981)
4. L. F. Errea, L. Méndez and A. Riera: J. Chem. Phys., 79, 4221 (1983)
5. L. F. Errea, L. Méndez and A. Riera: Chem. Phys. Lett., 104, 401 (1984)
6. T. S. Ho, S. I. Chu and C. Laughlin: J. Chem. Phys., 82, 788 (1984)
7. S. Ganguly, K. Rai Dastidar and T. K. Rai Dastidar: Phys. Rev. A31, 1171 (1985)
8. Y. P. Hsu, M. Kimura and R. E. Olson: Phys. Rev. A31, 576 (1985)
9. L. F. Errea, L. Méndez and A. Riera: J. Chem. Phys., 82, 4536 (1985)
10. L. F. Errea, L. Méndez and A. Riera: J. Phys. B18, 4271 (1985)

11. A. Riera: in Electronic and Atomic Collisions, ed. by D. C. Lorents, W. E. Meyerhof and J. R. Peterson (Elsevier Science Publishers B.V. 1986), p. 319
12. E. Salzborn: Report IPPJ-AM-13, Nagoya University, Nagoya, Japan (1979)
13. D. A. Copeland and C. L. Tang: J. Chem. Phys., 65, 3161 (1976)
14. E. J. Shipsey, T. A. Green and J. C. Browne: Phys. Rev. A27, 821 (1983)
15. W. Fritsch and C. D. Lin: Phys. Rev. A29, 3039 (1984)
16. A. Macías and A. Riera: J. Phys. B11, L489 (1978)
17. L. F. Errea, L. Méndez and A. Riera: J. Chem. Phys., 85, 825 (1986)

# Collisional Ionization in Laser-Excited Barium Vapor

D. Cubaynes[1], J.L. Picqué[2], G. Spiess[3], J.M. Bizau[1], B. Carré[3], F. Roussel[3], and F.J. Wuilleumier[1]

[1]Laboratoire de Spectroscopie Atomique et Ionique and LURE, F-91405 Orsay, France
[2]Laboratoire Aimé Cotton, C.N.R.S., F-91405 Orsay, France
[3]Service de Physique des Atomes et des Surfaces, CEN-Saclay, F-91191 Gif-sur-Yvette, France

## 1. INTRODUCTION

Experimental study of atomic vapors resonantly excited by a laser is facilitated for alkali metals and alkaline earths because the transition to the first excited state lies in the visible range. Observation of a high fractional ionization, close to 100%, in dense alkaline vapors after a short time ($\approx$ 1 μs) and under modest laser illumination ($\approx$ 1 MW/cm$^2$) had become a challenge since the pioneer experiments of LUCATORTO and McILRATH [1,2]. During the last decade, studies of ionization processes taking place in the laser-excited medium have rapidly expanded [3-10]. More recently, electron spectrometry has proved to be a remarkable means of investigation offering a decisive insight into the ionization mechanisms in a Na vapor [11-13]. The many experimental data accumulated by several groups [14- 17] has made Na a model and a starting point to develop a better understanding of the intricate situation in Ba vapor studied in the present work.

It has been demonstrated through different experiments [1,8-10] that high fractional ionization in dense metallic vapors ($10^{14}$ -$10^{15}$ at/cm$^3$) results basically from the same chain mechanism:
i) seeding processes, to be identified, produce the first free electrons of low energy, a part of which is trapped in the medium by the plasma potential.
ii) the free electrons are "heated" up to several eV in successive superelastic collisions with the dominant excited species originating from the laser pumping effect [18].
iii) the "hot" electrons initiate electron-impact ionization of the vapor which combines with step ii) to give rise to an avalanche regime of ionization.

In vapors of lower density ($10^{11}$ -$10^{13}$ at/cm$^3$) the last step of avalanche does not develop, even for CW laser excitation, owing to the longer mean free path of the electrons. Such a medium is particularly well adapted to the detailed study of superelastic collisions and seeding processes. In Na it has been early established that Penning ionization and associative ionization, in collisions between excited atoms, are the dominant seeding processes for CW laser excitation [11,13]. The important role of energy-pooling collisions was also outlined as an intermediate step that transfers energy into the highly excited atoms likely to be further ionized [15,16].

In this work we have studied the energy distribution (0-2.5 eV) of the electrons produced in a Ba effusive beam resonantly excited by a CW laser tuned to the $6s^2\ ^1S_0$ - $6p\ ^1P_1$ optical transition. The observed spectrum results from seeding processes and superelastic effects which are identified following an analysis based on the Na model. Although the sequential seeding processes present a close analogy with the Na case, they exhibit several original features. In Ba, the dominant excited population is not in the laser pumped $6p\ ^1P_1$ state but almost entirely in the lower metastable states $5d\ ^1D$ and $5d\ ^3D$ radiatively populated from the $6p\ ^1P_1$. These metastable atoms are a key of the medium as they represent an important and available source of internal energy [19].

## 2. EXPERIMENT

The experimental set-up has been previously described in detail [17]. An effusive beam of Ba ($n = 10^{11} - 10^{13}$ at/cm$^3$) is resonantly excited by a perpendicular CW laser beam at the source volume of a cylindrical mirror analyser (CMA). A part of the electrons produced in the interaction volume diffuses out through the annular slits of the CMA and are energy-analyzed. The width of the slits is defined corresponding to a resolution $\Delta E/E$ of the order of 1% . The CW, single mode, ring laser excites the $6s^2\ {}^1S_0 - 6p\ {}^1P_1$ optical transition of Ba (2.23 eV) and is locked to the center of the Doppler absorption line in an auxiliary Ba beam. For a typical laser intensity of a few W/cm$^2$ up to 95% of the atoms are transferred in the low-lying excited states including the radiative $6p\ {}^1P_1$ state and the metastable $5d\ {}^1D$ and $5d\ {}^3D$ states.

Figure 1 shows a detailed view of the region around the interaction volume. As low-energy electron detection (0-2.5 eV) was the main purpose of this experiment, an improvement of the electron transmission should be obtained. Two conical extraction grids, in front of the CMA entrance slit, and a plane grid, behind the source region, have been mounted. They work as an accelerating system for the electrons coming from the interaction volume and as a shielding barrier against unwanted electrons emitted by different surfaces surrounding the source region. When Ba is heated, the nozzle of the oven at ground potential was found to be an intense source of spurious electrons that are rejected by applying a -2V bias potential to the inner conical grid. The outer grid potential has been optimized to +1V to compensate the contact potential appearing on the walls of the inter-

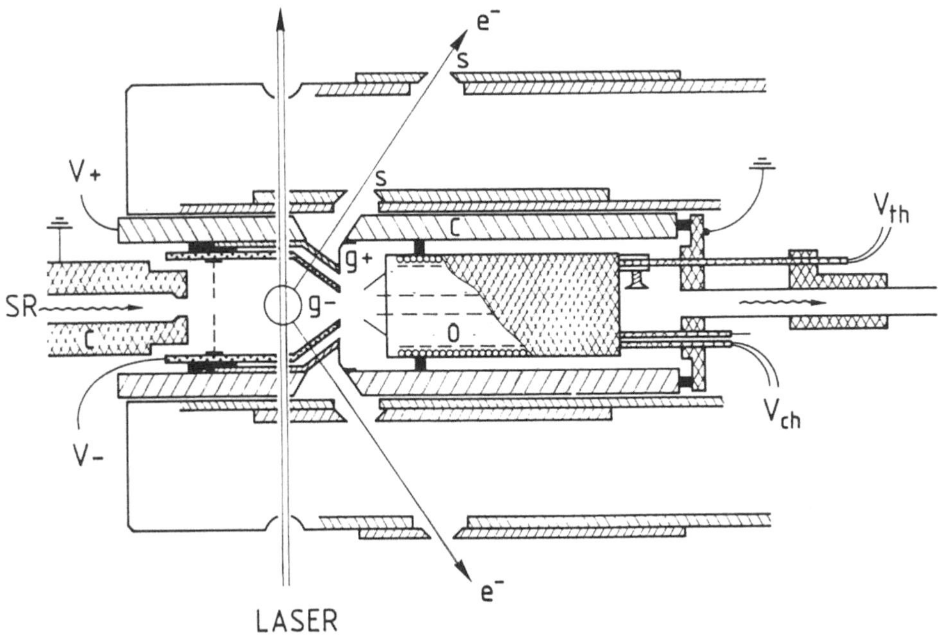


Fig. 1 - Schematic view of the experimental arrangement (cut in horiz. plane). Effusive Ba beam produced from the oven (O) (elements at ground potential are cross-hatched) is laser-excited in the source volume of a CMA (interaction region circled ; entrance slits (s) of the CMA). Shielding (grid $g^-$, $V^-$) and accelerating devices ($g^-$, $g^+$ ; elements at $V^+$ simply hatched) are installed around the interaction region. Synchrotron radiation (SR) is superimposed collinear with the Ba beam. Cooling elements are marked (C). Except for the laser, the system has axial symmetry around the SR-Ba beam axis.

action chamber. With a 3V accelerating potential between the two grids, electrons of near zero kinetic energy can be detected. However, the decreasing transmission of the CMA affects the intensity of the signal below 0.5 eV.

Twin microchannel plates mounted on the CMA axis detect and count the transmitted electrons.

## 3. CALIBRATION PROCEDURE

The experimental arrangement is also designed for the study of inner-shell excitation and ionization of atoms, either in the ground state or in laser-excited states, by synchrotron radiation emitted from the ACO storage ring at LURE [20]. In the present work, the photoionization by synchrotron radiation (SR) provided us with a very convenient method for electron energy calibration and for probing the densities of the different species present in the laser-excited medium. For each low-energy electron spectrum due to the collisional processes under study (left

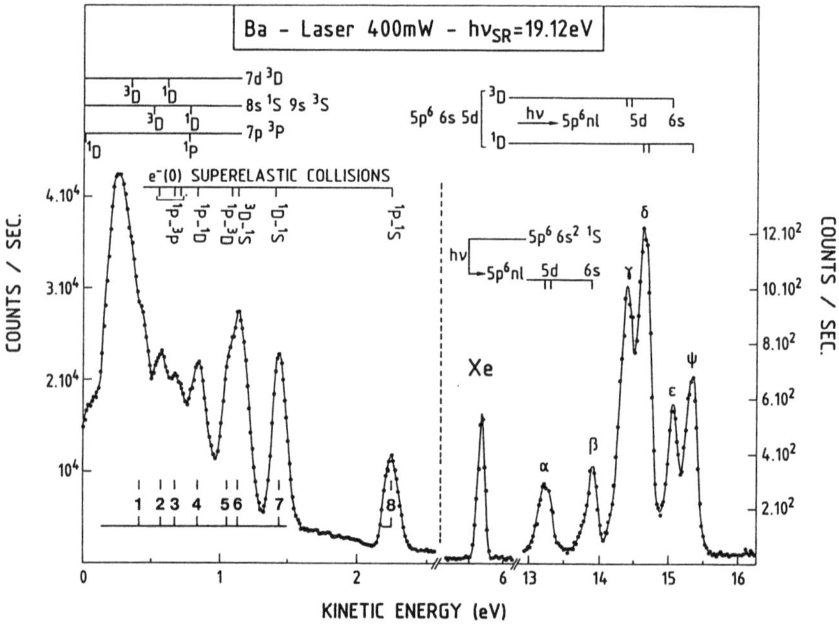


Fig. 2 - Electron energy spectrum measured in ($6s^2\ ^1S_0$-$6s\ 6p\ ^1P_1$) laser-excited Ba vapor, in the presence of SR.
i) high-energy range [13-16 eV]: photoionization of Ba by SR at 19.12 eV. Peaks $\alpha$, $\beta$ ; photoionization of the ground state atom in the 6s subshell (principal and satellite lines). Peaks $\gamma,\varepsilon$ ; photoionization of Ba ($5p^6\ 6s\ 5d\ ^3D$) in, respectively, the 6s and the 5d subshells, enhanced by resonant inner-shell excitation of a 5p electron - autoionization - see text. Peaks $\delta,\psi$ ; same as $\gamma,\varepsilon$ for Ba ($5p^6\ 6s\ 5d\ ^1D$). Densities in ground state and $5d\ ^3D$ , $^1D$, $6p\ ^1P_1$ excited states can be measured from the peaks $\alpha$-$\psi$ intensities. Additional photoionization $Xe(5p^6)$ + $h\nu_{SR}$ → $Xe(5p^5)$ (peak Xe at ≈ 5.7 eV) provides absolute calibration of the low-energy scale.
ii) low-energy range [0-2.5 eV]: sequential collisional mechanisms. Peaks 4-8 result from superelastic heating (3) of near-zero energy electrons, in collision with $Ba^*$($6p\ ^1P_1$ ,$5d\ ^1D$ , $5d\ ^3D$) de-excited to either the metastable or the ground state. The primary near-zero energy electrons (highest peak strongly affected by the decreasing transmission of the CMA) are produced in energy-pooling process (4) populating $Ba^{**}$($7p\ ^3P$) and Penning ionization (5) with $Ba^*$ the $5d\ ^1D$ state. Analogous sequential processes contribute to peaks 1-4.

part of Fig. 2), SR produced photoelectrons of higher energy, coming from either autoionization of core-excited Ba or photoionization of an inert gas (Xe), are simultaneously observed (right part of Fig. 2). With laser on, SR of selected energy (19.12 eV in Fig. 2) induces inner-shell excitation of a 5p electron either from the ground state $5p^6\,6s^2\;^1S_0$ (peaks $\alpha,\beta$) or from the metastable states $5p^6\,6s\,5d\;^{1,3}D$ (peaks $\gamma$-$\psi$ between 14 and 16 eV).

Doubly inner and outer-shell excited states decay by autoionization to one of the ion states $Ba^+(5p^6\,6s)^2S_{1/2}$ or $Ba^+(5p^6\,5d)\;^2D_{3/2,\,5/2}$ ; the ejected electrons give rise to the inner-shell spectrum in Fig. 2. Excitation from the laser-pumped $5p^6\,6s\,6p\;^1P_1$ state produces a negligible signal under the conditions of Fig. 2. Peak at 5.7 eV from photoionization of Xe [19] assures an accurate calibration of the electron kinetic energy i.e. determines the retarding potentials in the laser-excited medium (plasma potential and contact potential). The inner-shell spectrum is very sensitive to the SR energy and to the Ba density. Comparative measurements of the spectrum with laser on and with laser off make possible the determination of the atomic densities in the ground state and in the low-lying excited states.

## 4. POPULATION IN LOW-LYING EXCITED STATES

At the maximum laser power (400 mW) up to 95% of the Ba atoms are excited. A large fraction of them are in the metastable states $5d\;^1D$ (≈ 60%) and $5d\;^3D$ (≈35%) whereas only a few percent are in the laser-pumped radiative state $6p\;^1P_1$ . This feature underlines the dominant role of the metastable atoms as a source of energy in the laser-excited Ba. No trace of the excited $6p\;^3P$ states has been detected in contrast with other experiments [10] where pulsed laser excitation was used.

The ratio of the two metastable populations $n(^1D)/n(^3D)$, determined from the inner-shell spectra, is equal to $1.7 \pm 0.1$ and remains fairly constant over a large range of density ($10^{11}$ - $10^{13}$ $cm^{-3}$) showing that the metastable states are mainly populated by radiative decay from the resonant $6p\;^1P_1$ state. Spontaneous decay rates from $6p\;^1P_1$ to $5d\;^1D_2$ and $5d\;^3D_2$ are of the order of $5 \times 10^5$ $s^{-1}$ [21-22]. Decay rates to $5d\;^3D_1$ and $5d\;^3D_3$ are much smaller [22]. From simple rate analysis for the different species in the medium we found that the previous orders of magnitude for decay rates are consistent with the measured relative populations in the $6p\,^1P_1$ , $5d\;^1D_2$ and $5d\;^3D_2$ states.

## 5. RESULTS AND DISCUSSION

The electron spectrum between 0 and 2.5 eV (left part of Fig. 2) results from ionization processes in Ba vapor resonantly excited by the laser. A reproducible structure, including 9 peaks, is clearly visible. The observed distribution between 0 and .4 eV reflects the rapidly decreasing transmission of the CMA as well as the trapping in the medium of the lowest energy electrons by the plasma potential. In fact a large amount of electrons is produced at a near-zero energy and constitutes the dominant electron population in the laser-excited medium. The energy of each peak, apart the peak at ≈ 0 eV, is indicated in Table I within an uncertainty of 0.02 eV. These energies have been determined by averaging over a series of 25 spectra. One notices that peaks 1 and 5 are only shoulders located on more intense peaks whereas peak 8 at 2.25 eV is well separated from the others.

Figure 3 shows the behavior of the spectrum when the excited atom density varies from $2.7 \times 10^{11}$ to $5.3 \times 10^{11}$ at/$cm^3$. Intensity of peaks 4, 5, 6, 7 and 8 increases approximately as the $4^{th}$ power of the density suggesting that the electrons originate from a sequential collisional mechanism between Ba excited atoms. By analogy with the well-studied case of a Na vapor resonantly excited [12,19] in which such sequences have been demonstrated, one is led to consider a succession of processes of the type:

Table I - Electron energy (eV) of the peaks in spectrum of Fig. 2

| Peak number | 1 | 2 | 3 | 4 | 5 | 6 | 7 | 8 |
|---|---|---|---|---|---|---|---|---|
| Peak energy (eV) | 0.41 | 0.56 | 0.67 | 0.83 | 1.05 | 1.15 | 1.43 | 2.25 |

$$Ba^{*} + Ba^{*} \rightarrow Ba^{**} + Ba + \Delta E \qquad (1)$$

$$Ba^{**} + Ba^{*} \rightarrow Ba^{+} + Ba + e(\varepsilon) \qquad (2)$$

$$e(\varepsilon) + Ba^{*} \rightarrow e(\varepsilon') + Ba' \qquad (3)$$

The energy-pooling collision (1) between two low-lying excited $Ba^*(6p\ ^1P_1,\ 5d\ ^1D_2,\ 5d\ ^3D_2)$ can populate highly excited states $Ba^{**}$ with an energy close to the energy of the ingoing colliding system [15]. In the case of Na (Na(3p) + Na(3p)), for energy defects $\Delta E$ of the order of kT, where T is the vapor temperature, energy-pooling processes cross-sections are of the order of $10^{-15}$ $cm^2$ [16,23]. In Ba, parity conservation during the collision [24] and energy consideration fix severe limitations to the number of $Ba^{**}$ states which can be populated by energy-pooling process of type (2). Penning ionization cross-sections for two colliding excited atoms may be quite large, of the order of $10^{-13}$-$10^{-12}$ $cm^2$ [25]. Photoionization of the $Ba^{**}$ states is expected to be negligible under the present low laser inten-

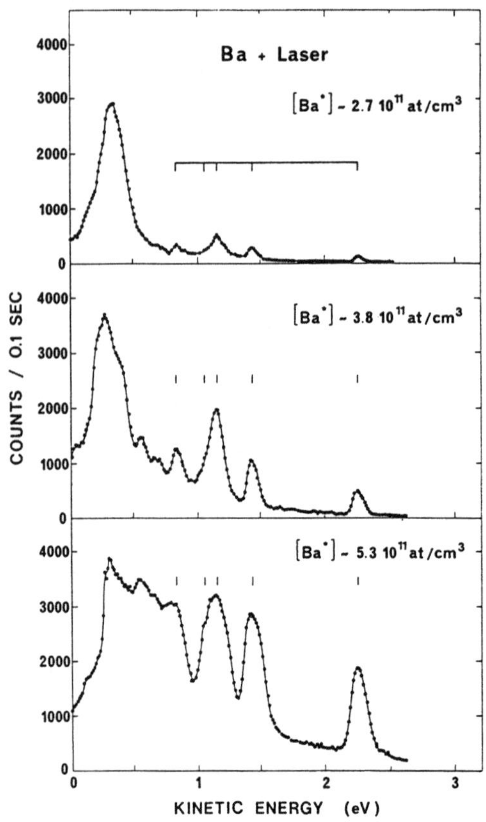


Fig. 3 - Variation of the electron energy spectrum (0-2.5 eV) with the total excited atoms density [Ba*]. Peaks are labelled and attributed as in Fig. 2. Variation of peaks 4-8 intensities is compatible with a quartic dependence on [Ba*] characterizing the collisional sequence.

sity conditions. Associative ionization seems to be ruled out in the studied Ba vapor as it was not possible to detect $Ba_2^+$ ions in another experiment performed at SACLAY [26]. The last step (3) is a superelastic process [18] in which the primary or "seeding" electrons e(ε) gain an energy quantity ε'-ε equal to the energy difference between Ba* and Ba'. The state Ba' may be either the ground state or any excited state below Ba*.

In the spectrum of Fig. 2, it comes out that peaks 6, 7 and 8 exhibit a remarkable feature, suggesting that superelastic collisions are contributing: within the experimental uncertainty (.02 eV) their energies are precisely those of the respective Ba* excited states 5d $^3D_2$ (1.14 eV), 5d $^1D_2$ (1.41 eV) and 6p $^1P_1$ (2.23 eV). A direct interpretation is that the dominant primary electron population of near-zero energy is partially trapped in the medium by the plasma potential and thus efficiently heated in superelastic collisions with low-lying Ba* atoms present in the vapor and de-excited to the ground state. In the same way, superelastic de-excitations of the 6p $^1P_1$ to the 5d $^1D_2$ and 5d $^3D_2$ states produce electrons of respective energies 1.09 and 0.82 eV which correspond to peaks 4 and 5 in Fig. 2. All these superelastic processes, except for 5d $^3D_2$ - $6s^2\ ^1S_0$ de-excitation, have been already observed [27] in a laser-excited Ba beam impinged by 30 eV and 100 eV electrons. In agreement with this interpretation, the branching ratio between peaks 4 and 9 is found constant when the atomic density is varied. In conclusion, a series of superelastic processes, involving electrons of near-zero energy and the low-lying Ba* excited states, may interpret peaks 4 to 8 as indicated in the upper part of Fig. 2. The de-excitation to the triplet 6p $^1P_1$ - 6p $^3P$ is also indicated but is probably not observed.

The next task consists in assigning the Ba** highly excited states contributing in processes (1) and (2) to the primary electron spectrum, i.e., the dominant peak at near-zero energy and the neighbour peaks below 0.9 eV in Fig. 2. The simultaneous population of the radiative 6p $^1P_1$ and of the two metastable 5d $^{1,3}D_2$ states in the vapor makes the analysis more intricate than in the Na case where only the resonant 3p population is active. Examination of the 3a energy levels (Fig. 4) shows that the colliding systems 5d $^3D_2$ + 5d $^3D_2$ and 5d $^3D_2$ + 5d $^1D_2$ bring respectively 2.28 and 2.54 eV to be eventually transferred to Ba**. This is far below the minimum energy required (2.98 eV) in any Penning ionization process of type (2). The system 5d $^1D_2$ + 5d $^1D_2$, with 2.81 eV, is also ruled out. Excitation of the closest 7s $^3S$ state at 3.24 eV has a too large energy defect of 5.5 kT and should be negligible. The formation of the doubly-excited triplet 6p' $^3D$, although resonant, is forbidden by the parity conservation (assuming the parity of the atomic state is given by $(-1)^{\ell_1+\ell_2}$ for weakly correlated electrons of momentum $\ell_1$ and $\ell_2$ [24]). Thus, contributing collisional sequences require that at least one 6p $^1P_1$ atom is involved in the energy-pooling process. Energy consideration shows that only one sequence results in producing near-zero energy electrons. It proceeds via the highly excited 7p $^3P$ states:

$$Ba(5d\ ^1D_2) + Ba(6p\ ^1P_1) \rightarrow Ba(7p\ ^3P) + Ba(6s^2\ ^1S_0) + 2kT \tag{4}$$

$$Ba(7p\ ^3P) + Ba(5d\ ^1D_2) \rightarrow Ba(6s^2\ ^1S_0) + Ba^+ + e(0.015\text{-}0.045\ eV) \tag{5}$$

Energy-pooling process (4) has an energy defect of only 2 kT and results from two atomic transitions that are dipole-allowed. Penning process (5) is direct when the metastable atom is ionized. In this sequence the electron yield is proportional to the density product $n^2(^1D_2)\ n\ (^1P_1)$, it is therefore relatively large on account of the dominant 5d $^1D_2$ population in the medium.

Energy-pooling formation of the 6d $^{1,3}D$ states, although at energy levels comparable to those of 7p $^3P$, should be forbbiden by parity conservation. Under pulsed laser excitation the 6d $^3D_1$ state has been observed [10,28] to contribute strongly to the ionization of the medium. But in that case the 6d $^3D_1$ state was populated by laser pumping effect from the 6p $^3P_1$ state which is not detected in the present experiment.

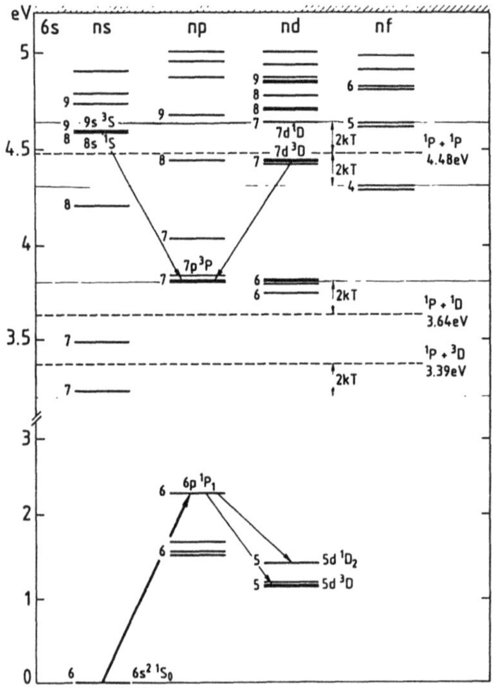


Fig. 4 - Energy diagram of Ba illustrating population of the excited states in the medium. Energy-pooling (1) in Ba* + Ba* colliding system (dashed lines) populates the 7p $^3P$ (process (4)) or the 7d $^3D$, 8s $^1S$, 9s $^3S$ highly excited states. Ionization proceeds by subsequent Penning ionization (2) of Ba**.

Another direct Penning process is possible for Ba(7p $^3P$) but is less favoured than (5) on account of the low n($^1P_1$) density:

$$Ba(7p\ ^3P) + Ba(6p\ ^1P_1) \rightarrow Ba^+ + Ba(6s^2\ ^1S_0) + e(0.83\text{-}0.87\ eV) \quad (6)$$

The electrons have energies between 0.83 and 0.87 eV according to the J sublevel of 7p $^3P$ and can contribute to the observed peak 4.

The remaining part of the primary electron spectrum (peaks 1-3) may be interpreted by collisional sequences starting with the energy-pooling system 6p $^1P_1$ + 6p $^1P_1$ at 4.48 eV. Again from parity conservation we assume that only the following highly excited Ba** states may be formed with small energy defects (Fig. 4): 7d $^3D$(+0.5 kT), 8s $^1S$/9s $^3S$(-0.5 kT) and 7d $^1D$(-2 kT). They can decay radiatively to the 7p $^3P$ states and contribute indirectly to the peak at near-zero energy. Since the product density $n^2(^1P_1)$ is not favourable in the energy-pooling process, Penning ionization by metastable atoms 5d $^{1,3}D$ should be considered as the subsequent step (2). It has been observed in Rb [29] that cross-sections for Penning ionization between excited atoms are of the same order of magnitude in either the case of dipole-allowed transitions (direct process) or of dipole-forbidden transitions (electron exchange process). Penning ionization for the abovementioned Ba** atoms in collisions with Ba (6p $^1P_1$) is a direct process that would produce electrons at energies 1.46, 1.62 and 1.67 eV. These peaks are not observed, which shows that the corresponding cross-sections are insufficient to compensate the unfavorable n($^1P_1$) density. But, on the other hand, exchange Penning processes involving metastable atoms give rise to electron peaks that are in general agreement with the observed primary spectrum, as indicated in the upper part of Fig. 2. Thus the following Penning processes are assumed to contribute to the observed peaks 1-4.

$$Ba(7d\ ^3D) + Ba(5d\ ^{3,1}D_2) \rightarrow Ba^+ + Ba + e(0.37\ eV,\ 0.63\ eV) \quad (7)$$

$$Ba(8s\ ^1S/9s\ ^3S) + Ba(5d\ ^{3,1}D_2) \rightarrow Ba^+ + Ba + e(0.52\ eV,\ 0.79\ eV) \quad (8)$$

$$Ba(7d\ ^1D) + Ba(5d\ ^{3,1}D_2) \rightarrow Ba^+ + Ba + e(0.57\ eV,\ 0.84\ eV) \quad (9)$$

The experimental resolution is not sufficient to distinguish between 8s $^1S$ and 9s $^3S$ contributions, the two levels being separated by only 0.01 eV. We remark a systematic shift of about + 0.04 eV in the experimental peak energies for processes (7) and (8) and no significant shift for process (9). Although a definite conclusion would need more experimental data, energy shifts of a few hundredths of eV have been already observed in Penning exchange processes with metastable helium [30,31]. They are attributed to transitions at interatomic distances where the potential energy of the quasi-molecule formed by the colliding atoms departs slightly from its asymptotic value. The absence of shift in peaks 2 and 4 may reflect that the state 7d $^1D$ has a negligible contribution.

Doubly-excited states of Ba have to be mentioned since they are eventually formed in energy-pooling processes when the energy defect is of the order of kT. Several doubly-excited states have been observed in a Ba vapor [10] under pulsed laser-resonant excitation. From the known energy levels of doubly-excited states, it is deduced that 6p' $^1F$ may be formed in the vicinity of 3.39 eV that is the energy of the 6p $^1P_1$ + 5d $^3D_2$ system, 6p' $^1P$ in the vicinity of 3.64 eV (6p $^1P_1$ + 5d $^1D_2$ system) and a group of states: 6d' $^3D$, $^3G$, $^1F$ and 6p$^2$ $^1D$ in the vicinity of 4.48 eV (6p $^1P_1$ + 6p $^1P_1$ system). Penning ionization for these doubly-excited states is energetically possible but the cross-sections for such processes are expected to be small as 3 electrons are rearranging. Even in the case of almost resonant energy-pooling processes such as

$$Ba(6p\ ^1P_1) + Ba(6p\ ^1P_1) \rightarrow Ba(6d'\ ^3D_3 \text{ or } ^3G_5) + Ba, \quad (10)$$

where the energy defect is smaller than 0.01 eV, the Penning ionization peaks expected at 0.74 eV and 0.77 eV, for collisions with 5d $^1D_2$ atoms, are not detected.

## 6. CONCLUSION

We have investigated the low-energy electron spectrum produced in a low density Ba vapor resonantly excited by a CW laser. Electron spectrometry allows a selective analysis of the different ionization channels. The atomic populations in the laser-excited medium are measured using synchrotron radiation-produced photoionization. A few percent of the atoms are in the laser-pumped 6p $^1P_1$ state whereas a large fraction is stored in the metastable 5d $^1D_2$ and 5d $^3D_2$ states mainly coming from radiative decay of the 6p $^1P_1$. The ionization is attributed to collisional sequences developing in the medium through a limited number of intermediate highly excited states Ba**: mainly the 7p $^3P$ triplet and, to a lesser extent, the 7d $^3D$, 8s $^1S$ and 9s $^3S$ states. These Ba** states are selectively populated by quasi-resonant energy-pooling in collisions between the low-lying excited atoms produced from the laser. Subsequent Penning ionization of Ba** in a second collision with the same low excited species is at the origin of the observed primary electron spectrum between 0 and 0.9 eV. A dominant electron population at near-zero energy is due to a favourable collisional sequence via the excited 7p $^3P$ states and involving the large 5d $^1D_2$ population. In consequence, the spectrum exhibits five additional peaks between 0.9 and 2.5 eV resulting from superelastic heating of the low-energy electrons which are partially trapped by the plasma potential in the ionized medium.

## LITERATURE


1. T.B. LUCATORTO and T.J. McILRATH: Phys. Rev. Lett. 37, 428 (1976).
2. T.J. McILRATH and T.B. LUCATORTO: Phys. Rev. Lett. 38, 1390 (1977).
3. A.N. KLUCHAREV, A.Z. DEVDARIANI, V. Yu. SEPMAN and V. VUJNOVIC: X ICPEAC Paris (1977) Abstract of papers, 1150.

4. G.H. BEARMAN and J.J. LEVENTHAL: Phys. Rev. Lett. 41, 1227 (1978).
5. J. WEINER and P. POLAK-DINGELS: J. Chem. Phys. 74, 508 (1981).
6. B. CARRÉ, F. ROUSSEL, P. BREGER and G. SPIESS: J. Phys. B: At. Mol. Phys. 14, 4271 (1981).
7. M. CHERET and L. BARBIER: Proc. of the Colloque International du CNRS "Collisions in a laser field", J. Phys., Série des Colloques, 46, C1-193 (1985).
8. C.H. SKINNER: J. Phys. B: At. Mol. Phys. 13, 55 (1980).
9. I.A. BACHOR and M. KOCK: J. Phys. B: At. Mol. Phys. 14, 2793 (1981).
10. L. JAHREISS and M.C.E. HUBER: Phys. Rev. A 28, 3382 (1983).
11. J.L. Le GOUËT, J.L. PICQUÉ, F. WUILLEUMIER, J.M. BIZAU, P. DHEZ, P.M. KOCH and D.L. EDERER: Phys. Rev. Lett. 48, 600 (1982).
12. B. CARRÉ, J.M. BIZAU, P. DHEZ, D.L. EDERER, P. GERARD, J.C. KELLER, P.M. KOCH, J.L. Le GOUËT, J.L. PICQUÉ, G. SPIESS and F. WUILLEUMIER: Optics Comm. 52, 29 (1984).
13. B. CARRÉ et al.: Proc. of the Colloque International du CNRS "Collisions in a laser field", J. Phys., Série des Colloques, 46, C1-163 (1985).
14. A. de JONG and F. van der VALK: J. Phys. B: At. Mol. Phys. 12, L561 (1979).
15. M. ALLEGRINI, G. ALZETTA, A. KOPYSTYŃSKA, L. MOI and G. ORRIOLS: Optics Comm. 19, 96 (1976).
16. J. HUENNEKENS and A. GALLAGHER: Phys. Rev. A 28, 1276 (1983).
17. B. CARRÉ, F. ROUSSEL, G. SPIESS, J.M. BIZAU, P. GERARD and F. WUILLEUMIER: Z. Phys. D - At., Mol. and Clusters 1, 79 (1986).
18. R.M. MEASURES: J. Appl. Phys. 48, 2847 (1977).
19. B. CARRÉ et al.: XIV ICPEAC Palo Alto (1985) Invited papers, p. 493.
20. F. WUILLEUMIER: J. Physique 43, C2-347 (1982).
21. P.McCAVERT and E. TREFFTZ: J. Phys. B: At. Mol. Phys. 7, 1270 (1974).
22. P. HAFNER and W.H.E. SCHWARZ: J. Phys. B: At. Mol. Phys. 11, 2975 (1978).
23. P. KOWALCZYK: J. Phys. B: At. Mol. Phys. 17, 817 (1984).
24. L. LANDAU and E. LIFCHITZ: Ed. MIR, Moscow 1967, p. 119, Mécanique Quantique.
25. L. BARBIER and M. CHÉRET: J. Phys. B: At. Mol. Phys. 16, 3213 (1983).
26. F. ROUSSEL, P. BREGER and G. SPIESS: private communication.
27. D.F. REGISTER, S. TRAJMAR, S.W. JENSEN and R.T. POE: Phys. Rev. Lett. 41, 749 (1978).
28. R. KUNNEMEYER and M. KOCK: J. Phys. B: At. Mol. Phys. 16, L607 (1983).
29. L. BARBIER: Thesis. Université de Paris-Sud, 1984.
30. A. LE NADAN, G. LE COZ, P. TUFFIN and J. PERESSE: J. Physique 43, 1607 (1982).
31. C. BRION, C. McDOWEL and W. STEWART: J. Electron. Spectrosc. 1, 113 (1972/1973).

# Laser-Assisted Ionization in the He*($2\,^1S$, $2\,^3S$) + He($1\,^1S$) Collision System

*P. Monchicourt, P. Pradel, D. Dubreuil, and J.J. Laucagne*

Service de Physique des Atomes et des Surfaces, Institut de Recherche Fondamentale, Commissariat à l'Energie Atomique, Centre d'Etudes Nucléaires de Saclay, F-91191 Gif-sur-Yvette Cedex, France

## 1. DEFINITION OF THE LASER-ASSISTED IONIZING COLLISION

Laser-assisted atomic collisions are processes where photon absorption (or emission) is made possible by the collisional interaction between two atoms, whereas it is not allowed when these two atoms are far apart. Since the work of GODZENKO and YAKOVLENKO [1], there has been a great interest for such processes, as shown by the large number of theoretical studies [2], and also by the less numerous experimental works [3].

Here we present an experimental study of a laser-assisted collision involving two helium atoms and a moderately intense laser field. In this process, a one-photon absorption event induces an electron transition from a bound level to a continuum of states. The ionizing collision under study is [4]:

$$He^*(2^1S,2^3S) + He(1^1S) + \hbar\omega \rightarrow He^+(1^2S) + He(1^1S) + e^- \qquad (1)$$

where $He^*(2^1S,2^3S)$ is a very low density metastable beam (density less than $10^3$ atoms/cm$^3$), whose kinetic energy is adjustable, typically between 25 and 150 eV in the center-of-mass system, that is, twice these values in the laboratory. This beam includes both singlet and triplet species in proportion depending on kinetic energy. In our energy range, the two concentrations are roughly of the same order. This beam crosses a He target, that is, a region filled with helium gas whose density is such that only single collision events are observed, as checked experimentally. The collision occurs in the field of a pulsed laser beam, whose frequency corresponds to a photon energy of 3.5 eV. This is below the photoionization threshold of the metastable, and non resonant with any atomic transition in the system. The laser intensity can be varied up to 10 Megawatts/cm$^2$. In these experiments, the "fast" $He^+$ ions are detected, that is, the atomic ions having nearly the kinetic energy of the parent metastable beam. This assisted ionizing collision (1) produces ions via a single electron transition, and can be pictured as a photoionization process occurring at short internuclear distances. This is shown by the potential curves relevant to the system [5,6] in Fig. 1: At large internuclear distances, the atomic structure is not perturbed by the collision, and the photon energy is too low to ionize the system. Below a threshold distance $R_0 \approx 2$ Å, and as a result of the collisional interaction, the two initial field-dressed curves (dashed lines) enter inside the $He^+$ + He + $e^-$ continuum. For $R < R_0$, a photon can be absorbed. The heavy particles come out on the ion curve (dashed-dotted line), while the electron is ejected with a kinetic energy equal to the difference between the dressed curve and the ion curve at the distance where the transition occurs, if the Franck-Condon principle holds. The molecular ion cannot be observed in our conditions, since its potential well of 2.5 eV is much smaller than the working kinetic energies of a few tens of eV's.

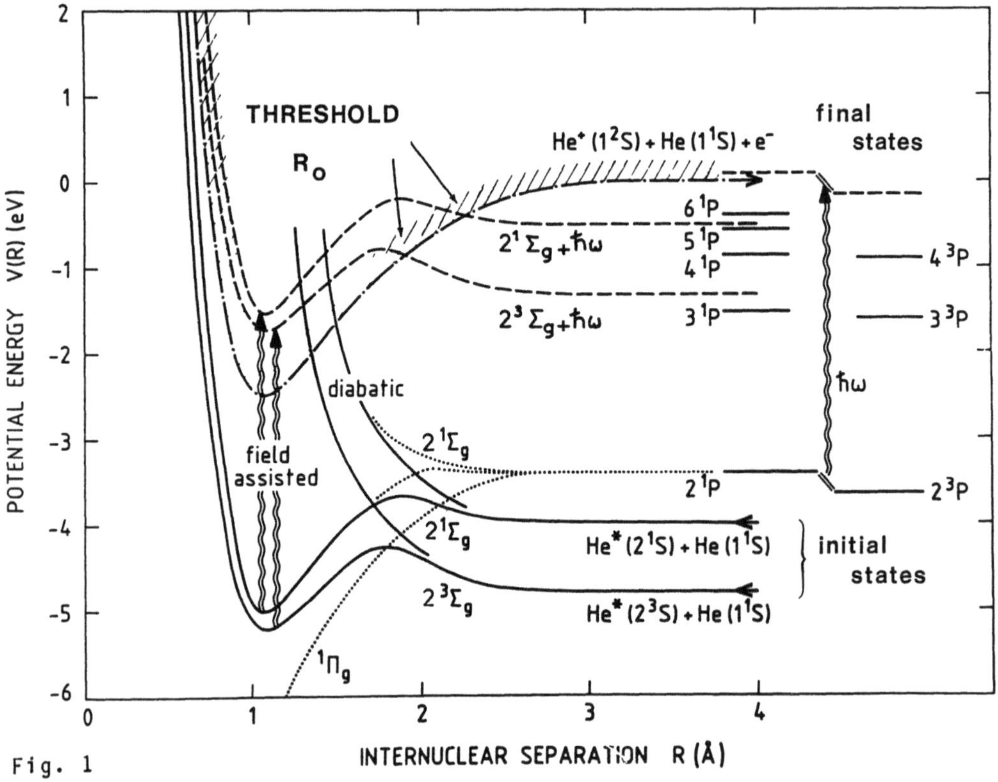


Fig. 1

The assisted process (1) is characterized by its cross-section $\sigma_a$ which gives the ion production rate from the metastable and ground state concentrations $n^*$ and n, and from their relative velocity v:

$$\frac{dn^+}{dt} = \sigma_a \, n^* \, n \, v \tag{2}$$

In $\sigma_a$ is included the effect of the field. This effect becomes explicit by using the photoionization probability per unit time w(R) given by the Fermi Golden Rule:

$$w(R) = \frac{2\pi}{\hbar} \rho \, | < f \, | \, e \, \vec{r} . \vec{\mathcal{E}} | i > |^2 \tag{3}$$

where $\rho$ is the density of continuum states and the brackets define the dipole matrix element between the initial and the final continuum states, which are R-dependent. The cross-section $\sigma_a$ is obtained classically by integrating w over the trajectory and over all impact parameters b as:

$$\sigma_a = 2\pi \int_0^{R_0} b \, db \int_0^{R_0} 2 \, w(R) \frac{dR}{v_R(R)} \tag{4}$$

where $v_R(R) \simeq v_R(\infty) \left(1 - \frac{b^2}{R^2}\right)^{\frac{1}{2}}$ is the radial component of the relative velocity.

By assuming:

$w(R) = 0 \qquad$ for $R > R_0$

$$w(R) = \sigma_I \mathcal{J} \quad \text{for } R < R_0$$

where $\sigma$ is an averaged photoionization cross-section and $\mathcal{J}$ the photon flux, expression (4) is easily integrated to give:

$$\sigma_a = \frac{2}{3}\pi R_0^2\, \sigma\varphi \mathcal{J}\, t_c = \frac{4}{3}\pi R_0^3 \frac{\sigma\varphi\mathcal{J}}{v_R(\infty)} \tag{5}$$

which shows that $\sigma_a$ is 2/3 of the geometrical cross-section $\pi R_0^2$ times the photoionization probability $\sigma\varphi\ \mathcal{J}\ t_c$ during the time available for photon absorption $t_c = 2\ R_0/v_R(\infty)$. The second expression of (5) displays the characteristic dependence of $\sigma_a$, i.e., $\sigma_a \approx \mathcal{J}/\sqrt{E_{CM}}$, where $E_{CM} = \frac{1}{2}\ m\ v_R^2(\infty)$. This dependence is used in the following to distinguish the assisted process among the others. Expression (5) shows also that from the measurement of $\sigma_a$, it is possible to get information about $\sigma_a$, that is, about the dipole moment of the bound-free transition during the collision. This will be discussed in section 5.

## 2. COMPETING PROCESSES

A careful examination of the possible competing processes has been performed [4]. For instance:

- Atomic photoionization is not allowed for a 3.5 eV photon.
- Two-photon ionization does not contribute. This is checked by its $\mathcal{J}^2$ dependence and also by removing the He target: in this case, the ion signal disappears.
- Two-step processes such as photoexcitation of $He^*$ followed by collisional ionization, and collisional excitation of $He^*$ on high excited states followed by photoionization give a negligible contribution to the ion signal.

Actually, two competing processes contribute in our conditions:

. Field-Free ionization is an inelastic ionizing collision via a so-called diabatic channel [7]. Because of the kinetic energy, when following the initial curves (see Fig. 1), the system has a finite probability to jump across the adiabatically avoided 2S-2P crossing, to follow the repulsive "diabatic" curve, and to enter into the continuum where autoionization occurs. This process is written as:

$$He^*(2^1S, 2^3S) + He(1^1S) \rightarrow He^+(1^2S) + He(1^1S) + e^- \tag{6}$$

and has already been studied. A diabatic cross-section $\sigma_d$ has been calculated for the triplet system [8]. This field-free channel which produces ions in all our experimental conditions is used to calibrate our results and deduce the absolute magnitude for the assisted cross-section.

. Collisional excitation of the $2^1P$ level, followed by photoionization. In this two-step process (see Fig. 1), the collisional step occurs either through the same crossing as above [8], or through the $2^1\pi$-$2^1\Sigma$ crossing jumped by angular coupling, yielding cross-sections as high as $10^{-18}$ cm$^2$ [9]. The $2^1P$ level only contributes, since the $2^3P$ cannot be photoionized by a 3.5 eV photon. This process is efficient because the $2^1P$ is the nearest neighbour of the initial $2^1S$ state: the small energy defect of 0.6 eV allows thus high cross-sections. Higher levels are not significantly populated, because of the much higher energy defects of the inelastic step.

One should emphasize that these two competing processes involve an inelastic transition, and have therefore cross-sections which are increasing functions of kinetic energy. The opposite behaviour is expected for the assisted cross-section $\sigma_a$, which decreases as $1/\sqrt{E_{CM}}$. Hence, the experimental arrangement allows to vary the kinetic energy, in order to distinguish these processes from process (1).

## 3. EXPERIMENTAL SET-UP

A schematic view of the experiment is shown on Fig. 2.

. The metastable beam is obtained in a neutralization cell by a near-resonant charge exchange of a primary $He^+$ beam on Cesium. The remaining ions are removed from the beam by a transverse electric field.

. The beam enters into the interaction cell through a diaphragm 2mm in diameter. The interaction cell is 12 mm long, and is filled with helium gas at a pressure monitored by a Westinghouse gauge. This region is illuminated by a parallel beam of light coming from either one of these two lasers: (i) the $3^{rd}$ harmonic of a Yag MOLECTRON laser, or (ii) a LAMBDA PHYSICS excimer laser using XeF as active medium, yielding nearly the same irradiation conditions: $\hbar\omega$ = 3.5 eV; light intensity up to 10 $MW.cm^{-2}$ and pulse duration ≈ 20 nsec.

. The ions produced in the direction of the beam are analyzed as a function of their time arrival on the detector (channeltron). This analysis is performed using a multichannel analyzer, which stores, as a function of the time delay from the laser pulse, the ions produced over a large number of laser shots. Since the laser-induced ions are produced only during the laser pulse duration $\tau$ ≈ 20 nsec, while the field-free ions are continuously produced in the interaction region, one can expect to measure a field-free background and an extra ion peak of assisted ions whose time width is the direct time transform of the interaction length ($\ell$ = 12 mm) translated by the time-of-

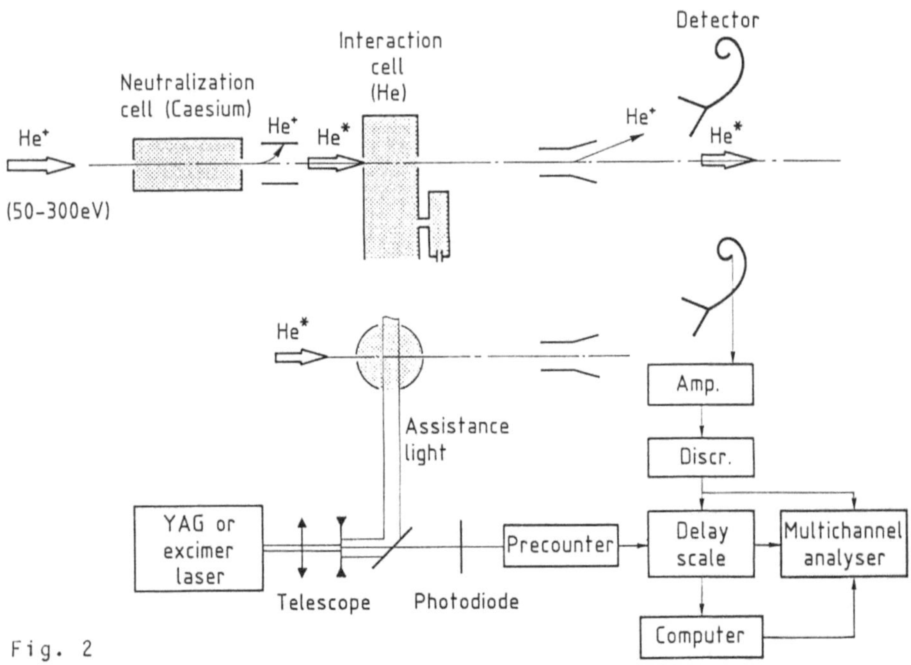


Fig. 2

flight of the ions. This time width is $\approx \ell/v + \tau$. The multichannel analyzer is triggered on before this time arrival, and turned off after this time window, in order to store the two signal amplitudes, each being proportional to the corresponding cross-sections $\sigma_a$ and $\sigma_d$.

## 4. EXPERIMENTAL RESULTS

A typical histogram obtained after $6\times10^5$ laser shots, for $\mathcal{J} \approx 10$ MW cm$^{-2}$ is shown in Fig. 3. The average value of the continuous field-free background $\overline{S_d}$ is seen to be around 80 ions per channel (each dot represents the number of ions contained in one channel, each 50 nsec in duration). The assisted ion peak $S_a$ is observed in the expected time window (vertical dashed lines), i.e., at the expected time delay and with the calculated timewidth.

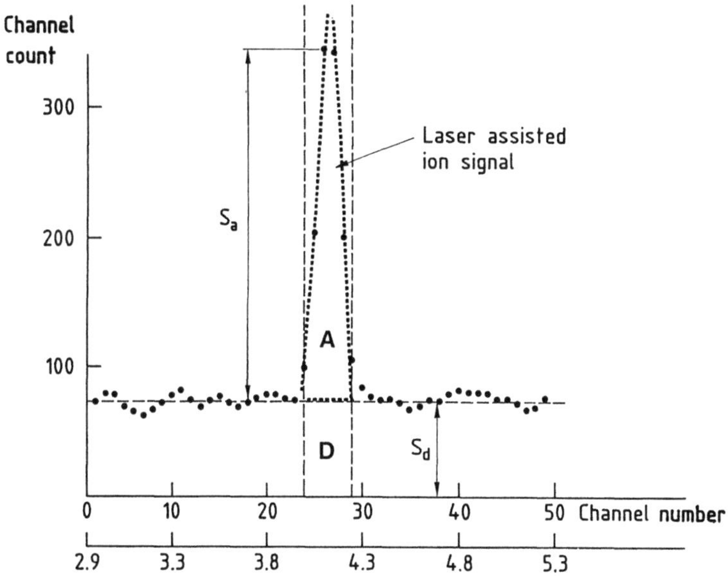


Fig. 3 DELAY TIME FROM LASER PULSE (μsec)

In order to compare the two ion signals produced in the whole interaction volume $\mathcal{V}$, one has to integrate these signals over the whole time window. Each of these two areas, A or D, is proportional to the cross-section times the interaction time, i.e.:

$$A \propto \sigma_a \mathcal{V} \tau$$
$$D \propto \sigma_d \mathcal{V} \ell/v$$

The cross-sections $\sigma_a$ and $\sigma_d$ as a function of $E_{CM}$ and $\mathcal{J}$ are obtained from the experimental values of A and D; absolute values are deduced by normalizing $\sigma_d$ to a published theoretical value [8].

The field-free cross-section $\sigma_d$ obtained this way is plotted in Fig. 4 as a function of kinetic energy. The black dots are our experimental values obtained from D and from the theoretical calculation of EVANS, COHEN, and LANE [8] at $E_{CM}$ = 50 eV for the triplet system (large black dot). The two other theoretical values calcula-

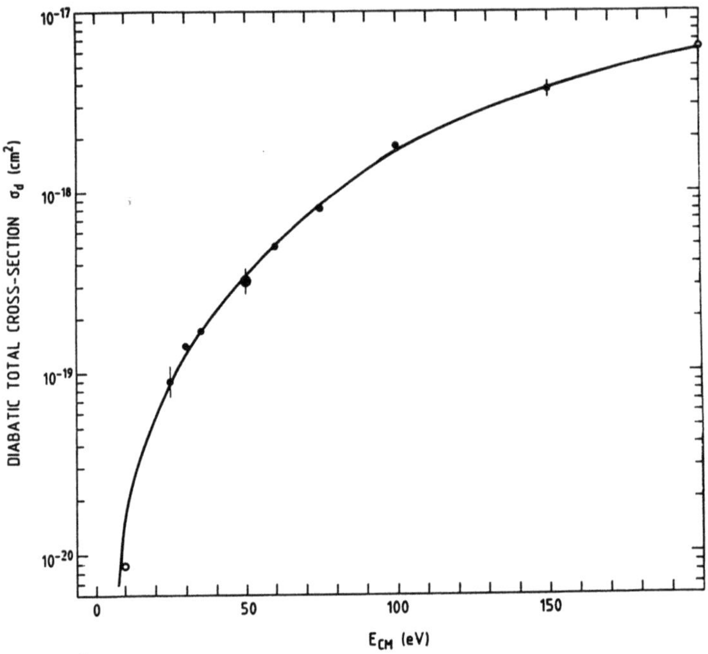
DIABATIC TOTAL CROSS-SECTION σd (cm²)
10⁻¹⁷
10⁻¹⁸
10⁻¹⁹
10⁻²⁰
0
50
100
150
ECM (eV)

Fig. 4

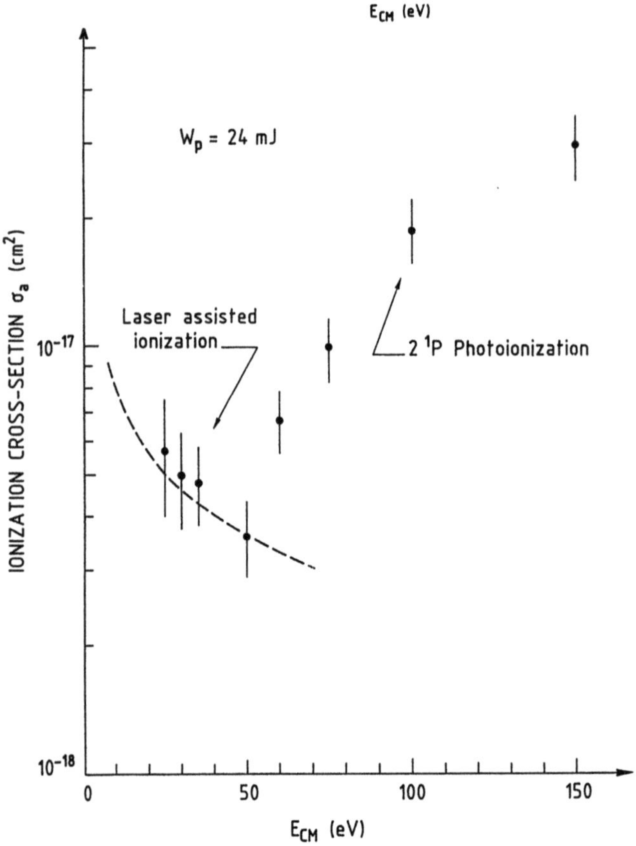
IONIZATION CROSS-SECTION σa (cm²)
Wp = 24 mJ
Laser assisted ionization
2 ¹P Photoionization
10⁻¹⁷
10⁻¹⁸
0
50
100
150
ECM (eV)

Fig. 5

ted by the same authors [8] at 10 and 200 eV (circles) are in excellent agreement with our measurements. The values of $\sigma_d$ are found to grow nearly as $E_{CM}^{5/2}$.

The field-induced ionization cross-section $\sigma_a$ obtained from the signal A measured at different kinetic energies and from the normalization of $\sigma_d$, is shown in Fig. 5, for $\mathcal{J} \approx 5$ MW.cm$^{-2}$, in a semi-log plot, as a function of $E_{CM}$. The error bars represent the statistical uncertainty on each experimental point. Measurements below 25 eV have not been possible, because the metastable density in the beam was too low to give acceptable statistics for the signal. A very different behaviour is obtained below and above 50 eV. This is interpreted as follows:

. Below 50 eV, the experimental values follow closely the characteristic dependence $\propto 1/\sqrt{E_{CM}}$ (dashed line), that is, correspond to the laser-assisted cross-section given by (5).

. Above 50 eV, the laser-assisted signal becomes smaller than the two-step process involving the $2^1P$ level. What is measured in this range is not $\sigma_a$, but the $2^1S$-$2^1P$ inelastic cross-section times the photoionization probability of the $2^1P$ level. This inelastic cross-section is found to follow an $E_{CM}^2$ variation law.

One should note here that all these values are lower limit values, because they are normalized to a pure triplet value [8], whereas our beam is a triplet + singlet mixed one. Since the singlet is characterized by a smaller energy defect for field-free ionization, one may expect higher cross-sections for it.

Another important characteristic of $\sigma_a$ is the behaviour as a function of laser intensity. This is shown in Fig. 6, where $\sigma_a$ is seen to follow (5), i.e., a linear variation law versus laser energy per pulse, which is directly proportional to laser intensity.

## 5. CONCLUSION

Two important features can be emphasized from this experimental study of the laser-assisted process (1):

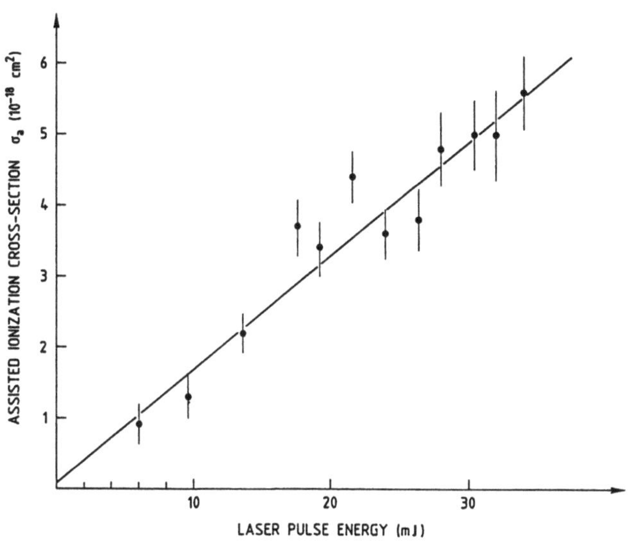


Fig. 6

The first point concerns the experimental evidence for laser-assisted ionization. In our conditions, which are among the simplest one can get, we have three different ionizing processes:

- Field-free ionization, whose cross-section varies as $\approx E_{CM}^{5/2}$.
- Laser-assisted ionization, which follows the characteristic dependence $\mathcal{J}/\sqrt{E_{CM}}$ given by (5).
- Two-step ionization via the $2^1P$ level, which is found to grow as $E_{CM}^2$.

The two former processes have been discriminated by the time-of-flight technique, while the two latter processes have been distinguished by their different dependence with kinetic energy. Without these two diagnostics, one should emphasize that it would have been impossible to isolate the assisted process. This process is found to be predominant only below $E_{CM}$ = 50 eV, for a laser intensity of 5 MW.cm$^{-2}$.

The second point is the magnitude of $\sigma_a$, which is found to be larger than $10^{-18}$ cm$^2$ at 5 MW.cm$^{-2}$. If one uses the simple analytical expression (5), one finds that the dipole moment during the collision is roughly 30 times the unperturbed atomic value, that is, the value corresponding to the photoionization of the atomic metastable [10]. Hence, one can conclude that the 2S orbital of the optical electron is strongly modified during the collision. Such a dipole enhancement should be expected since, during the collision, the optical electron suffers a 2s-2p hybridization precisely in the range of internuclear distances where photoionization becomes allowed [5]. During this hybridization, the optical electron is pushed away from its nuclear center, and has therefore a much higher probability to be photoionized. Calculations are in progress to check this simple interpretation, or will indicate if more sophisticated interpretations are needed.

## REFERENCES


1. L.I. Gudzenko and S.I. Yakovlenko, Sov. Phys. JETP 35, 877 (1972).
2. See for instance J.C. Bellum and T.F. George, J. Chem. Phys. 68, 134 (1978) with references therein.
3. R.W. Falcone, W.R. Green, J.C. White, J.F. Young, and S.E. Harris, Phys. Rev. A15, 1333 (1977).
   J. Boulmer and J. Weiner, Phys. Rev. A27, 2817 (1983) and references therein.
4. P. Pradel, P. Monchicourt, D. Dubreuil, J. Heuzé, J.J. Laucagne, and G. Spiess, Phys. Rev. Lett. 54, 2600 (1985).
   P. Monchicourt, P. Pradel, F. Roussel, and J.J. Laucagne, Phys. Rev. A33, 3515 (1986).
   P. Pradel, P. Monchicourt, D. Dubreuil, and J.J. Laucagne, to be published.
5. S.L. Guberman and W.A. Goddard III, Phys. Rev. A12, 1203 (1975).
6. R.P. Saxon, K.T. Gillen, and B. Liu, Phys. Rev. A15, 543 (1977).
7. K.T. Gillen, J.R. Peterson, and R.E. Olson, Phys. Rev. A15, 527 (1977).
8. S.A. Evans, J.S. Cohen, and N.F. Lane, Phys. Rev. A4, 2235 (1971).
9. E.J. Shipsey, J.C. Browne, and R.E. Olson, Phys. Rev. A11, 1334 (1975).
10. R.F. Stebbings, F.B. Dunning, F.K. Tittel, and R.D. Rundel, Phys. Rev. Lett. 30, 815 (1973).

# The Collision Dynamics of Associative Ionization

*J. Weiner*

Department of Chemistry, University of Maryland, College Park, MD 20742, USA

The common question motivating my research program continues to be: What dynamical factors influence the outcome of an inelastic or reactive collision? The key word is dynamical because even if all the potential curves of all the relevant states were known accurately, the outcome of a collisional encounter would still be impossible to characterize or predict. Dynamical factors such as collision velocity, reactant state excitation and polarization, spin-orbit coupling, and Coriolis (electronic-rotational) coupling profoundly alter the final probability distribution of inelastic channels.

The associative ionization process (AI),

$$A + B \rightarrow AB^+ + e \tag{1}$$

is a good choice to study because it is a simple, chemical-bond-forming mechanism, one (and sometimes both) collision partners can be state-prepared, aligned or oriented by laser excitation, and the charged particle products lend themselves to efficient and sensitive detection.

## 1. Absolute Rate Constants

My research group first undertook to measure accurately AI rate constants in alkali systems so as to gauge their importance relative to other inelastic processes. The results of these studies [1-4] demonstrate three important points: (1) rate constants for

$$Na(3s) + Na^*(np) \rightarrow Na_2^+ + e;\ 5 < n < 15 \tag{2}$$

are large, ranging from $3\times10^{-10}$ to a maximum of $2\times10^{-9}$ $cm^3$ $sec^{-1}$ at n=11. (2) rate constants increase rapidly with principal quantum number to a maximum between n=9-11, after which they gradually decline. Reference 1 compares these findings to a theory advanced by Janev [5] and Duman [6]. Agreement was fair, but theory and experiment appeared to be diverging at higher n. Recently Boulmer and Weiner [4] extended these results to n=27 and measured rate constants involving s and d as well as p excited orbitals. These new results (see Fig. 1) clearly demonstrate that the dynamical picture implicit in the theory proposed by Janev and Duman is inadequate to account for the observations. (3) The rate constant for AI between two excited Na atoms,

$$Na(3p^2P_{3/2}) + Na(3p^2P_{3/2}) \rightarrow Na_2^+ + e \tag{3}$$

is surprisingly small ($1.1\times10^{-11}$ $cm^3$ $sec^{-1}$) compared to those of process (2). Reaction (3) is important because of the role it plays in initiation of laser-triggered alkali vapor plasmas and because earlier studies had reported wildly different numbers for the AI rate coefficient. The work by Gallagher and Huennekens [7] in a high-pressure cell and our own work in crossed beams [2,3] are the only two studies to properly consider the important effects of radiation trapping. These two determinations are in agreement to within about 25%.

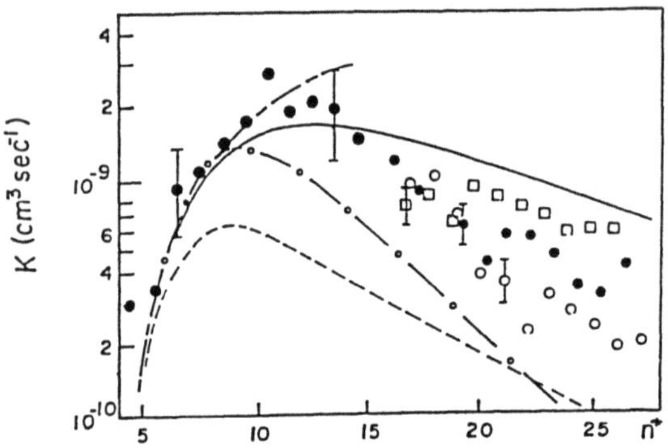


Fig. 1 Associative Ionization Rate Constant vs. Effective Quantum Number. Data symbols: □ s levels, ● p levels, o $\ell \geq 2$ levels. Theoretical curves: --- s levels, —— p levels, -o- $\ell \geq 2$ levels.

## 2. Velocity Dependence of AI Rate Constant

The surprisingly small rate constant for (3) immediately raised the possibility of a potential barrier and suggested a study of AI as a function of collision energy. We decided to attack this problem by initiating a velocity-selected crossed beams experiment. A work station was established at the Regional Laser Lab in Philadelphia in order to take advantage of the available single-frequency ring dye laser needed to excite narrow "δ-function" velocity groups from the global velocity distribution of each Na beam. Scanning the laser frequency and measuring the $Na_2^+$ production from AI leads to a direct determination of the AI "excitation function" (probability vs. collision energy). Reference 8 reports the initial results of the experiment. We found a strong dependence of cross-section on collision velocity with a minimum at about $1.6 \times 10^3$ m $sec^{-1}$ (0.15 eV) and a steeply rising curve on both sides of the minimum (see Fig. 2). It appeared that a single potential curve could not possibly account for this behavior, and our tentative guess was that at least two potentials, one attractive and the other repulsive, contribute significantly to the final associative ionization state. The curious shape of the excitation function was quite unexpected, and it was clear from these initial results that the range of available collision velocities had to be greatly expanded, especially toward the energy region below 80 meV. Reference 9 describes recent new results where we report the deter-

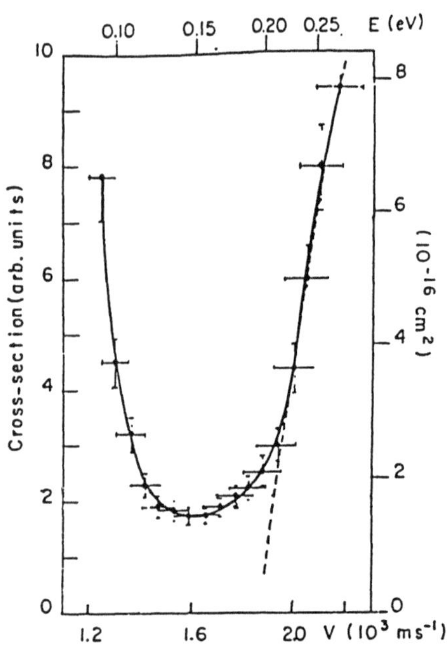


Fig. 2 Associative ionization cross-section vs. relative energy (upper abscissa) and relative velocity (lower abscissa) for process (3). The steeply ascending portion of the curve may be due to a repulsive potential and the steeply descending portion to an attractive potential.

mination of separate excitation functions for singlet and triplet manifolds over a collision energy range from 2.4 meV to 290 meV (see Fig. 2). Despite these more complete results the interpretation of these excitation functions in terms of adiabatic potential curves and nonadiabatic coupling is still far from complete and is the object of continuing research in my laboratory.

## 3. Internal State Distributions

The distribution of rotational and vibrational states of the product ions provide important information on the collision process in two ways: (1) the rotational energy is directly related to the amount of orbital angular momentum available in the collision, and (2) the vibrational energy is related to the shape and position of the electronic potential curves involved in the transition [10]. Reference 11 reports the results of an experiment in which we have measured the internal state distributions of process (3) and process (2) with n=6. As an alternate approach to accurately measuring the energy distribution of low-energy electrons (a very difficult task below about 0.5 eV), we have chosen photofragment translational spectroscopy [11,12]. This technique involves photodissociation of the product ions and measurement of fragment kinetic energies by time-of-flight (TOF). The nascent $Na_2^+$ internal energy distribution follows directly from the fragment recoil kinetic energy distribution. Figure 3 shows the fragment TOF spectrum for process (2) together with calculated spectra obtained for various choices of model internal state distributions. The conclusion of a rather extensive analysis based on Monte Carlo simulations is that: (1) the associative ionization probability (opacity function) falls rapidly with internuclear distance after a separation of about 2Å; and (2) the distribution of populated vibrational states is nonthermal, increasing approximately linearly with vibrational level. We find essentially the same results for process (3). In fact the behavior of the opacity function is consistent with our earlier study of the velocity dependence (section 2) where, in order to fit the high-velocity side of the excitation function, we assumed an exponentially decreasing form for the autoionization width and found that the opacity function dropped essentially to zero beyond b≈2.3Å. It is interesting to note that although processes (2) and (3) have similar opacity functions, the overall rate constant of (3) is only 3% of (2). Whether the disparity in rate constants is due to widely differing Franck-Condon overlap factors or autoionizing widths is a question for continuing research.

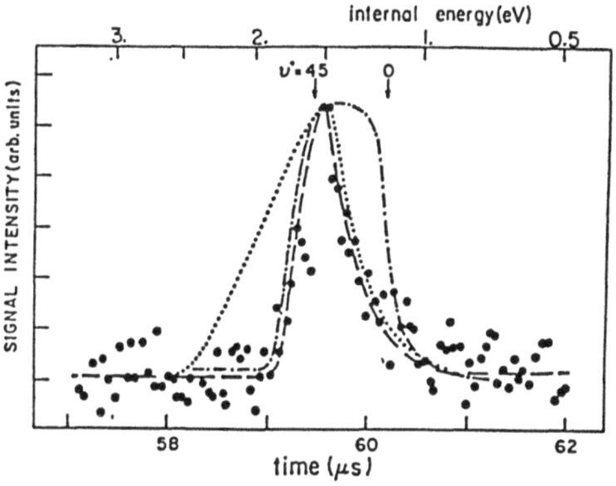


Fig. 3 Photofragment TOF spectrum from experiment 3 (Internal State Distribution). Profile shape reflects nascent distribution of rotational and vibrational energy in $Na_2^+$ ions produced by associative ionization via process (2) with n=6. The figure shows data points (large filled circles) together with three different Monte Carlo simulations: (1) impact parameter b = 5Å and only v"=0 populated ..., (2) impact parameter b = 0Å and all vibrational levels equally populated, ·-·-·, (3) impact parameter b = 2Å and vibrational population linear with v", ---. Latter choice is best fit to measured spectrum.

## 4. Polarization Dependence of AI Rate Constant

One of the first questions one seeks to answer when considering the nature of process (3) is: Do the p-orbitals of the collision partners prefer an axial ($\sigma^2$) or transverse ($\pi^2$) configuration. Axial and transverse approach are illustrated in Fig. 4. A clear-cut preference would greatly simplify the task of analysis since only a subset of the bewildering number (12) of possible potential curves arising

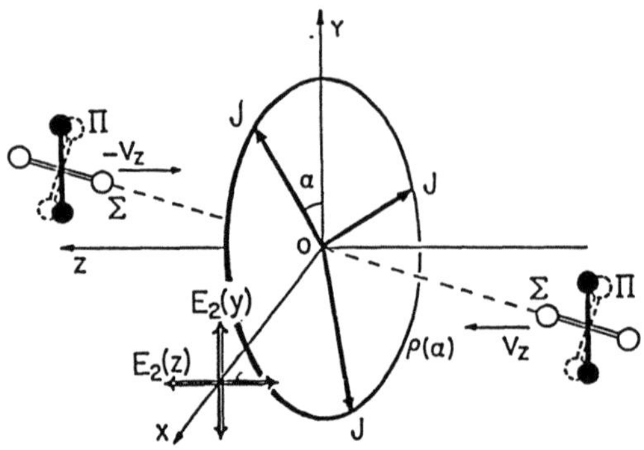


Fig. 4 Figure shows initial vertical polarization of 3p orbitals (parallel to Y) by filled-in dumbells. Projection onto molecular axes shown by open dumbells. Axial approach is labelled Σ and transverse approach is labeled π. Distribution of rotational J vectors indicated by disc centered at the origin and confined to XY plane. Polarization of $E_2$ samples the actual J distribution.

from the $^2P$ + $^2P$ asymptote need be considered. For example if σ approach is preferred, then only Σ states of the collision complex play a significant role. Earlier studies [14,15] showed that the rate constant for process (3) is sensitive to polarized excitation of the reactants in space-fixed laboratory coordinates, but a straight forward interpretation of the results in the molecular collision frame in terms of molecular orbitals remains unsettled [16,17].

In reference 18 we report the results of an effusive beam experiment which shows that the rate dependence on polarization persists even when all hyperfine states of the $3p^2P_{3/2}$ level are populated by broad-band pulsed laser excitation. Secondly we analyze the dimer ion products of process (3), using polarized photodissociation, to show that the spatial distribution of $Na_2^+$ rotational angular momentum is also sensitive to initial-state polarization. The experiment is essentially a TOF photofragment analysis of the nascent $Na_2^+$ ions produced by process (3) with the reactants polarized alternately parallel and perpendicular to the atomic beam axis. A schematic diagram of the experiment is shown in Fig. 5. Two factors enable us to relate the experimental results to collision symmetry: (1) the well-known angular dependence of dimer photofragmentation probability and (2) the conversion of orbital to rotational angular momentum during the course of the collision. In order to interpret the results in terms of molecular orbital symmetries we developed [19] a semi-

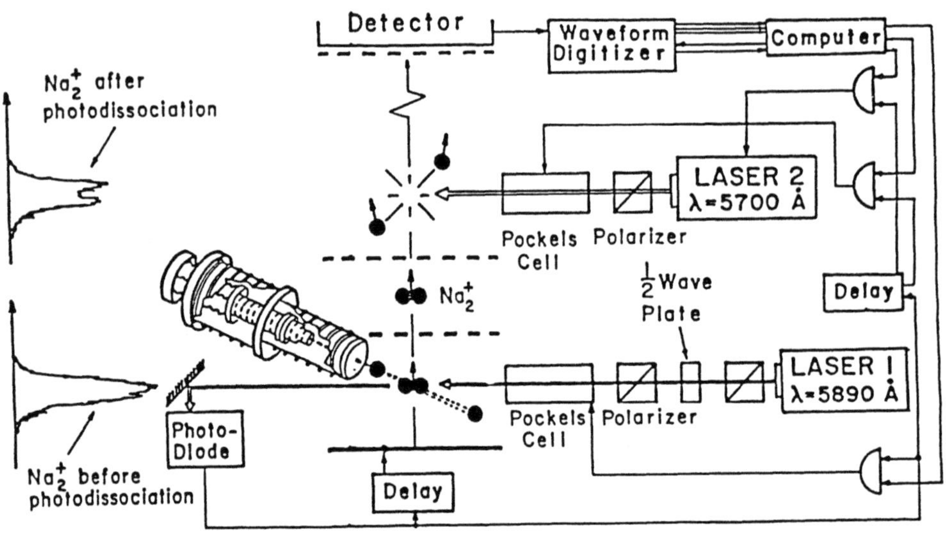


Fig. 5 Schematic diagram of experimental apparatus used in polarization dependence experiment 4.

classical model which relates laboratory-fixed polarization to the molecular collision axis, averages over all collision angles and impact parameters, assigns a ratio of AI rate constants, $k_\sigma/k_\pi$, for axial and transverse p-lobe orientation, and averages over all nuclear spin orientations. From the polarization dependence of the AI rate constant and $Na_2^+$ rotational angular momentum emerge a clear physical picture of the collision symmetries involved in process (3). The results show that $\sigma^2$ configuration is favored by more than a factor of six in the rate constant and that the "locking radius", where the quantization axis transforms from laboratory-fixed to molecular-fixed, occurs at an internuclear separation greater than 25Å.

## 5. Summary

We know much more now about associative ionization than we did only two years ago. At least in the case of process (3), we know Σ states are strongly favored; and that AI occurs at small impact parameter with relatively little rotational excitation. In principle we should be able to explain the vibrational distribution by applying the Franck-Condon principle to the (as yet unknown) shapes of the relevant potential curves. We anticipate new developments in the area from theorists [20] and from further development of the velocity-selected technique described in section 2. The results of section 2, while tantalizing, are incomplete. The problem is that technical constraints obliged us to perform the experiment in a suprathermal jet, and therefore these results cannot be related directly to the other effusive experiments. It is possible, for example, that the internal state distributions and sensitivity to p-orbital alignment may be entirely altered by the collision energy regime in which AI takes place. The velocity dependence study (section 2) indicates that both a high energy and low energy regime contribute significantly to the cross-section. Experiments in sections (2) and (3) were carried out in thermal beams, and therefore reflect the relative weighting of the velocity-dependent cross-section according to the Maxwell-Boltzmann distribution. If these same experiments were carried out under velocity-selected conditions, the results might show vibrational distributions and orbital alignment dramatically different at high and low collision energy. In the future we plan to explore the internal state distributions and alignment of AI as a function of collision energy, using the laser velocity-selected technique.

1. R. Bonanno, J. Boulmer, and J. Weiner: J. Phys. B 16, 3015 (1983).
2. R. Bonanno, and J. Weiner: Phys. Rev. A 28 604 (1983).
3. R. Bonanno, J. Boulmer, and J. Weiner: Commts. At. Mol. Phys. XVI, 109 (1985).
4. J. Weiner, and J. Boulmer: J. Phys. B 19, 599 (1986).
5. R. Janev and A. Mihajlov: Phys. Rev. A 21, 819 (1980).
6. E. Duman, and I. Shmatov: Sov. Phys. JETP 51, 1061 (1980).
7. J. Huennekens, and A. Gallagher: Phys. Rev. A 28, 1276 (1983).
8. M-X. Wang, M.S. deVries, J. Keller, and J. Weiner: Phys. Rev. A 32, 681 (1985).
9. M-X. Wang, J. Keller, J. Boulmer, and J. Weiner: Phys. Rev. A (in press).
10. J. Lorenzen, H. Morgner, W. Bussert, M.W. Ruf, and H. Hotop: Z. Phys. A 310, 141 (1983) and references cited therein.
11. J. Keller, R. Bonnano, M-X. Wang, M.S. deVries, and J. Weiner: Phys. Rev. A 33, 1612 (1986).
12. H. Helm, and R. Möller: Phys. Rev. A 27, 2493 (1983) and references cited therein.
13. C. Brechignac and Ph. Cahuzac: Chem. Phys. Lett. 112, 20 (1984).
14. J.C. Kircz, R. Morgenstern, and G. Nienhuis: Phys. Rev. Lett. 48, 610 (1982).
15. E.W. Rothe, R. Theyunni, G.P. Reck, and C.C. Tung; Phys. Rev. A 31, 1362 (1985).
16. G. Nienhuis: Phys. Rev. A 26, 3137 (1982).
17. D.M. Jones, and J.S. Dahler: Phys. Rev. A 31, 210 (1985).
18. M-X. Wang, M.S. deVries, and J. Weiner: Phys. Rev. A 33, 765 (1986).
19. M-X. Wang, M.S. deVries, and J. Weiner: Phys. Rev. A 34, 1869 (1986).
20. A. Henriet, F. Masnou-Seeuws, and C. Le Sech: Chem. Phys. Lett. 118, 507 (1985) and private communication.
21. H.A.J. Meijer, H.P.v.d. Meulen, H.G.M. Heideman, G. Nienhuis, and R. Morgenstern: XIV ICPEAC, Palo Alto (1985).
22. R.N. Zare and D.R. Herschbach: In Proc. IEEE 51, 173 (1963).

# Production of Atomic Rydberg States in Optical Collisions

*C.E. Burkhardt*[1], *M. Ciocca*[1], *W.P. Garver*[1], *J.J. Leventhal*[1], *and J.D. Kelley*[2]

[1]Department of Physics, University of Missouri-St. Louis, St. Louis, MO 63121, USA
[2]McDonnell Douglas Research Laboratories, St. Louis, MO 63166, USA

The advent of tunable lasers has provided the means to study many processes of fundamental importance in atomic and molecular physics. Among the phenomena of current interest are "laser-assisted" or "laser-induced" collisional processes[1]. In this context these terms refer to processes which result in atomic or molecular final states that are only accessible to irradiated systems undergoing collisional interaction[2];

$$X + Y + \hbar\omega \rightarrow X^* + Y. \qquad (1)$$

In a process described by (1), X must simultaneously interact with both the collision partner Y and the photon for $X^*$ to be produced. In the experiments described below, X represents Na(3p), Y is either Na(3p) or Na(3s), and $X^*$ is Na(np) where $n \geq 20$. Thus, (1) represents collision-induced absorption from Na(3p) to produce Na(np) Rydberg states, a transition that is dipole-forbidden for isolated Na(3p) atoms. Because this transition is dipole-forbidden, (1) does not precisely represent the traditional definition of an optical collision[1], i.e. broadening of an allowed transition. Nevertheless, the nature of the process makes the term "optical collision" appropriate.

Most previous studies of laser-assisted collision processes[3-6] employed laser power densities greater than 1 $MW/cm^2$. The work reported here required only modest laser power densities (~1 $kw/cm^2$), primarily because the final state is a Rydberg atom with an electron distribution of such large spatial extent that collisional interactions with a perturber occur at very long range[7]. This long-range interaction leads to large effective collision cross-sections and to long collision durations. The latter characteristic results in unusually narrow collision-induced absorption linewidths.

The apparatus used in these experiments has been described previously[7]. An electrically shielded stainless steel cell containing sodium atoms is irradiated by two coincident counterpropagating pulsed laser beams, one (yellow) tuned to a D-line to effect the 3s→3p excitation and the other (blue and tunable) to produce Na(nℓ). Both dye lasers are pumped by the same YAG laser with laser powers adjusted so that the 3s→3p transition is barely saturated, but the 3p→nd and 3p→ns transitions are not. As the wavelength of the blue laser beam, $\lambda_B$, is scanned, the most abundantly-produced Rydberg atoms are Na(ns) and Na(nd). These are dipole-allowed final states from Na(3p) and do not require a collision for their production.

Because blackbody radiation and heavy-body collisions effectively ionize a significant fraction of the Rydberg atoms[7,8] their production can be verified by detecting mass-selected $Na^+$. These $Na^+$ are produced by some combination of blackbody photoionization, Na(3s)/Na(nℓ) collisions, Na(nℓ)/Na(nℓ) collisions, and even e/Na(nℓ) collisions[8-10]. The measured ion-production rates[7] show that at $N \simeq 10^{14} cm^{-3}$, the approximate density at which we observe the "forbidden" transitions, virtually all of the the initially produced Rydberg atoms are ionized.

The spectra obtained in this experiment are plots of $Na^+$ signal intensities as a function of $\lambda_B$ with the yellow laser wavelength $\lambda_Y$ tuned to the Na $3s_{1/2}$-$3p_{1/2}$ (D1) or $3s_{1/2}$-$3p_{3/2}$ (D2) line. At lower Na(3s) density ($\sim 5 \times 10^{13} cm^{-3}$) only peaks corresponding to ionization from laser-produced ns or nd states are observed. As the Na(3s) density is increased, additional peaks appear; some of these at $\lambda_B$ corresponding to the 3p-np energy separation. Figure 1 shows a partial spectrum with $\lambda_Y$ fixed at the D2

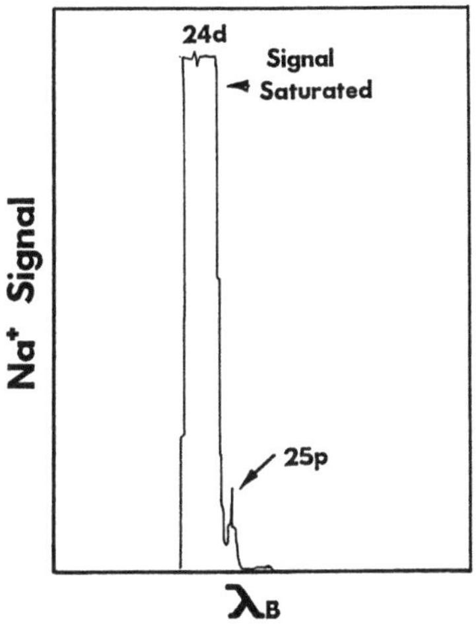


Figure 1 $Na^+$ signal as a function of blue laser wavelength $\lambda_B$. The signal from the allowed transition to Na(24d) is off the scale to allow observation of the peak corresponding to the forbidden transition to 25p.

line. The dipole-forbidden 25p peak appears adjacent to the more intense 24d peak. The latter is truncated because the ion-detection system was saturated. Within the resolution of the experiment ($\sim 0.25 cm^{-1}$) the allowed and forbidden peaks have the same half-width and their intensity ratio is about 200:1 at the highest density.

Before the np peaks can be attributed to collision-induced absorption, other mechanisms for their production must be ruled out. One possibility is Stark mixing of the np levels with the nearby (n-1)d levels. The apparatus employs an electric field to steer the $Na^+$ ions into a quadrupole mass spectrometer for identification and detection. Under typical conditions this field is only ~1 V/cm, which is much too small to account for the observed 1/200 np/(n-1)d ratio. Moreover, the applied steering field is larger than the actual field inside the laser-Na-vapor interaction region, which was electrically shielded to provide a field-free environment. No effect on the observed spectra was observed when the external steering field was increased to ~100 V/cm, so that Stark mixing cannot be the source of the peaks.

Electric quadrupole transitions can produce np states from Na(3p), but the quadrupole transition strength is less than $10^{-5}$ that of the dipole transition[11] and thus leads to np/(n-1)d ratios much smaller than those observed.

Both Stark mixing and electric quadrupole absorption would lead to signals from the laser-produced np levels that would vary linearly with the total Na number density N. Collision-induced absorption as in (1) would produce a signal with a quadratic N dependence. If the np signal could be measured over a sufficiently wide range of accurately known N values, observation of a quadratic N dependence would establish collisional origin. At the values of N at which the effect is observed however, $\sim 10^{14} cm^{-3}$, there is too much uncertainty in the determination of the initial N values to use a direct signal vs. density measurement. This difficulty can be avoided by comparing the np signal under varying experimental conditions to that from a process of known collisional origin. If the signal intensities from the two processes maintain a constant ratio as their magnitudes change, then a quadratic density dependence for the np signal is established.

Fortunately, a feature suitable for application of this technique appears in the data. At low density only one set of ns or nd peaks appear as $\lambda_B$ is varied. These represent dipole-allowed transitions from the particular fine-structure level, either $3p_{1/2}$ or $3p_{3/2}$, excited by the appropriate fixed value of $\lambda_y$. As N is increased to higher values, the dipole-forbidden np peaks appear, but the spectra also show a second set of ns and nd peaks at $\lambda_B$ values corresponding to excitation from the 3p fine-structure level that was not laser excited. At the N-values used in this work, $10^{11}cm^{-3} < N < 10^{15}cm^{-3}$, this fine structure mixing clearly originates from binary collisions[12, 13] between Na(3p) and either Na(3s) or Na(3p). The rate constant for this process is about $2 \times 10^{-9} cm^3 s^{-1}$. The only other process which can lead to fine structure mixing is spontaneous Raman scattering from one 3p fine structure level to the other. This effect has been observed[14] in laser-irradiated Na vapor, but that work employed N values of about $10^{10}cm^{-3}$ and power densities of ~0.5 MW/cm², 500 times greater than the power densities used in our studies. Under our conditions, collisional fine structure mixing is orders of magnitude more rapid than the Raman process.

Figure 2 shows two spectra taken at different values of N, both of which show the effects of fine structure mixing as well as those of optical collisions. In these spectra, the yellow laser was tuned to the $3s \rightarrow 3p_{3/2}$ transition. At the lower N the forbidden np peaks and the second set of ns and nd peaks originating from $Na(3p_{1/2})$ are easily identified. At the higher N both np peaks, and the ns and nd peaks from $Na(3p_{1/2})$ increase in constant ratio. The np peaks must therefore have a quadratic density dependence and result from the effects of binary collisions.

The ratio of the nd fine structure peaks originating from $3p_{3/2}$ and $3p_{1/2}$ can be used to provide a rough estimate of the total Na number density. For the experimental conditions under which the highest density data in Fig. 2 were obtained, the $3s$-$3p_{3/2}$ transition was saturated by the yellow laser and the dipole-allowed ns and nd signals were independent of a further increase in power at $\lambda_Y$. These signals depended linearly on power at $\lambda_B$, and were well below the saturation level. Under these conditions, the total number density N is approximately given by the sum of the $3s_{1/2}$ and $3p_{3/2}$ densities, denoted $N_0$ and $N_{3/2}$. Moreover, $N_0$ and $N_{3/2}$ can be taken as constant throughout most of the laser pulse duration $\tau$, and they will be close to their statistically weighted saturation values. Therefore $N_0$ is approximately 0.33 N, and $N_{3/2}$ is 0.66 N during the pulse. The $3p_{1/2}$ density $N_{1/2}$ produced by $3p_{3/2}$ collisions with $3s_{1/2}$ or $3p_{3/2}$ atoms evolves according to

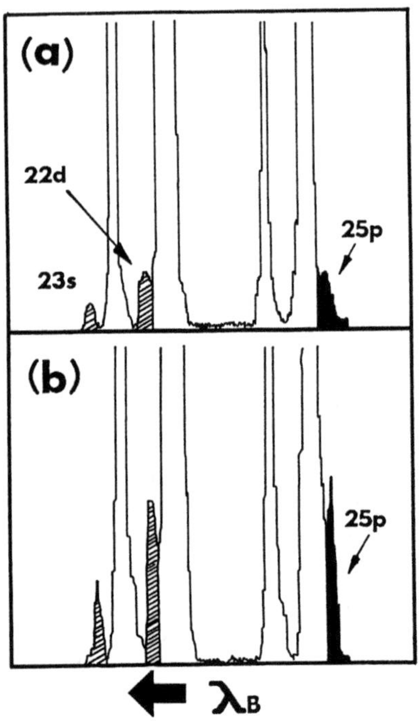


Figure 2 $Na^+$ signal as a function of blue laser wavelength, $\lambda_B$. a) $\lambda_Y$ set to the wavelength of the $D_2$ line at $N \approx 10^{14}$ cm$^{-3}$. The peaks that are off-scale are from the $3p_{3/2} \rightarrow$ ns and nd excitations. The cross-hatched peaks are $3p_{1/2} \rightarrow$ ns and nd excitations. The peak labeled 25p results from the forbidden $3p_{3/2} \rightarrow$ 25p transition. b) Same as a) except N is increased. The rate of growth of the on-scale peaks is the same showing that the 25p signal has the same dependence on N as do those due to fine-structure mixing of the 3p level, quadratic.

$$\frac{d N_{1/2}}{dt} = (k_0 N_0 + k_{3/2} N_{3/2})N_{3/2} , \qquad (1)$$

where $k_0$ and $k_{3/2}$ are the fine structure mixing rate constants for collisions with $3s_{1/2}$ and $3p_{3/2}$ atoms. Integrating and substituting for $N_0$ and $N_{3/2}$ as discussed above one obtains

$$N_{1/2}(t) = (0.22\ k_0 + 0.44\ k_{3/2})N^2t \qquad (2)$$

The rates of formation of $N_{d1}$ and $N_{d2}$, the two nd levels originating from $3p_{3/2}$ and $3p_{1/2}$ respectively, are given by

$$\frac{dN_{d1}}{dt} = \sigma I_B(t)\ N_{3/2} \qquad (3)$$

and

$$\frac{dN_{d2}}{dt} \sigma I_B(t)\ N_{1/2} \qquad (4)$$

The optical pumping cross-section $\sigma$ is the same for both processes because the fine structure components of the Rydberg nd levels are degenerate with respect to the 0.25 $cm^{-1}$ bandwidth of the blue laser. The average power of the blue laser beam varies by only ~10% over the bandwidth used to obtain Fig. 2. The time dependence of the blue laser pulse must be explicitly considered; to a good approximation these pulses can be taken as triangular with a temporal width $\tau$ and a peak intensity $I_0$. Substituting for $N_{3/2}$ and $N_{1/2}$ from (2) and integrating over the laser pulse duration, one obtains

$$\frac{N_{d1}}{N_{d2}} \simeq (4/3)[(0.22\ k_0 + 0.44\ k_{3/2})N\tau]^{-1} \qquad (5)$$

The ratio $N_{d1}/N_{d2}$ represents the total production of the two nd states during the laser pulse and therefore is the ratio of the signals from these two states. The observed value of $N_{d1}/N_{d2}$ at the highest density shown in Fig. 2 is about 200. The fine-structure mixing rate constant for $3p_{3/2}$-$3s_{1/2}$ collisions $k_0$ is about[12,13] $2 \times 10^{-9}\ cm^3\ s^{-1}$. If $k_{3/2}$ is taken to be the same as $k_0$, and a $\tau$ value of 10 ns is used, (5) can be solved to obtain $N \simeq 5 \times 10^{14}\ cm^{-3}$. This value may be an overestimate, since $k_{3/2}$ may well be greater than $k_0$, but the actual N value is unlikely to be less than $2 \times 10^{14} cm^{-3}$.

The interaction responsible for collision-induced absorption to the np levels is the same as that which produces $\ell$ mixing in state-selected Rydberg atoms. The very large radial extent of Rydberg states ($r_{max} \simeq 2n^2$ for $n >> \ell$) allows Rydberg electron-perturber interactions to occur at large internuclear separation. As a consequence, a Rydberg state with an initially well-defined $\ell$-value acquires a mixed $\ell$ character when the separation between two atoms is comparable with n less than $2n^2$ atomic units. In the experiments discussed here, the forbidden np states arise primarily from the fractional (n-1)d character acquired during the long-range collision. An estimate of the extent of this (n-1)d character can be obtained by assuming that it can be added to the final state np wavefunction only when an Na(3p)/perturber pair are within $2n^2$ atomic units of one another. This procedure leads to

$$R \simeq (4\pi/3)(2n^2)^3 N f_d \qquad (6)$$

where R is the observed ratio of forbidden np to allowed nd signals and N is the total Na density in $a.u.^{-3}$. R is about $5 \times 10^{-3}$ and N is about $5 \times 10^{14} cm^{-3}$ or $7 \times 10^{-11}\ a.u.^{-3}$ as discussed above. Therefore $f_d$ is approximately 0.01, a qualitatively reasonable value.

The long-range $\ell$-mixing collision picture is obviously not valid when the Na-Na collision system separation is small enough so that the final state is more properly considered an $Na_2$ Rydberg molecule. For $n > 20$, the $\ell$-mixing interaction extends to $R > 400$ a.u.; the fact that impact parameters less than 20 a.u. or so are not properly treated by such a model is not important in this context. If one uses a straight-line trajectory for the Na-Na system with an average relative velocity of $10^5$ cm $s^{-1}$, and averages over impact parameter from $b = 0$ to $b = 2n^2$, the average collision duration for $n = 25$ is about $2.5 \times 10^{-11}$s. This value leads to an Uncertainty Principle linewidth of about 0.2 $cm^{-1}$, consistent with the data and comparable with the instrumental resolution.

A collisional absorption effect related to that reported here has been observed[15] in Mg vapor at considerably higher number density, $\sim 10^{17}$ $cm^{-3}$. In that experiment focused 330 nm laser light was used to produce Mg ($3s3p^1p°$) atoms by two-photon excitation. This process is "two-photon forbidden" and has been interpreted as arising from excitation of a quasimolecule, Mg($3s^{21}S$) - Mg($3s^{21}S$) with large internuclear separation, a description similar to that presented here.

We have also recently learned that Compton and coworkers [16] have observed $3p \rightarrow np$ transitions in sodium using two laser excitations of Na vapor confined in a heat pipe at about 50 torr. The spectra are very similar to those of Lu et al. in that the peaks corresponding to the "forbidden" transitions are ~1/10 the height of those of the allowed atomic transitions.

Acknowledgements

This work was supported by the National Science Foundation under Grant No. PHY 8418075, by U.S. DOE, Division of Chemical Sciences, under Grant No. DE-FG02-84ER1327, and by the McDonnell Douglas Independent Research and Development program. One of us (M.C.) wishes to thank the Fondazione Angelo Della Riccia, Firenze for a fellowship. The authors would like to thank Dr. R. N. Compton for useful comments and Dr. K. T. Lu for making the data of Reference 15 available to us and for valuable discussions of the effects.

References

1. See for example:
   a. L. I. Gudzenko and S. I. Yakovlenko, Sov. Phys. JETP 39, 759 (1974).
   b. P. R. Berman and E. J. Robinson, in Photon-Assisted Collisions and Related Topics, edited by N. K. Rahman and C. Guidotti (Harwood Academic, Chur, 1982).
2. See for example:
   a. K. Duffy, M. Hutchinson, J-M. Yuan and T. F. George, J. de Phys. C146, 1 (1985).
   b. A. Gallagher and T. Holstein, Phys. Rev. A 16, 2413 (1977) and references cited therein.
3. T. Sizer and M. G. Raymer, Phys. Rev. Lett. 56, 123 (1986).
4. P. D. Kleiber, A. M. Lgyra, K. M. Sando, S. P. Heneghan and W. C. Stwalley, Phys. Rev. Lett. 54, 2003 (1985).
5. P. Pradel, P. Monchicourt, D. Dubveuil, J. Heuze, J. J. Laucagne and G. Spiess, Phys. Rev. Lett. 54, 2600 (1985).
6. P. Monchicourt, P. Pradel, F. Roussel and J. J. Laucagne, Phys. Rev. A 33, 3515 (1986).
7. See for example:
   C. E. Burkhardt, R. L. Corey, W. P. Garver, J. J. Leventhal, M. Allegrini, and L. Moi, Phys. Rev. A 34, 80 (1986).
8. C. E. Burkhardt, W. P. Garver, V. S. Kushawaha and J. J. Leventhal, Phys. Rev. A 30, 652 (1984).
9. G. Vitrant, J. M. Raimond, M. Gross and S. Haroche, J. Phys. B 15, L49 (1982).
10. E. F. Worden, J. A. Paisner and J. G. Conway, Opt. Lett. 3, 156 (1978).
11. E. U. Condon and G. H. Shortley, Theory of Atomic Spectra (Cambridge, London, 1970).

12. J. Huennekens and A. Gallagher, Phys. Rev. A 28, 238 (1983).
13. B. P. Kibble, G. Copley and L. Krause, Phys. Rev. 153, 9 (1967).
14. P. B. Hogan, J. L. Carlsten, F. M. J. Pichanick, S. J. Smith, and W. W. Smith, Phys. Lett. 37, 990 (1976).
15. a. J-Y Zhang, H-T Zhou, J. Yang, Q-R Li and K. T. Lu submitted to Tenth International Conference on Atomic Physics.
    b. K. T. Lu, private communication.
16. R. N. Compton, Oak Ridge National Laboratory, private communication.

# Energy Transfer Collisions in Laser-Excited Mixtures of Alkali Vapors

S. Abdullah*, M. Allegrini, S. Gozzini, and L. Moi

Istituto di Fisica Atomica e Molecolare del C.N.R., Via del Giardino 7, I-56100 Pisa, Italy

*ICTP Fellow, Permanent Address: Iraqi Atomic Energy Commission, Tawaitha, Baghdad, P.O. Box 765, Iraq

ABSTRACT Experiment on energy transfer in collisions between two laser excited alkali atoms of different species are reported and rate coefficients have been measured in a sodium-potassium mixture for the Na(3D) and K(7S,5D) levels. These values are of same order of magnitude as rate coefficients for sodium 3P/3P collisions.

## 1. Introduction

Since the discovery of electronic energy transfer in collisions between two excited atoms [1] numerous experiments on pure alkali vapors have been reported and some absolute rate coefficients have been measured [2]. The common method of determining these rate coefficients is to optically excite the alkali atoms to one of the first P-resonance states and detect the fluorescence emitted by the highly excited nL levels populated in the collisional transfer process. Results from different laboratories show that, as expected, in these thermal collisions only nL states close in energy (within a few kT) to the sum energy of the collision reactants are significantly populated.

To our knowledge no experiments have been performed with alkali atom mixtures, except for those performed in our laboratory [3,4] on sodium and potassium. The aim of those experiments was to observe and possibly to measure rate coefficients for the process

$$\begin{aligned} Na^{\star}(3P) + K^{\star}(4P) &\rightarrow Na^{\star\star}(nL) + K(4S) \\ &\rightarrow Na(3S) + K^{\star\star}(nL) \end{aligned} \quad (1)$$

where $Na^{\star\star}(nL)$ and $K^{\star\star}(nL)$ indicate highly excited states, lying close to the sum energy of $N^{\star}(3P) + K^{\star}(4P)$ (see fig.1). However in [3] fluorescence was observed only from potassium 5P, 6S and 4D states which have an energy close to twice $E[K^{\star}(4P)]$, therefore direct proof of reaction (1) was not possible. Despite this negative result it was clear that in a mixture of alkali atoms there should be more exit channels than for pure vapor experiments[4]. Little data are available for heteronuclear diatomic alkali molecules and very little is known about long-range interactions in these systems, therefore studies of process (1) and similar processes for all alkali pair combinations and measurements of their rate coefficients should provide knowledge about quasi-molecular potentials at internuclear distances for which other experimental methods are insensitive.

A list of levels which are expected to be populated by energy transfer collisions between two different atoms excited to the first P-resonance level has been made [4] and experiments are under way in our laboratory. Considering all possible combinations of alkali pairs there are about sixty high-lying levels in which the atoms may be excited through collisions. We have first investigated a mixture of sodium and potassium. The experimental method we used (essentially intermodulated spectroscopy) allowed the observation of reaction (1) for all expected Na**(nL) and K**(nL) levels but one. Here we briefly describe our experiment and report our measurements of the relative cross-section for levels 3D of sodium, 7S and 5D of potassium.

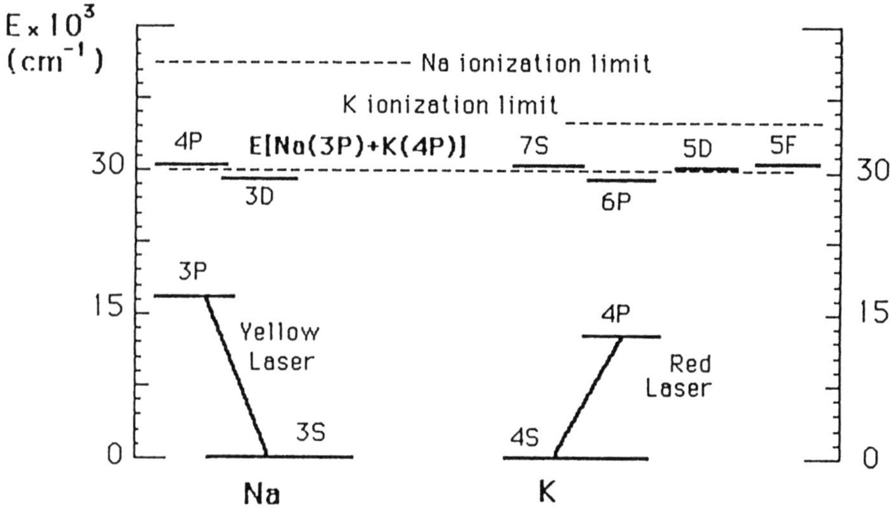


Fig. 1 Diagram of the Na and K levels of interest in the experiment.

## 2. Sodium-Potassium Experiment

Figure 1 shows the levels of interest in this experiment which is intended to observe reaction (1) by looking at the fluorescence emitted by levels 5D,5F,6P,7S of potassium and 3D, 4P of sodium.

The experiment was carried in a pyrex capillary cell of cylindrical shape (0.15cm internal diameter and 5 cm long). Preparation procedure of this kind of cell has already been described [5] . The cell was sealed under vacuum with an amalgam of approximately 90% pure sodium and 10% pure potassium and placed in a temperature-controlled oven. Confident in the validity of the empirical Raoult law, this particular amalgam has been chosen in order to make Na and K atom densities nearly equal. This assumption has been indeed verified in the experiment, as we will describe later on. Measurements have been made over a wide temperature range, however cross-sections have been evaluated for T = 240°C. Sodium and potassium atoms were excited to the first $P_{1/2}$ or $P_{3/2}$ resonance states by a Rhodamine 6G and an Oxazine 750 cw dye laser respectively. The two lasers were carefully superimposed on the entrance window of the cell; alignment in this experiment is very critical because of the small radial dimensions of the cell. Fluorescence emission was detected and analyzed in the visible and infrared regions with the usual apparatus for intermodulated spectroscopy. Although we tested the i.r. apparatus for proper operation, emission from 5F ⟶ 3D at 1.1μm or 5F ⟶ 4D at 3.1 μm was not observed, probably because of the poor sensitivity of the i.r. detection apparatus. Therefore, the cross-section for the 5F level has not been measured and in the following this aspect of the experiment will not be considered any more.

Under combined laser excitation of both sodium and potassium pure sodium 3P/3P and potassium 4P/4P collisions are present and produce very excited atoms; moreover, these pure collisions, either directly or by decay from higher states, also populate the levels produced by process (1). In order to measure cross-sections for process (1) we must discriminate it against these unwanted contributions. With this aim we introduce in our experimental apparatus (see figure 2) the intermodulation technique. The two laser beams exciting Na and K atoms respectively were modulated at two different frequencies $\omega_1$ and $\omega_2$. Fluorescence was detected at the sum frequency $\omega = \omega_1+\omega_2$. This assured that signals from the lock-in amplifier were due only to levels produced by Na(3P)/K(4P) collisions.

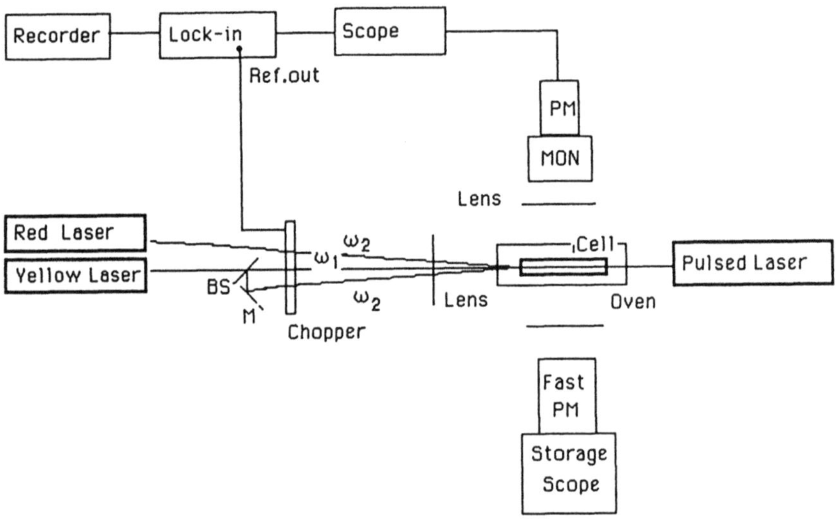


Fig. 2 Sketch of the experimental apparatus
BS beam splitter; M mirror. The lasers are shown as acting simultaneously. In fact, only one of the two $\omega_2$-modulated laser beams was present at a time.

With this apparatus we have observed fluorescence from all expected levels except 5F, namely the following transitions:

Table 1. Transitions observed from levels populated by process(1).

| Atom | Level | Transition | Wavelength (nm) |
|---|---|---|---|
| Na | 4P | 4P-3S | 330 |
| | 3D | 3D-3P | 819 |
| K | 5D | 5D-4P | 582 |
| | 7S | 7S-4P | 579 |
| | 6P | 6P-3S | 345 |

Instead of measuring absolute cross-sections of process (1) we took advantage of the fact that absolute cross-sections for energy transfer have been already measured in different laboratories [5,6] for the following process in pure sodium vapor.

$$Na^{\star}(3P) + Na^{\star}(3P) \longrightarrow Na^{\star\star}(5S,4D) + Na(3S) \qquad (2)$$

In our case we chose the 5S-3P transition at 616nm and thus our cross-sections for reaction (1) are relative to $\sigma_{5S}$ of process(2). In order to measure I(5S-3P) we replaced the red laser with a second yellow beam, still at the frequency $\omega_2$. In this way we could then measure 5S-3P fluorescence intensity with the same density of Na $\omega_1$-excited atoms: as we will see later, this will permit simplification of the formulas. It was convenient to choose the 3D level of sodium as the one for which we measure cross-section relative to the 5S state. Other level cross-sections are then simply deduced from intensity ratios relative to 3D. In the experiment we have also measured the intensity ratio between the fundamental resonances in sodium and potassium, $I(3P-3S)/I(4P-4S)$ excited by the $\omega_2$-modulated laser beams; we will see that we need this ratio for the final evaluation of cross-sections, unless absolute atom densities for both sodium and potassium were known.

A second part of the experiment involved measurement of the effective lifetime of sodium 3P and potassium 4P levels that again enter the cross-section determination. In this case pulsed dye lasers, pumped by a nitrogen laser (see figure 2), were used for excitation to first $P_{1/2}$ or $P_{2/3}$ states of sodium and potassium; fluorescence was detected by a fast photomultiplier and a storage oscilloscope. Measured lifetimes were roughly ten times natural lifetimes for both sodium and potassium, which indirectly proved that our empirical choice of the amalgam to produce comparable atom densities of sodium and potassium at a given temperature, was correct.

## 3. Data Analysis for Cross-Section Determination

Let us work first with the 3D level of sodium Upon combined excitation of sodium and potassium atoms, solution of rate equations at steady-state gives its population - $[Na_{3D}]$ as a function of $[Na_{3P}]_1$ and $[K_{4P}]_2$ where subscripts 1 and 2 indicate that 3P atoms are excited by laser modulated at frequency $\omega_1$ and 4P atoms are excited by laser modulated at frequency $\omega_2$

$$[Na_{3D}] = \frac{k_{3D}\,[Na_{3P}]_1\,[K_{4P}]_2}{A_{3D-3P}} \tag{3}$$

where $k_{3D}$ is the rate coefficient for production of 3D atoms by (1) and $A_{3D-3P}$ is the transition probability from 3D to 3P, which is the only one because the 3D level decays 100% to 3P. Dealing with thermal collisions, rate coefficients and cross-sections are related simply by the mean interatomic velocity $k_{nL} = v\sigma_{nL}$ with $v = [8kT/\pi\mu]^{\frac{1}{2}}$. Since in the experiment we measure fluorescence intensities it is convenient to express relation (3) as

$$I_{3D-3P} = K_{3D}\,\alpha_{3D-3P}\hbar\omega_{3D-3P}\,(V/4\pi)\,[Na_{3P}]_1\,[K_{4P}]_2 \tag{4}$$

where $I_{3D-3P}$ is the intensity of 3D-3P transition, $\alpha_{3D-3P}$ is a correction factor for the response of the apparatus, $\omega_{3D-3P}$ is the transition frequency and V the fluorescence volume.

We consider now the case when only sodium excitation is present at two different modulation frequencies with the second laser beam replacing geometrically the potassium laser beam of the previous case. The population of collision-produced levels $[N_{nL}]$ is then

$$[N_{nL}] = \frac{k_{nL}\,[Na_{3P}]_1\,[Na_{3P}]_2}{\Sigma_{n'L'}\,A_{nL,n'L'}} \tag{5}$$

where $[Na_{3P}]$ is modulated at frequency $\omega_1$ from one beam and at $\omega_2$ from the other beam. Writing explicitly expression (5) for level 5S, our reference level, we get

$$[N_{5S}] = \frac{k_{5S}\,[Na_{3P}]_1\,[Na_{3P}]_2}{A_{5S-4P} + A_{5S-3P}} \tag{6}$$

where $k_{5S}$ is known /5,6/. $I_{5S-3P}$ has been measured, then

$$I_{5S-3P} = k_{5S}\ \gamma_{5S-3P}\ \alpha_{5S-3P}\ \hbar\omega_{5S-3P}\ (V/4\pi)\ [Na_{3P}]_1\ [Na_{3P}]_2 \qquad (7)$$

where $\gamma_{5S-3P}$ is the branching ratio of the transition.

Assuming that fluorescence is coming from the same volume (a good approximation with the geometry of our cell) and making the ratio we get

$$\frac{I_{3D-3P}}{I_{5S-3P}} = \frac{k_{3D}}{K_{5S}}\ \frac{\alpha_{3D-3P}}{\alpha_{5S-3P}}\ \frac{1}{\gamma_{5S-3P}}\ \frac{\omega_{3D-3P}}{\omega_{5S-3P}}\ \frac{[K_{4P}]_2}{[Na_{3P}]_2} \qquad (8)$$

Densities of sodium and potassium are unknown, however we have determined their ratio by measuring the intensity of resonance fluorescence lines and the effective lifetimes of sodium 3P ($\tau_{3P}$) and potassium 4P ($\tau_{4P}$) levels as shown in (9).

$$\frac{[K_{4P}]}{[Na_{3P}]} = \frac{\alpha_{3P-3S}}{\alpha_{4P-4S}}\ \frac{\tau_{4P}}{\tau_{3P}}\ \frac{\omega_{3P-3S}}{\omega_{4P-4S}}\ \frac{I_{4P-4S}}{I_{3P-3S}} \qquad (9)$$

We have assumed that correction factors are essentially due to monochromator and photomultiplier spectral response. Making all substitutions with quantities measured at T = 240°C, where effective lifetimes were roughly ten times longer than natural lifetimes, we obtain $k_{3D} \cong 0.6 k_{5S}$. $k_{5S}$ was determined in refs.[5,6] with an uncertainty of 35%.
$k_{3D}$ as derived quantity has higher uncertainty, especially as related to measurements of effective lifetimes. A resonable estimate of overall uncertainty may rise to 50%, giving $\sigma_{3D} = (1.4 \pm 0.7) \times 10^{-15}$ cm-2.

Cross-sections for others levels can be determined in a similar way, or, more simply by using rate coefficient ratios already determined in [4] . However, for the two transitions in the u.v., 4P-3S in sodium and 6P-3S in potassium, we could not correct fluorescence intensity for glass absorption precisely enough to give their cross-sections with meaningful uncertainty. Therefore final results of the present work are $k_{7S}$, $k_{5D}$ for potassium and $k_{3D}$ for sodium; they are summarized in table 2.

Table 2. $Na^*(3P) + K^*(4P) \longrightarrow Na^{**}(nL) + K(4S)$ cross-sections.
$\longrightarrow Na(3S) + K^{**}(nL)$

$\Delta E$ is the difference between the energy of $Na^*(3P) + K^*(4P)$ and the energy of highly excited level nL.

| Atom | Level | $\Delta E(cm^{-1})$ | $\sigma(cm^{-2})$ |
|---|---|---|---|
| Na | 3D | + 768 | $(1.4 \pm 0.7) \times 10^{-15}$ |
| K | 5D | - 169 | $(1.9 \pm 1.0) \times 10^{-15}$ |
| | 7S | - 258 | $(0.9 \pm 0.5) \times 10^{-15}$ |

## 4. Conclusions

We have started a series of experiments to investigate production of highly excited states by energy transfer in collisions between two alkali atoms of different species, laser excited to the first P-resonance states. The experiment is based on an intermodulation technique, which distinguishes between the desired signals and those from collisions between atoms of the same species. First results have been obtained for a sodium-potassium mixture. Fluorescence has been observed only from expected levels (except for one which emits only in the infrared), proving once again that only levels with energy defects (positive or negative) within a few kT are significantly produced. Cross-section values have been estimated for three such levels; D-levels seems to be favourite compared with S-levels. These values are comparable to those for energy transfer in sodium/sodium collisions, suggesting the same exchange mechanism in pure as well as mixed alkali vapors. However, calculation of the corresponding molecular adiabatic potentials and nonadiabatic energy transfer at the avoided crossing would provide a check on our cross-section measurements. Experiments are now in progress in our laboratory on a potassium/rubidium mixture for which only three levels are expected to be populated by energy transfer collisions.

## REFERENCES


1. M.Allegrini, G.Alzetta, A.Kopystynska, L.Moi and G. Orriols: Opt.Commun. 19, 96 (1976)
2. Results up to 1984 are reviewed in: M.Allegrini, C.Gabbanini and L.Moi: J.Physique Colloq. 46, C1-61 (1985); more recent results for sodium are in: M.Allegrini, C.Gabbanini, L.Moi and R.Colle: Phys. Rev. A32, 2068 (1985); S.A.Davidson, J.F.Kelly and A.Gallagher: Phys. Rev. A33, 3756 (1986)
3. M.Allegrini, P.Bicchi, S.Gozzini and P.Savino: Opt. Commun. 36, 449 (1981)
4. A.Cremoncini, S.A.Abdullah, M.Allegrini, S.Gozzini and L.Moi: Proc. Intern. Conf. LASERS'85, in press.
5. M.Allegrini, P.Bicchi and L.Moi : Phys. Rev. A28, 1338 (1983)
6. J.Huennekes and A.Gallagher: Phys. Rev. A27, 771 (1983)

# The Far Wing Behaviour of the LICET Lineshape

*A. Bambini*[1] *and P.R. Berman*[2]

[1]Istituto di Elettronica Quantistica, Consiglio Nazionale delle Ricerche, Via Panciatichi 56/30, I-50127 Firenze, Italy

[2]New York University, 4 Washington Place, New York, NY 10003, USA

Abstract

In this paper we analyse the behaviour of the quasistatic wing of the LICET lineshape. It is found that, under several experimental conditions, the simple two-level model is inadequate to describe the fall off of the spectral wing at large detunings between the LICET transition and the laser frequency. The failure comes from the fact that other levels, lying nearby the initial or the final level of the transition, do not act as true virtual states in the process and affect the lineshape in its quasistatic region.

As a result, the wing of the spectrum does not fall off to zero, as the detuning increases, with a constant slope (0.5 for van der Waals collisional interaction), but exhibits a more complicated dependence on the detuning, with a local slope ranging between 0.5 and 2, depending on the energy level configuration of the colliding partners.

Light-Induced Collisional Energy Transfer (LICET) has been studied extensively in the past decade, following its theoretical prediction /1/ and its experimental observation /2/. References to recent work in this field can be found in /3/.

The process can be described as a second order process, in which excitation energy is transferred from one atom to another during a collision, with the simultaneous absorption of a photon from a laser field, whose frequency $\Omega$ is chosen to match the frequency difference between the initial and the final states:

$$A^{*} + B + \hbar\Omega \rightarrow A + B^{*} \qquad (1)$$

The energy level scheme shown in Fig.1 is one possible scheme for the transition. In Fig.1a, the energy level scheme is shown for the two atoms undergoing the transition, while in Fig.1b the corresponding levels are represented in the separated atom basis. The LICET process transfers energy between level 2 and level 3 in Fig.1b, and it occurs only if the laser field is present during the collisional interaction. The level labelled by 1 act as a virtual level, close to the state where the system is prepared at the initial time. It enhances the transition cross-section but it is not populated by the collisional interaction.

The LICET excitation spectrum of the final level 3, which is monotored at various laser wavelengths, can be roughly divided into two regions: a central region (core), characterised by a peak in the cross-section, which occurs when the laser frequency is tuned at the transition frequency between levels 2 and 3, and a wing on one side of the core, characterised by a long tail which extends far beyond the Doppler width of the atomic transitions. On the other side, the excitation spectrum falls off very rapidly.

In the wing, contributions arise from collisions at short impact parameters, where the energy levels are strongly affected by the collisional interaction. In this region the quasistatic approximation is expected to be valid: the transition occurs at the interatomic separation where the instantaneous frequency difference between the two levels, distorted by the collisional interaction, is matched by the laser frequency.

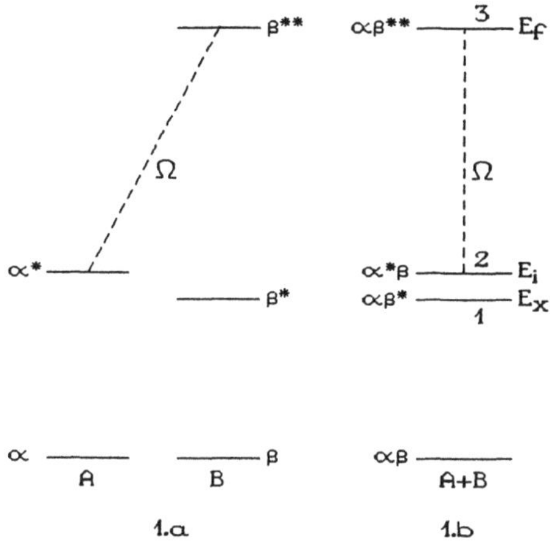


Figure 1 - Energy level scheme for a LICET reaction. On the left side the separate atoms' energy levels are shown. The LICET transition occurs from state $\alpha^*$ of atom A to state $\beta^{**}$ of atom B. The laser frequency $\Omega$ is tuned at the frequency difference between these two states. On the right side, the compound states are shown. The relevant levels of the transition are labelled by $E_x$, $E_i$ and $E_f$.

A simple calculation performed within this scheme of approximation shows that the quasistatic wing should fall off, at large frequency detunings, with the power law $|\Delta|^{-0.5}$, where $\Delta$ is the frequency detuning of the laser field from the LICET transition. However, experiments carried out for several colliding partners /4,5/ showed that the quasistatic wing was falling off with a larger slope. In the experiment performed on Europium and Strontium atoms /4/ the slope in the tail was estimated to be 0.85, while in the experiment on Sodium and Calcium atoms /5/ the estimated slope was found to be 0.80.

It isthe purpose of this paper to explain this discrepancy. To do so, we refer to the energy level scheme of Fig.1. The levels of atom A are indicated as $\alpha$ (the ground state) and $\alpha^*$ (the excited state where atoms are prepared at the initial time). The levels of atom B are indicated as $\beta$ (the ground state), $\beta^*$ (the intermediate state, which acts as a virtual state), and $\beta^{**}$ (the final level of the LICET transition). We disregard magnetic degeneracies in the atomic level configuration.

Out of the six compound states which can be formed from these atomic states, we choose only the three states which are relevant to the transition, namely state 1,(formed by $\alpha\beta^*$), state 2 (formed by $\alpha^*\beta$) and state 3 (formed by $\alpha\beta^{**}$).

Dipole-dipole interaction connects states 1 and 2 whereas atom-field interaction connects state 1 to state 3.

We write down the Hamiltonian for the atomic system, treating the laser field as a classical, unperturbed field:

$$H = H_\alpha + H_\beta + V + W_\alpha + W_\beta \tag{2}$$

Here, $H_\alpha$ and $H_\beta$ represent the unperturbed energies of the two atoms, V their collisional interaction, which we will assume for simplicity as a dipole-dipole interaction. The next two terms represent the interaction of the laser field with atom A and atom B respectively.

The collisional operator is written as

$$V = \frac{1}{R_{AB}^3} \left[ \hat{\vec{P}}_A \cdot \hat{\vec{P}}_B - 3(\hat{\vec{P}}_A \cdot \vec{\varepsilon})(\hat{\vec{P}}_B \cdot \vec{\varepsilon}) \right] \tag{3}$$

where $\hat{\vec{P}}_A$ and $\hat{\vec{P}}_B$ represent the dipole moment operators of atom A and atom B respectively, and $\vec{\varepsilon}$ is the unit vector along the interatomic axis.

Since we disregard the magnetic sublevels of the system, we adopt a simplified version of the dipole-dipole interaction, which does not depend on the relative orientations of the two atoms, but only on their separation $R_{AB}$

$$V = \hat{P}_{\alpha}\hat{P}_{\beta}/R_{AB}^{3} \tag{4}$$

where $\hat{P}_{\alpha}$ and $\hat{P}_{\beta}$ are effective dipole moments of atom A and atom B. We make the assumption of classical, straight line collisional trajectories. Thus, the interatomic separation $R_{AB}$ takes the simple form

$$R_{AB} = (b^2 + v^2t^2)^{\frac{1}{2}} \tag{5}$$

where b is the impact parameter of the collision and v is the relative speed of the two atoms.

Furthermore, we will assume that the laser field amplitude is constant during each collision. This is a good approximation, since the collision, under usual experimental conditions, takes place in a few picoseconds, whereas the electric field component of the laser field varies with a time scale typical of the light pulse duration used in these experiments, of the order of a few nanoseconds. The Rotating Wave Approximation (RWA) is also assumed in the derivation of our results.

Renaming states 1,2 and 3 as x (intermediate), i (initial) and f (final) respectively, we can now write down the Schrödinger equation for the three levels under the approximations discussed above ($\hbar = 1$)

$$i\dot{c}_i = V_{ix}e^{-i(\omega_x - \omega_i)t}c_x \tag{6a}$$

$$i\dot{c}_x = V_{xi}e^{-i(\omega_i - \omega_x)t}c_i - \chi e^{-i(\omega_f - \omega_x - \Omega)t}c_f \tag{6b}$$

$$i\dot{c}_f = -\chi e^{-i(\omega_x - \omega_f + \Omega)t}c_x \tag{6c}$$

where

$$V_{ix} = \frac{1}{R_{AB}^3} \langle i\,|\hat{P}_{\alpha}\hat{P}_{\beta}|\,x\rangle \qquad (V_{ix} = V_{xi}\ ,\ \text{real})$$

and

$$\chi = \langle x\,|P_{\beta}|\,f\rangle E_0/2 \qquad (E_0 \text{ is the laser field amplitude})$$

The usual approximation which is done at this point is to consider the level x as a virtual state. There are two conditions to be satisfied for level x to be a true virtual state, namely:

1) The collisional interaction V must be small enough in order not to mix state i and state x. This in turn requires that the matrix element of V taken between those two states must be small when compared to their energy difference.

2) The rate of variation of V must be small enough to prevent real population transfer from state i to state x. This would occur if there is a non vanishing component at $(E_i - E_x)/\hbar$ in the Fourier transform of the matrix element of V.

If these two conditions are matched, then the probability amplitude $c_x$ can be eliminated from our system of equations, to find

$$i\dot{c}_i = \frac{V_{ix}^2}{\omega_i - \omega_x} c_i - \frac{\chi V_{ix}}{\omega_f - \omega_x - \Omega} e^{-i(\omega_f - \omega_i - \Omega)t} c_f \tag{7}$$

$$i\dot{c}_f = \frac{\chi^2}{\omega_f - \omega_x - \Omega} c_f - \frac{V_{xi}\chi}{\omega_i - \omega_x} e^{-i(\omega_i - \omega_f + \Omega)t} c_i \tag{8}$$

The first term in eq.(7) is a collisional (van der Waals in our case) shift, while the first term in eq.(8) is a Stark (or light) shift. They are both induced by the interaction with the virtual level. The light shift in eq.(8), in our approximation, is constant during each collision (although it may vary from collision to collision due to the variation of the pulse amplitude), and can therefore be accounted for (if significant) by redefining the energy of the final level f.

Other virtual levels, not included in our original equations, may affect the energy of the initial or the final level. We assume that the shift of the final level is negligible, and that the initial level undergoes a van der Waals shift $S_i(t)$ when the collision takes place.

The transition term which appears in both equations (7) and (8) is a product of a field term and a collisional term, and is the effective matrix element which induces the LICET transition.

To evaluate the cross-section of the process, we need to integrate the system of differential equations for given b and v, starting from the initial conditions $c_i = 1$ and $c_f = 0$, and then average the square modulus of $c_f(+\infty)$ over the distribution of b and v. We evaluate the solution at first order perturbation theory, which amounts to neglecting the flow of population from level f back into level i, induced by the last term in eq.(7). We find

$$ic_f = -\int_{-\infty}^{t} T(t') \exp\left[-i\Delta t' + i\int_{-\infty}^{t'} S_i(t'')dt''\right] dt' \tag{9}$$

where the time dependence of $S_i$ comes from the time dependence of $R_{AB}$, eq.(5). In eq. (9) we have indicated by T the transition element

$$T(t) = \frac{V_{xi}(t)\,\chi}{\omega_i - \omega_x}$$

To evaluate the square modulus of $c_f(\infty)$ in the wing of the spectrum, we can use the method of the stationary phase, to obtain

$$|c_f(\infty)|^2 = \frac{\pi}{3v} |T(t_\pm)|^2 \frac{R_s^7}{|Q|(1 - b^2/R_s^2)^{\frac{1}{2}}} \left| e^{i(\frac{\pi}{4} + \phi_+)} + e^{-i(\frac{\pi}{4} - \phi_-)} \right|^2 \tag{10}$$

In eq.(10), $R_s$ is the interatomic separation at which the phase is stationary, and is given by

$$R_s^6 = -\frac{Q}{\Delta} \tag{11}$$

where Q is the van der Waals coefficient in the collisional shift $S_i$

$$Q = \frac{|\langle i|\hat{P}_\alpha \hat{P}_\beta|\,\chi\rangle|^2}{\omega_i - \omega_x}$$

and $\phi_-$ , $\phi_+$ are the values of the phase in the integrand of (9) evaluated at the two points $t_-$ and $t_+$ where it is stationary. The sign of the collisional shift $S_i(t)$ determines the side of the spectrum on which the phase can take stationary values. For the scheme shown in Fig.1, $S_i(t)$ is negative and the excitation spectrum extends towards the "red" side ($\Delta < 0$). On the other side of the spectrum, where there is no point of stationary phase, the excitation spectrum falls off exponentially.

If we neglect the oscillations that come from the interferential term in (10) we can easily average Eq.(10) over the distribution of the impact parameter b to find

$$<|c_f(\infty)|^2>_b = \text{const.}\ \frac{\chi^2 R_s^3}{v|\omega_i - \omega_x|} \tag{12}$$

Substituting the value given by (11) for $R_s$ in (12), we arrive at the following expression for the cross-section in the quasistatic wing

$$\sigma(|\Delta|) = \text{const.}\ \frac{\chi^2 |V_0|}{v|\omega_i - \omega_x|^{3/2}} \cdot \frac{1}{|\Delta|^{\frac{1}{2}}} \tag{13}$$

where $V_0$ is the matrix element of the atomic dipoles

$$V_0 = <i|\hat{P}_\alpha \hat{P}_\beta|x>$$

Equation (13) shows that the wing in the LICET spectrum falls off to zero as $|\Delta|^{-\frac{1}{2}}$ (for van der Waals collisional interaction). As discussed above, in several experiments this behavior has not been confirmed. The reason for such discrepancy lies in the fact that the intermediate elvel, level x in our notations, does not behave like a true virtual level. When the interaction energy between the initial and the intermediate states becomes comparable with their energy separation, the first condition stated above (see discussion before equation (7) does not hold true: the two states are mixed by their interaction.

In the Europium-Strontium experiment, for instance, the collisional interaction energy between the initial and the intermediate levels was shown in Ref./6/ to be comparable with their energy separation. This can also be shown from our derivation. Indeed, if we evaluate the collisional interaction energy at the points of stationary phase

$$V_{ix}(R_s) = \frac{V_0}{R_s^3} = \frac{\sqrt{|Q.(\omega_i - \omega_x)|}}{R_s^3} \tag{14}$$

and substitute the expression for $R_s$ as given by (11), we see that it becomes comparable with the energy separation between the two levels of the detuning $\Delta$ is of the same order of magnitude than $\omega_i - \omega_x$ :

$$V_{ix}(R_s) = \sqrt{|\Delta.(\omega_i - \omega_x)|} \tag{15}$$

In the Europium-Strontium case, the LICET excitation spectrum was measured up to 50 inverse cm, while the frequency separation between the initial and the intermediate levels is 63 inverse cm (transition $^8P_{9/2}$ - $^1D_2$). In this range of measurement, one should not expect that the intermediate level acts like a virtual level.

To calculate, at least approximately, the cross-section for the process in these new conditions, we need to abandon the virtual level approximation used so far, and to include the effects brought into the process by the intermediate level. To this end, we consider again the system of equations (6a)-(6c). In first order perturbation theory, the last term in eq.(6b) can be neglected, and the first two equations are decoupled from the third. These equations can be solved in an approximate way since the second condition stated above can still be considered valid. Namely, we assume that the characteristic time scale b/v of variation of the collisional interaction energy is much longer than the period associated with the frequency difference $\omega_i - \omega_x$. Thus, an instantaneous diagonalisation of these two equations can be done.

We find that the amplitudes $c_i$ and $c_x$ can be combined in a linear superposition

$$\vec{b} \equiv \begin{matrix} b_1 \\ b_2 \end{matrix} = \underset{\sim}{T} \cdot \begin{pmatrix} c_x e^{-i(\omega_x - \omega_i)t} \\ c_i \end{pmatrix} \; ; \quad \underset{\sim}{T} = \begin{pmatrix} \cos\theta & -\sin\theta \\ \sin\theta & \cos\theta \end{pmatrix} \tag{16}$$

(the factor $\exp[-i(\omega_x - \omega_i)t]$ has been introduced here for convenience) which evolves with time as

$$\dot{\vec{b}} = \underset{\sim}{D}\vec{b} \quad ; \quad \underset{\sim}{D} = \begin{pmatrix} \lambda_1 & 0 \\ 0 & \lambda_2 \end{pmatrix} \tag{17}$$

where ($V = V_{ix} = V_{xi}$):

$$\cos\theta = \frac{1}{\sqrt{2}}\left(1 + \frac{\delta}{(\delta^2 + V^2)^{\frac{1}{2}}}\right)^{\frac{1}{2}} \qquad \sin\theta = -\frac{1}{\sqrt{2}}\left(1 - \frac{\delta}{(\delta^2 + V^2)^{\frac{1}{2}}}\right)^{\frac{1}{2}} \tag{18a}$$

$$\delta = \frac{\omega_i - \omega_x}{2} \qquad \lambda_{1,2} = -\delta \pm \sqrt{\delta^2 + V^2} \tag{18b}$$

These equations can be easily solved. Inserting $c_x$ found from these equations into eq.(6c) and integrating, we find

$$|c_f(\infty)|^2 = \left|\int_{-\infty}^{+\infty} \chi \cdot \sin\theta \exp\left[-i\Delta t + i\int_{-\infty}^{t}\{\delta - \sqrt{\delta^2 + V^2}\}dt'\right]dt\right|^2 \tag{19}$$

In the quasistatic wing, this integral can be evaluated again by using the method of the stationary phase, as was done above for the two level case. Now we find that the (two) points at which stationary phase occurs are given by

$$R_s^6 = -\frac{V_0^2}{\Delta\cdot(\omega_i - \omega_x + |\Delta|)} \tag{20}$$

so that we find

$$|c_f(\infty)|^2 = \text{const.}\ \frac{\chi^2|\Delta|\ R_s^7}{vV_0^2(1 - \frac{b^2}{R_s^2})^{\frac{1}{2}}} \cdot \left|\exp\left(\frac{i\pi}{4} + i\phi_+\right) + \exp\left(-\frac{i\pi}{4} + i\phi_-\right)\right|^2 \tag{21}$$

Again, if we ignore the oscillations which come from the interferential term in (21), we obtain for the cross section in the quasistatic wing

$$\sigma(|\Delta|) = \text{const.}\ \frac{\chi^2|V_0|}{(\omega_i - \omega_x + |\Delta|)^{3/2}|\Delta|^{\frac{1}{2}}} \tag{22}$$

If we compare this result with eq.(13), we see that the two expressions are similar, but now a new dependence on the detuning has appeared, namely the term $(\omega_i - \omega_x)$ in eq.(13) has changed into $(\omega_i - \omega_x + |\Delta|)$ in eq.(22). If the detuning is not negligible with respect to the frequency difference between the initial and the intermediate states, the lineshape changes appreciably on its wing and the cross-section goes to zero with a different behaviour.

It is to be noted that now one cannot define an overall slope of the cross-section in the wing. In different regions of the spectrum the local slope may vary continuously from 0.5 to 2, where the last value can be found only for those cases in which one can measure the lineshape at detunings from resonance very large compared with $\omega_i - \omega_x$. We note also that, at very large detunings from resonance, the main contributions to the lineshape come from collisions which occur at very small impact parameters. In these conditions, v/b may become comparable with $\omega_i - \omega_x$, and the approximation used to derive eq.(22) may be not valid.

In the experiment on Europium and Strontium the measurement was made in the range 5-50 inverse cm. We have compared the cross-section as evaluated from (22) for the Eu-Sr case with the curve $\sigma(|\Delta|) \simeq |\Delta|^{-0.85}$, which was estimated by the authors with a best fit procedure. In Fig.2 we report the log-log plot of the cross-section (22) and the straight line with the constant slope of 0.85. The two curves are very similar, showing that there is a good agreement between the experimental data reported in /4/ and our results.

Detailed calculations and discussion of several cases of interest will be reported elsewhere.

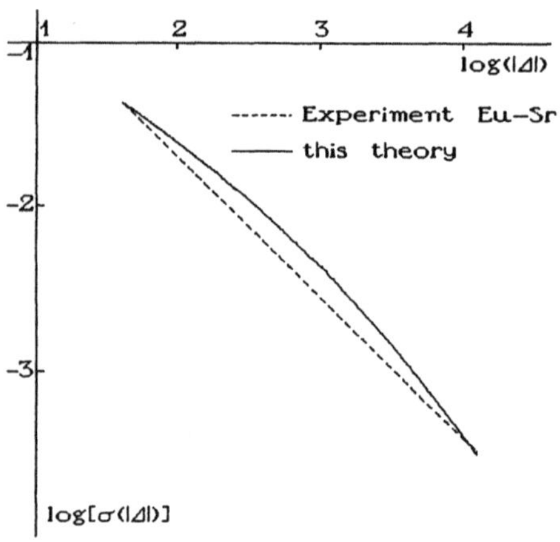


Figure 2 - The logarithmic plot of the cross-section of the LICET spectrum in the quasistatic wing, for the Europium-Strontium experiment. The full line represents the plot from eq.(22), with $\omega$= 63 cm$^{-1}$ (transition $^8P_{9/2}$-$^1D_2$). The broken line represents the plot of the curve $\sigma = |\Delta|^{-p}$, with p = 0.85. This value of the slope was estimated in ref./4/ to be the best fit to the experimental data. Both plots are shown in the range from 5 to 60 cm$^{-1}$. As can be seen from the graph, the plot of eq.(22) in this range displays an everage slope very close to the estimated one.

## References


1. L.I.Gudzenko, S.I.Yakovlenko: Sov.Phys.JETP 39, 759 (1974)
2. S.E.Harris, R.W.Falcone, W.R.Green, D.B.Lidow, J.C.White, J.F.Young: In Tunable Lasers and Applications, ed.by A.Mooradian, T.Jaeger and P.Stokseth (Springer Verlag, New York 1976)
3. S.I.Yakovlenko: Sov.J.of Quant.Elec. 8, 151 (1978) ; P.R.Berman, E.J.Robinson: In Photon-Assisted Collisions and Related Topics, ed.by N.K.Rahman and C.Guidotti, (Harwood Academic Publisher, Chur, Switzerland 1982)
4. C.Brechignac, Ph.Cahuzac, P.E.Toschek: Phys.Rev. A21, 1969 (1980)
5. C.Brechignac, Ph.Cahuzac, A.Debarre: J.de Physique 46, C1-107 (1985)
6. K.Niemax: Phys.Rev.Lett. 55, 56 (1985)

# High-Resolution Lineshape Study of a LICET Process in the Far-Wing

*M. Matera*[1], *M. Mazzoni*[1], *R. Buffa*[2], *S. Cavalieri*[2,*], *and E. Arimondo*[3]

[1]Istituto di Elettronica Quantistica del CNR, I-50127 Florence, Italy
[2]Dipartimento di Fisica, Università di Firenze, I-50125 Florence, Italy
[3]Dipartimento di Fisica, Università di Pisa, I-56100 Pisa, Italy

An experimental study of laser-induced collisional energy transfer between europium and strontium atoms is reported. The analysis of the lineshape shows evidence that the cross-section does not follow a simple power law in the quasistatic wing. The experimental results are in agreement with a new theoretical model predicting that this laser-induced excitation cannot be characterised by a universal profile.

## 1. Introduction

Since the pioneering work of YAKOVLENKO et al./1/ the atomic transitions induced by the simultaneous action of a laser field and a collision have been the subject of a large number of experimental and theoretical investigations. The processes are described as a reaction of the general form:

$$A_i + B_i + \hbar\omega \rightarrow A_f + B_f \qquad (1)$$

where $A_{i,f}$ and $B_{i,f}$ are the internal states of the atoms A and B undergoing a collision and $\omega$ is the frequency of an applied laser field. In the Laser-Induced Collisional Energy Transfer (LICET), also known as Radiatively-Assisted Inelastic Collision (RAIC), the atoms prepared in initial states $A_i$ and $B_i$, as a consequence of the combined atom-atom and atom-field interactions, emerge in new final states $A_f$ and $B_f$. Since the first experimental demonstration by HARRIS and coworkers in a strontium-calcium mixture /2-3/, the laser-induced collisions have been investigated in various binary atomic systems, as Eu-Sr /4-6/, Rb-K /7/, Na-Ca /8/, Li-Sr /9/. Determination of quasi-molecular transient states of the colliding atoms, excitation of high-lying states unaccessible by direct photon absorption and switching of chemical reactions represent the main results obtainable from LICET experiments.

Laser-induced inelastic collisions are mainly studied through the spectral profiles obtained by varying the laser frequency around the frequency $\omega_0$ corresponding to the interatomic transition for infinitely separated atoms. The strongly asymmetric shape of the LICET spectrum, as a function of the detuning $\Delta\omega = \omega-\omega_0$, exhibits an extended wing, where the process can be described as a transition between quasi-molecular adiabatic states (quasistatic approximation /10/). From the quasistatic approximation as well as from a two-level atomic model, in the case of dipole-dipole interaction potentials, an asymptotic far-wing law $|\Delta\omega|^{-1/2}$ has been predicted for the cross section of the LICET process. High-resolution studies of the LICET spectrum,performed by BRECHIGNAC et al. on the Eu-Sr system /5/ and by DEBARRE on the Na-Ca system /8/, have shown that the detuning dependence of the cross-section in the far-wing can be fitted by a power law $|\Delta\omega|^{\alpha}$ with $\alpha = -0.85$ and $\alpha = -0.8$ respectively. This discrepan-

* Present address: Imperial College of Science and Technology, The Blackett Laboratory, Prince Consort Road, London SW7 2BZ

cy with the theoretical predictions was ascribed to the influence of short-range interactions, including a breakdown of the linear trajectory hypothesis, or to a role played by near-resonant atomic levels.

Very recently a new analysis of laser-induced collisions for the Eu-Sr system performed by BAMBINI and BERMAN /11/ has shown that the far-wing profile is strongly affected by near-resonant europium and strontium levels, providing for the cross-section the asymptotic law:

$$\sigma \propto |\Delta\omega|^{-1/2} (\Delta + |\Delta\omega|)^{-3/2} \qquad (2)$$

where $\Delta = 63\ cm^{-1}$ represents the Eu-Sr interatomic energy defect in the transfer process. This law reduces to the -1/2 power law when $|\Delta\omega| \ll \Delta$ and approaches the -2 power law when $|\Delta\omega| \gg \Delta$.

In this work a new high-resolution measurement of the far-wing LICET profile for the Eu-Sr system is reported. The lineshape has been measured with a signal-to-noise ratio allowing a quantitative comparison with the theoretical predictions in a detuning range $\approx 50\ cm^{-1}$. As main result of the present experiment, the far-wing profile has been found to deviate from the power law $|\Delta\omega|^{\alpha}$ showing evidence of a detuning dependence of the slope $\alpha$. Furthermore, in the explored detuning range, the experimental data are in agreement with the Bambini and Berman law (2).

## 2. Experimental

The Eu and Sr energy levels involved in the collisional reaction are schematically represented in Fig.1. Once prepared the initial state by a laser pulse of intensity $I_p$ (pump laser), resonant with the transition:

$$Eu(6s^2\ ^8S_{7/2}) \rightarrow Eu(6s6p\ ^8P_{9/2}) \qquad (3)$$

at $\lambda = 459.4$ nm, the reaction is induced by a second laser pulse of intensity $I_t$ (transfer laser), tuned to near the interatomic transition:

$$Eu(6s6p\ ^8P_{9/2}) \rightarrow Sr(5p^2\ ^1D_2) \qquad (4)$$

resonant at $\lambda = 657.7$ nm for infinite interatomic separation. Introducing the LICET

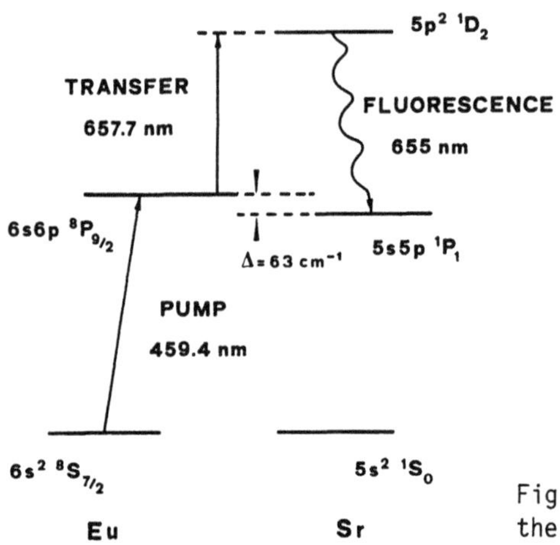


Fig.1 - Energy level diagram for the Eu-Sr laser-induced process

cross-section $\sigma(\Delta\omega, I_t)$, the density of events induced by the transfer laser may be expressed as /3/:

$$N_{ev} = \sigma(\Delta\omega, I_t)\ \overline{V}\ N(Eu^*)\ N(Sr)\ \tau \qquad (5)$$

where $\overline{V}$ is the mean relative speed of the colliding atoms, $\tau$ is the pulse width and N(Sr) is the ground-state number density of strontium. At the atomic densities of the experiment, the Eu(6s6p $^8P_{9/2}$) population $N(Eu^*)$, produced by the pump laser, is strongly affected by radiation trapping. Since at power levels of MW/cm$^2$ the europium transition is saturated, a weak dependence of $N_{ev}$ on the intensity of the pump laser results. In the weak-field regime we have for the LICET cross-section:

$$\sigma(\Delta\omega, I_t) = \sigma_0(\Delta\omega)\ I_t \qquad (6)$$

where $\sigma_0(\Delta\omega)$ is the lineshape function, which at the line center has been evaluated to be ≈500 Å$^2$/(MW/cm$^2$)/5/. The LICET events are monitored by observing the fluorescence from the Sr(5p$^2$ $^1D_2$) level at $\lambda$ = 655 nm.

A schematic of the experimental apparatus is shown in Fig.2. The laser pulses were provided by two narrowband, tunable dye lasers pumped by the same nitrogen laser. The dye lasers, described in /12/, provided 10 ÷ 50 μJ, 2 nsec pulses with a bandwidth ≈0.1 cm$^{-1}$. Using a variable optical delay line, the transfer laser pulse was delayed by T ≈ 16 nsec with respect to the pump laser. This arrangement, also used by previous authors /5/, was necessary to reduce the direct two-photon excitation of the final strontium state. The laser beams, mixed through a dichroic mirror, were focused into a heat-pipe oven up to a $10^{-3}$ cm$^2$ area with a transfer laser power density in the range 5 ÷ 10 MW/cm$^2$. The oven heated to 800 ÷ 850 °C provided Eu number densities near $10^{16}$ atoms/cm$^2$ and Sr number densities near $10^{17}$ atoms/cm$^2$. To protect the windows the cell also contained a buffer gas (argon) at a pressure of about 10 Torr. The backward fluorescence from the Sr(5p$^2$ $^1D_2$) level at $\lambda$ = 655 nm was collected and focused onto the entrance slit of a 0.85 m double monochromator. The frequency-selected fluorescence signal, detected by a photomultiplier, was measured by a gated integrator

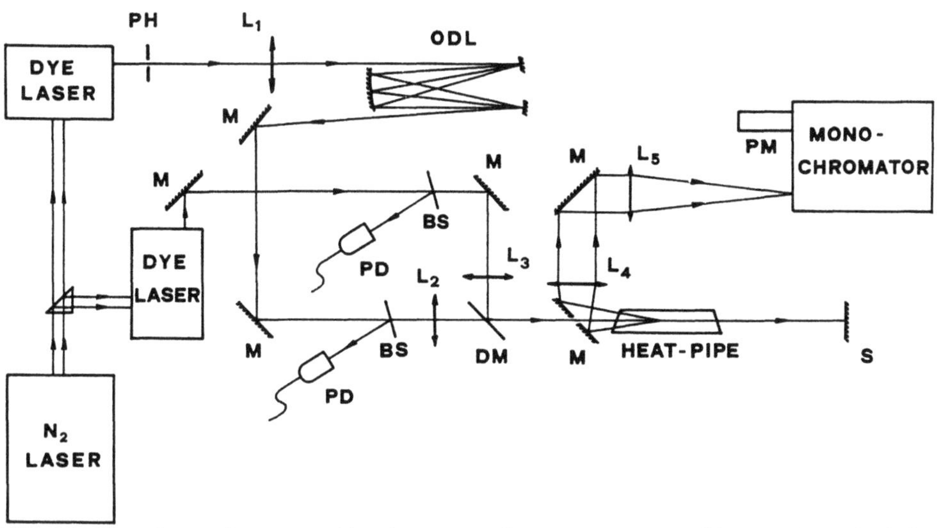


Fig.2 - Experimental set-up. M: mirror; L: lens; BS: beam splitter; PH: pin-hole; P: prism; ODL: optical delay line; DM: dichroic mirror; PD: photodetector; PM: photomultiplier; S: screen.

(gate width ≈ 0.5 μsec) interfaced to a microcomputer through a 10 bits bipolar analog to digital converter. Subtraction of baseline drift and gate feedthrough, monitor of the pulse-to-pulse laser energy and overall system control were performed by the computer.

A careful control of systematic error sources was required in order to make an accurate measurement of the far-wing lineshape. Particular care was taken to get a good long-term stability of the emission of the dye lasers. By using a 0.5 gr/lt solution of DCM dye in dimethylsulfoxide, the wavelength of the energy of the transfer laser in the scanning range was greatly reduced. The fine tuning system, with a resolution better than 0.1 $cm^{-1}$, allowed the determination of the detuning from line center with an accuracy of ±0.3 $cm^{-1}$.

## 3. Results

A typical result of the LICET lineshape measurement performed at T = 850 °C is shown in Fig.3. The line-core shape was found to be strongly dependent on the intensity of the transfer laser, proving that the stimulated emission on the $Sr(5p^2\ ^1D_2) \rightarrow$ $\rightarrow Sr(5s5p\ ^1P_1)$ transition played an important role in the final state decay. However, the comparison between the lineshapes measured at different laser intensities showed that, at the atomic densities of our experiment, stimulated emission effects were negligible for detunings larger than 5 $cm^{-1}$.

The fluorescence signal measured on the antistatic (blue) side of the resonance revealed the presence of a background process, dependent on the pump laser alone. This background signal was ascribed to a pump laser excitation of high-lying europium levels followed by collisional transfer to strontium, populating the $Sr(5p^2\ ^1D_2)$ state.

As our aim was the study of the LICET process in the quasistatic wing, in the same operating conditions of Fig.3, a measurement of the lineshape in the detuning interval 6.3 ÷ 53.5 $cm^{-1}$ was performed, with an optimised detection system. The spectrum was measured by averaging the signal over 16 laser pulses for each value of laser frequen-

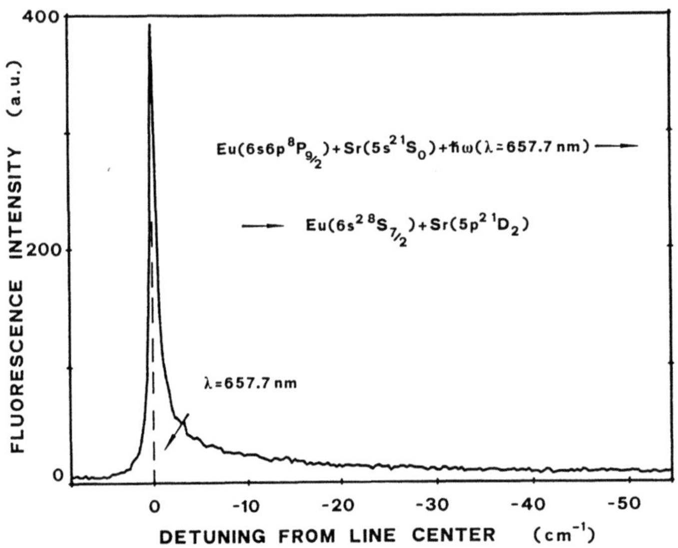


Fig.3 - Experimental lineshape of the Eu-Sr LICET process versus detuning $\Delta\omega = \omega - \omega_0$ of the transfer laser. The population of the final $Sr(5p^2\ ^1D_2)$ state is monitored by measuring the fluorescence emitted at $\lambda$ = 655 nm.

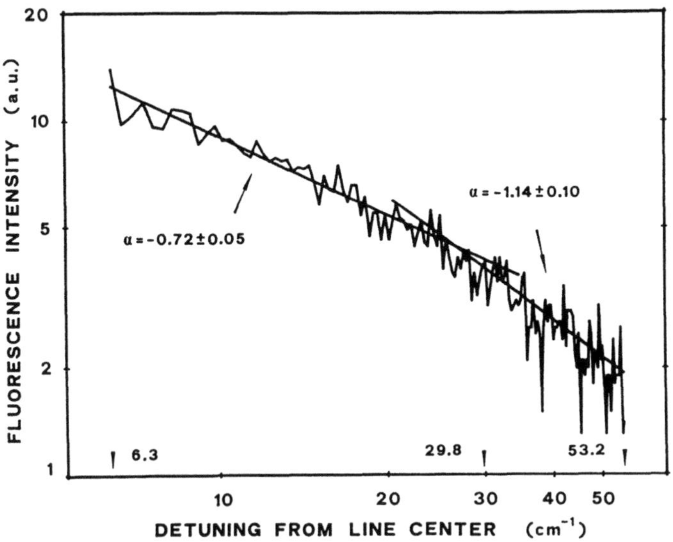


Fig.4 - Power law dependence of the Eu-Sr LICET profile in the far-wing. The straight lines of slope $\alpha = -0.72$ and $\alpha = -1.14$ reproduce the best fit to the experimental points in the detuning intervals 6.3 ÷ 29.8 $cm^{-1}$ and 29.8 ÷ 53.2 $cm^{-1}$ respectively.

cy. In order to compare our results with the predictions of the quasistatic theory, the recorded data in the detuning interval 6.3 ÷ 53.2 $cm^{-1}$ were fitted by a power law $|\Delta\omega|^{\alpha}$ . Looking for the best straight line fitting the log/log plot of the measured profile, a slope $\alpha = -0.79 \pm 0.05$ was found. This value, in agreement with the results of the previous high-resolution investigations, confirmed the discrepancy with the quasistatic law $|\Delta\omega|^{-1/2}$ predicted for a dipole-dipole interaction. An analysis of the far-wing profile was also performed by dividing the overall data into two groups, corresponding to the detuning intervals 6.3 ÷ 29.8 $cm^{-1}$ and 29.8 ÷ 53.2 $cm^{-1}$, and fitting each group with a different power law. A very good fit to the data in the small and large detuning intervals was obtained for $\alpha = -0.72 \pm 0.05$ and $\alpha = -1.14 \pm$ $\pm 0.10$ respectively, as shown by the straight lines in Fig.4. The uncertainties in the $\alpha$ values arise from a combination of the statistical errors with the systematic errors due to the inaccuracy of the detuning determination and to the background subtraction. The likelihood of our fit, checked by the $\chi^2$ test, proved that the measured profile can be more accurately described by a variable slope law.

A comparison with the Bambini and Berman theory was done by determining the average slope values predicted by (2) in the detuning intervals scanned in the experiment. The theoretical average slopes $\alpha = -0.76$ and $\alpha = -1.08$, for the detuning intervals 6.3 ÷ 29.8 $cm^{-1}$ and 29.8 ÷ 53.2 $cm^{-1}$ respectively, are in agreement with the measured average slopes.

## 4. Conclusion

The results reported here provide an experimental evidence that the LICET cross-section does not follow a simple power law in the quasistatic wing, since a significant slope change has been determined in the detuning interval 6.3 ÷ 53.2 $cm^{-1}$. Moreover, the experimental results are in agreement with the predictions of a new theoretical model developed by Bambini and Berman, taking into account relevant intermediate states of the quasi-molecular system. The present experiment demonstrates the feasibility of accurate measurements of a LICET lineshape in the far-wing. Modifications of the apparatus are in progress in order to improve the measurement accuracy up to the level required for a full test of the theoretically predicted behavior of the cross-section.

## 5. Acknowledgements

This work was performed at Istituto di Elettronica Quantistica del CNR and supported by Ministero della Pubblica Istruzione. The authors wish to thank Prof. R.Pratesi for his work devoted to the organization of the research and Dr. A.Bambini for arising the interest into the LICET problem and keeping it constant through stimulating discussions. They wish also to thank M.Neri for his valuable technical assistance.

## References

1. L.I.Gudzenko and S.I.Yakovlenko: Sov. Phys. JETP 35, 877 (1972); U.S.Lisitsa and S.I.Yakovlenko: Sov. Phys. JETP 39, 759 (1974)
2. D.B.Lidow, R.K.Falcone, T.F.Young and S.E.Harris: Phys. Rev. Lett. 36, 462 (1976); R.W.Falcone, W.R.Green, J.C.White, J.F.Young and S.E.Harris: Phys. Rev. A15, 1333 (1977)
3. W.R.Green, J.Lukasik, J.R.Willison, M.D.Wright, J.F.Young and S.E.Harris: Phys. Rev. Lett. 42, 470 (1979)
4. Ph.Cahuzac and P.E.Toscheck: Phys. Rev. Lett. 40, 1087 (1978)
5. C.Brechignac, Ph.Cahuzac and P.E.Toscheck: Phys. Rev. A21, 1969 (1980)
6. A.Debarre: J. Phys. B: At. Mol. Phys. 15, 1693 (1982)
7. B.Cheron and H.Lemery: Optics Comm. 42, 109 (1982)
8. A.Dèbarre: J. Phys. B: At. Mol. Phys. 16, 431 (1983)
9. D.Z.Zhang, B.Nikolaus and P.E.Toscheck: Appl. Phys. B28 2/3, 195 (1981)
10. A.Gallagher and T.Holstein: Phys. rev. A16, 2413 (1977)
11. A.Bambini and P.B.Berman: see previous paper in this volume
12. R.Buffa, S.Cavalieri, M.Matera and M.Mazzoni: Optics Comm. 58, 255 (1986)

# Collisionally Aided Coherent Emission at an Optical Frequency

*E. Giacobino and P.R. Berman**

Laboratoire de Spectroscopie Hertzienne de l'Ecole Normale Supérieure,
4 place Jussieu, F-75252 Paris Cedex 05, France

## I. Introduction

With the development of high-power laser sources, it has become possible to study a new class of collisional-radiative reactions. Theoretical and experimental studies of such collisionally-aided reactions, which require the action of a collision and an external radiation field simultaneously, were aimed initially at calculating or measuring the overall initial to final state cross-sections /1/. Subsequently, it became appreciated that such reactions can also produce coherence among degenerate magnetic sublevels, and that a study of the final state magnetic polarization yields additional information concerning the collisional interaction /2/. In a parallel development /3/, a series of experiments was carried out in which a collisionally-aided excitation of a coherence between the non-degenerate $3P_{1/2}$ and $3P_{3/2}$ states of Na was generated.

The above studies of collisionally-aided excitation of atomic coherence were limited to coherence between levels within the same electronic manifold. It was shown theoretically /4/ that one should also be able to produce collisionally-aided coherence between different electronic levels. If such a coherence is created between states of opposite parity, it can lead to coherent emission at the atomic transition frequency associated with these levels. Collisionally-aided generation of optical coherence can also provide a novel means for sum or difference frequency generation in a vapor.

In this work, we present experimental evidence for a collisionally-aided excitation of the 4D - $3P_{3/2}$ electronic state, optical coherence in Na produced via the reaction (Fig. 1):

$$Na(3S_{1/2}) + He + \hbar\Omega_1 + 2\hbar\Omega_2 \rightarrow Na(4D,3P_{3/2}) + He \qquad (1)$$

in which the notation (4D, $3P_{3/2}$) indicates a coherent superposition of the 4D and

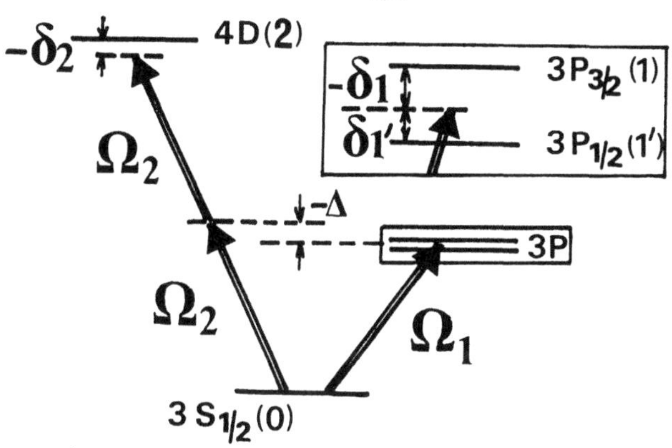


Fig. 1 Energy level diagram for Na ; the 4D fine structure and all the hyperfine structures are not resolved in the experiment

$3P_{3/2}$ states. The frequencies $\Omega_1$ and $2\Omega_2$ are detuned by several wave numbers from the $3S - 3P_{3/2}$ and $3S - 4D$ transitions, respectively. The frequency $\Omega_1$ is held fixed as $\Omega_2$ is varied. When the detuning for the two-photon (3S - 4D) transition is equal to that for the one-photon ($3S - 3P_{3/2}$) transition, a resonant structure is seen if one monitors emission from the sample at the $4D - 3P_{3/2}$ transition frequency (Figs. 1 and 2). This resonant structure is not present in the absence of collisions and, as is discussed below, is evidence for the creation of collisionally-aided $4D - 3P_{3/2}$ electronic state coherence in Na. It will also be seen that the coherence time associated with the pulsed laser sources plays an important role in the excitation process /5/.

## 2. Simplified Model of the Process

A qualitative interpretation of the results can be had by calculating the polarization of the sample in lowest order perturbation theory, assuming cw laser fields. The component of the polarization which is relevant to our discussion is the one that oscillates at frequency $(2\Omega_2 - \Omega_1)$. This component of the polarization is directly related to a density matrix element $\rho_{21}$, which, in lowest order perturbation theory, can be calculated from the optical Bloch equations as

$$\rho_{21}^{(3)}(2\Omega_2 - \Omega_1) = (\chi_2\chi_2'/\Delta)\,\chi_1^* \exp\{i\{(2\vec{k}_2 - \vec{k}_1)\vec{R} - (2\Omega_2 - \Omega_1)t\}\}$$

$$\times \frac{1}{\delta_1\delta_2}\{\frac{1}{f} + \frac{\Gamma}{\Gamma_{21} - i\delta_{21}}\} \tag{2}$$

where : $\Gamma = \Gamma_{20} + \Gamma_{01} - \Gamma_{21}$ (3)

$1/f = 1 + (\delta_1/2\delta_1')$ (4)

and : $\delta_{21} = \delta_2 - \delta_1$ (5)

The indices 0, 1, 1', 2 refer to states $3S_{1/2}$, $3P_{3/2}$, $3P_{1/2}$ and 4D respectively ; the $\chi$'s are the Rabi frequencies associated with the fields, $\chi_1 = \vec{E}_1.\vec{d}_{01}/2\hbar$, $\chi_2 = \vec{E}_2.\vec{d}_{01}/2\hbar$, $\chi_2' = \vec{E}_2.\vec{d}_{12}/2\hbar$ ($E_i$, $\Omega_i$, $\vec{k}_i$ = amplitude, frequency and propagation vector of field i ; $\vec{d}_{ij}$ is a dipole moment element) ; the detunings are defined as $\delta_1 = \Omega_1 - \omega_{10}$, $\delta_2 = \Omega_2 - \omega_{20}$, $\delta_1' = \Omega_1 - \omega_{1'0}$, $\Delta = \omega_{10} - \Omega_2$ ; and the $\Gamma_{ij}$'s are the (complex) collisional decay rates associated with the ij atomic-state coherence. The quantity $\chi_2^{TP} = \chi_2\chi_2'/\Delta$ plays the role of an effective two-photon Rabi frequency. In deriving Eq.(2), we have assumed that $|\Delta| >> |\delta_1|, |\delta_2| >> \Gamma_{ij}$ >> all Doppler widths >> all spontaneous decay rates. The last two inequalities, which are not valid at low perturber pressure ($P_{He} \sim 30$ Torr), explain why various Doppler shifts and spontaneous decay rates do not appear in Eq.(2).

Equation (2) is analogous to the one given by Bloembergen and coworkers /3/ for the creation of a coherence between two levels having the same parity. Two terms contribute to $\rho_{12}^{(3)}(2\Omega_2 - \Omega_1)$. The first term in brackets in Eq.(2) has as its origin a four-wave mixing process. This term is pressure independent (it exists even in the absence of collisions) and does not have a resonant structure in the region $\delta_2 \cong \delta_1$. Both the $3P_{3/2}$ and $3P_{1/2}$ states can serve as intermediate states in the four-wave mixing interaction, which is why both detunings $\delta_1$ and $\delta_1'$ appear in the factor f. The second term in brackets in Eq.(2) vanishes in the absence of collisions and does have a resonant structure when $\delta_2 \cong \delta_1$ ; the contribution from the $3P_{1/2}$ state in the vicinity of the resonance is negligible. It is the second term which represents the collisionally-aided excitation of the 2-1 coherence.

A fifth order contribution /7/ to $\rho_{21}$, arising from the collisionally aided excitation of the $3P_{3/2}$ level followed by a four-wave mixing process, can be neglected provided that $|\chi_1|^2 < \gamma_1|\delta_1|$, where $\gamma_1$ is the 3P spontaneous decay rate.

Since levels 1 and 2 are of opposite parity, an electric dipole coherence associated with $\rho_{21}$ is expected to give rise to emission in the direction

$$\vec{k}_s = 2\vec{k}_2 - \vec{k}_1 \tag{6}$$

with an intensity I proportional to

$$I \propto |\rho_{21}(2\Omega_2 - \Omega_1)|^2 = \frac{|\chi_2^{TP}\chi_1|^2}{\delta_1^2\delta_2^2 f^2}\left(1 + \frac{\{2yf + (yf)^2\}\Gamma_{21}^2}{\Gamma_{21}^2 + \delta_{21}^2}\right) \tag{7}$$

where

$$y = \Gamma/\Gamma_{21} \tag{8}$$

and we have kept only the third-order contribution to $\rho_{21}$. All collisional shifts have been neglected ($\Gamma_{ij}$'s taken as real). As can be seen in Eq.(7), the ratio of resonant signal (second term in Eq.(7)) to non-resonant background at $\delta_{21} = 0$ is given by $2yf + (yf)^2$. Both the resonant and non-resonant signals have a $\delta_2^{-2}$ dependence, which is readily seen in Fig. 2. Had we kept the fifth order term /6/, the signal would also contain a dispersive component, owing to the fact that the fifth order contribution is in phase quadrature with the third order one.

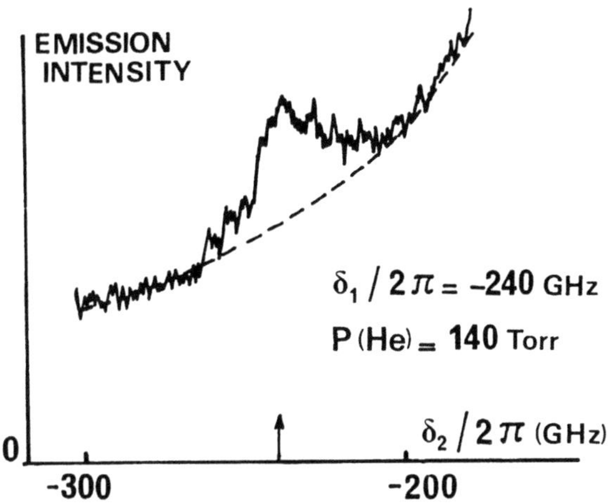


Fig. 2 Experimental curve of emission intensity (arbitrary units) at $\lambda$ = 568 nm in the direction $(2\vec{k}_2 - \vec{k}_1)$ as a function of $\delta_2/2\pi$ for fixed $\delta_1/2\pi$

## 3. Experiment

The experimental arrangement that was used to obtain Fig. 2 is shown schematically in Fig. 3. Two YAG pumped dye lasers with wavelengths approximately equal to 589 nm ($\Omega_1$) and 579 nm ($\Omega_2$), respectively, are sent with crossed polarizations into a heat-pipe containing sodium vapor and helium buffer gas. The sodium density is in the range $10^{13}$-$10^{14}$ at/cm$^3$, the helium pressure can be varied from one torr to several hundred torrs. The two laser beams overlap in the heat pipe, making an angle of a few mrd. A lens focuses the emerging signal on a pinhole located in the direction $\vec{k}_s$. This design eliminates most of the background lasers' light. Additional filtering of the signal is provided by a polarizer which eliminates the remaining light at frequency $\Omega_2$ (the polarization of the signal beam is expected to be parallel to that of the beam at $\Omega_1$) and by an interference filter centered on 568 nm. The signal is monitored with a photomultiplier followed by a box car integrator. Phase matching was achieved with a small angle between $k_1$ and $k_2$ and a detuning $\delta_1 < 0$ /8/.

A typical curve is shown in Fig. 2. The expected collisionally-aided Lorentzian emission profile centered at $\delta_2 = \delta_1$ can be seen superimposed on a background which can be reasonably well approximated with a $\delta_2^{-2}$ dependence. These curves were obtained with Rabi frequencies $\chi_1/2\pi \cong 0.7$ GHz and $\chi_2^{TP}/2\pi \cong 10$ MHz (the condition for neglect of the fifth order contribution is $\chi_1/2\pi < 2$ GHz). A dispersive structure,

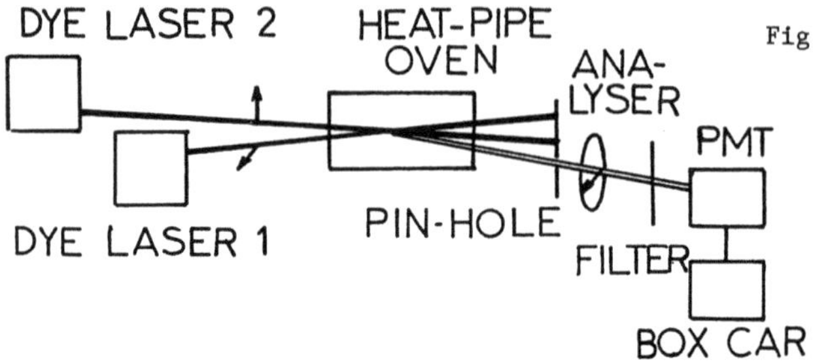


Fig. 3 Experimental set-up

characteristic of the higher order terms is seen when $\chi_1/2\pi$ is increased to 2.5 GHz. The absence of dispersive structure in the low power data is taken as a validity criterion for use of the third order theory.

The ratio r of the resonant signal component to the non-resonant background at $\delta_2 = \delta_1$ as a function of pressure is shown in Fig. 4. Although the data was taken with different detunings $\delta_1$, the results are corrected to an equivalent detuning $\delta_1/2\pi = -255$ GHz for each point. This ratio increases with pressure and then saturates. In attempting to compare the experimental results with the theoretical expression (7), one immediately encounters difficulties. Equation (7) predicts a constant value $r = 2yf + (yf)^2$ which is inconsistent with the low-pressure data. Even if the cw theory is modified to include effects of atomic motion (i.e. Doppler shifts), it cannot satisfactorily explain the low-pressure data (the curve labeled T = 5.0 ns is, effectively, the cw theory with inclusion of the Doppler shifts - see discussion below).

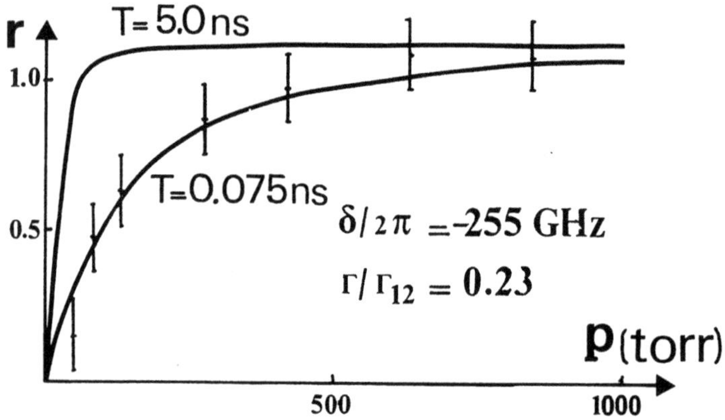


Fig. 4 Experimental points (with error bars) and theoretical curves of the ratio r of signal to background at $\delta_{21} = (\delta_2 - \delta_1) = 0$ as a function of helium pressure P. A ratio $y = \Gamma/\Gamma_{21} = 0.23$ is assumed for the theoretical curves.

## 4. Interpretation of the experimental results

In order to explain the differences between the cw theory and experiment, we have carried out a calculation assuming laser fields with gaussian temporal pulse envelope functions /8/.

A series of theoretical curves of Intensity versus $\delta_2$ is shown in Fig. 5 for $\delta_1/2\pi = -255$ GHz (f = 2), $k_S u/2\pi = 1.22$ GHz, T = 0.075 ns and several values of the pressure. In obtaining the curves in Fig. 5, we used collision parameters $\Gamma_{10}/2\pi = 11.3 \pm 0.4$ MHz/Torr, and $\Gamma_{20}/2\pi = 39.2 \pm 4.6$ MHz/Torr taken from the literature /9/ (collisional shifts are neglected and all values are quoted at 20°C). The value of $\Gamma_{21'}/2$ for the ($3P_{1/2}$ - 4D) transition is 36.2 ± 2.1 MHz/Torr /9/, but no

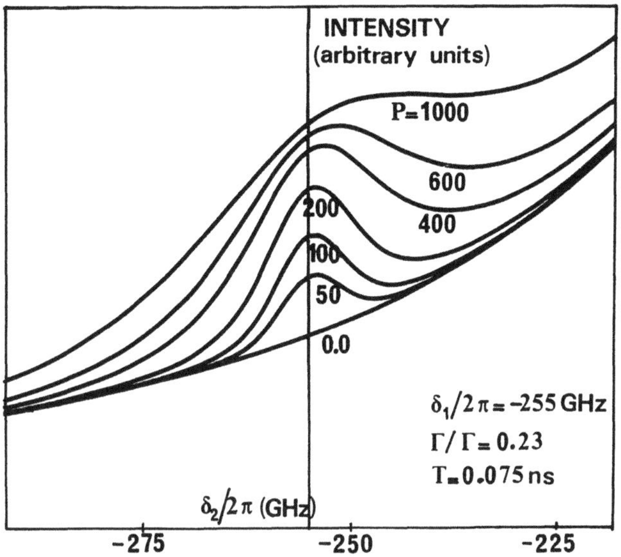


Fig. 5 Theoretical curves of emission intensity (arbitrary units) as a function of detuning $\delta_2/2\pi$ for $\delta_1/2\pi$ = - 255 GHz and T = 0.075 ns. Each curve is labelled by the helium pressure P in Torr.

data is available for $\Gamma_{21}$ for the ($3P_{3/2}$ - 4D) transition. To fit the high-pressure data we took $\Gamma_{21}/2\pi$ = 1.14, $\Gamma_{21}'/2\pi$ = 41.4 ± 2.4 MHz/Torr which leads to a value $y = \Gamma/\Gamma_{21}$ = 0.23 ± 0.17. In Figure 5, the collision-induced signal manifests itself as a resonant structure centered at $\delta_{21}$. The distortion caused by the $\delta_2^{-2}$ dependence is clearly seen at high pressures where $\Gamma_{21}/|\delta_1|$ is no longer negligibly small. The curves are normalized such that the non-resonant background contribution is unity at $\delta_2 = \delta_1$.

The theoretical ratio r of resonant to non-resonant signal amplitudes at $\delta_2 = \delta_1$ is shown in Fig. 4 as a function of pressure for T = 0.075 and 5.0 ns. The T = 0.5 ns curve approximates the cw limit. Its departure at low pressure from the constant value of r predicted by Eq.(7) is due to the fact that the effects of atomic motion cannot be neglected (as we assumed in deriving (7)) for pressures where $\Gamma_{21} < k_s u$. It is seen in Fig. 4 that the data is not consistent with the value T = 5.0 ns (and, consequently, not consistent with a cw theory). On the other hand, the data agrees rather well with a value T ≅ 0.075 ns. Since T serves as an effective laser coherence time in the problem, one can conclude that it is the laser coherence time T and not the laser pulse duration, $T_L$ ≅ 5.0 ns, which is the relevant time parameter for the laser field. Preliminary experiments on one- and two-photon absorption give linewidths that are consistent with pulse excitation with pulses whose temporal width is of order 0.075 ns.

In conclusion, we have observed a collisionally aided excitation of the 4D - $3P_{3/2}$ optical coherence in Na, which, in turn, leads to coherent emission. The laser coherence time plays an important role in the problem whenever it is the smallest relevant coherence time in the problem.

This work is supported by the U.S. Office of Naval Research and a C.N.R.S.-N.S.F. International Grant (Grant No INT 84-13300). We should like to thank G. Grynberg for helpful suggestions throughout the progress of this work.

* Permanent address : Physics Department, New York University, New York, NY 10003.


1. For a general review of this subject area, see Photon-Assisted Collisions and Related Topics, edited by N.K. Rahman and C. Guidotti (Harwood Academic, Chur, Switzerland, 1982) and Colloquium on Atomic and Molecular Collisions in a Laser Field, J. de Physique, volume 46, Colloque C1 (1985).
2. For a review of the use of so-called "optical" collisions to create magnetic-state coherences, see K. Burnett, Comment At. Mol. Phys. 13, 179 (1983) ; Phys. Rep. 118, 339 (1985). For creation of magnetic-state coherence in so-called radiative collisions, see P.R. Berman, Phys. Rev. A22, 1838, 1848 (1980) ; A. Debarre, J. Phys. B15, 1693 (1982).
3. N. Bloembergen, H. Lotem and R.T. Lynch, Jr., Indian J. Pure and Applied Phys. 16, 151 (1978) ; Y.A. Prior, A.R. Bogdan, M. Dagenais and N. Bloembergen, Phys. Rev. Lett. 46, 111 (1981) ; A.M. Bogdan, M. Downer and N. Bloembergen, Phys. Rev. A24, 623 (1981). For a recent excellent review of this subject, see L. Rothberg, to appear in Progress in Optics.
4. P.R. Berman and E. Giacobino, Phys.Rev. A28, 2900 (1983) ; P.R. Berman, Phys. Rev. B 29, 3234 (1984).
5. D. DeBeer, L.G. Van Wagenen, R. Beach and S.R. Hartmann, Phys. Rev. Lett. 56, 1128 (1986).
6. As the collision rates approach zero,the second term in Eq.(2) must be modified to include effects of atomic motion and spontaneous decay. When this is done, it is found that, even at $\delta_{21} = 0$, the second term vanishes as the collision rates go to zero.
7. The fifth order mentioned here is to be distinguished from pure field saturation effects (see, for example, H. Friedmann and A.D. Wilson-Gordon, Phys. Rev. A26, 2768 (1982) ; B. Dick and R.M. Hochstrasser, J. Chem. Phys. 75, 133 (1983) ; G.S. Agarwal and N. Nayak, Phys. Rev. A33, 391 (1986)) which are negligible in our experiment.
8. E. Giacobino and P.R. Berman, submitted to Phys. Rev. Lett..
9. Values are taken from the article of Allard and Kielkopf (N. Allard and J. Kielkopf, Rev. Mod. Phys. 54, 1103 (1980)).

# Part V

# Electronic Collisions in Laser Fields

# Close-Coupling Solution of Radiative e-H$^+$ Scattering in a Laser Field

*L. Dimou and F.H.M. Faisal*

Fakultät für Physik, Universität Bielefeld,
D-4800 Bielefeld 1, Fed. Rep. of Germany

We present the first results of application of the method of close-coupling equations to the fundamental radiative collision process of e + H$^+$ scattering in a laser field. Both elastic scattering and the photon absorption cross-sections are given. The prominent features of the cross-sections are the new "capture-escape resonances" involving the Rydberg states of the neutral H atom.

## Theory

The Schrödinger equation of the problem is ($e=\hbar=m=1$)

$$i\frac{\partial}{\partial t}\Psi(\vec{r},t) = \left[-\frac{1}{2}\nabla^2 - \frac{\vec{A}(t)}{c}\cdot\hat{p} + \frac{1}{2c^2}\vec{A}^2(t) - \frac{Z}{r}\right]\Psi(\vec{r},t) \tag{1}$$

where we have chosen a circularly polarized radiation field

$$\vec{A}(t) = A_0[\hat{\epsilon}_x\cos(\omega t+\delta) - \hat{\epsilon}_y\sin(\omega t+\delta)] \tag{2}$$

$\hat{\varepsilon}_x$ and $\hat{\varepsilon}_y$ are the orthogonal unit vectors, $A_0$ is the peak vector potential, $\omega$ is the frequency and $\delta$ is a constant phase; $\hat{p} = -i\nabla$. In the present form the interaction Hamiltonian in (1) does not fall off with increasing distance and is thus not convenient for direct application of the close-coupling asymptotics. We therefore change the representation by making the unitary transformation[1,2]

$$\Psi(\vec{r},t) = \exp\left\{-i\int^t dt\left[-\frac{\vec{A}(t)}{c}\cdot\hat{p} + \frac{1}{2c^2}\vec{A}(t)^2\right]\right\}\Phi(\vec{r},t) \tag{3}$$

in (1) to obtain

$$i\frac{\partial}{\partial t}\Phi(\vec{r},t) = \left[-\frac{1}{2}\nabla^2 - \frac{Z}{|\vec{r}-\vec{\alpha}_0(t)|}\right]\Phi(\vec{r},t) \tag{4}$$

where $\vec{\alpha}_0(t) = \frac{1}{c}\int^t dt\, \vec{A}(t)$

The interaction Hamiltonian in (4) has a Coulomb asymptotic behavior which can be conveniently treated within the well-known close-coupling method[3], extended for the strong-field radiative collisions[5]. We make a Floquet plus partial wave expansion of $\Phi(\vec{r},t)$

$$\Phi(\vec{r},t) = \sum_{n=-\infty}^{\infty}\sum_{lm} e^{-iEt+in(\omega t+\delta)}\,\frac{1}{r}F_{nlm}(r)Y_{lm}(\hat{r}) \tag{5}$$

and substitute it in (4), equate equal coefficients of $e^{in(\omega t+\delta)}$ and project on to $Y_{lm}^{*}(\hat{r})$ to obtain the Floquet representation of the radial close-coupling equations for the channel functions $F_{nlm}(r)$ in the channel $i = (nlm)$.

$$\left[\frac{d^2}{dr^2} + k_n^2 - \frac{l(l+1)}{r^2} + \frac{2Z}{\sqrt{r^2+\alpha_0^2}}\right]F_{nlm}(r) = \tag{6}$$

$$= -2Z \sum_{\lambda=1,p=0}^{\infty,\lambda} \sum_{l'm'} < Y_{lm}|V_{\lambda_p}(\theta,\phi)|Y_{l'm'} > F_{n-\lambda_p,l',m'}(r)$$

where the multipole-coupling potentials are

$$V_{\lambda_p}(\theta,\phi) = \frac{(2\lambda-1)!!}{2^\lambda \lambda!}\binom{\lambda}{p}\left(\frac{r\alpha_0\sin\theta}{r^2+\alpha_0^2}\right)^\lambda \frac{1}{\sqrt{r^2+\alpha_0^2}}\, e^{i\lambda_p\phi} \tag{7}$$

with $\lambda_p = \lambda-2p$. The asymptotic behavior of the $j^{\text{th}}$ solution in the $i^{\text{th}}$ open channel for the scattering problem is

$$F_i^{(j)}(r)_{r\to\infty} = \frac{1}{\sqrt{k_i}}\left[\delta_{ij}\sin(\theta_i) + K_{ij}\cos(\theta_i)\right] \quad i,j=1,2,\ldots,n_{op} \tag{8}$$

where $\theta_i = k_i r - \frac{1}{2}l_i\pi + \frac{Z}{k_i}\ln 2k_i r + \sigma_{l_i}$. $\sigma_{l_i}$ is the Coulomb phase shift and $n_{op}$ is the number of open channels included. $K = \{K_{ij}\}$ is the real K-matrix which is related to the (complex) S-matrix by $S_{ij} = \left(\frac{1+iK}{1-iK}\right)_{ij}$. Exponentially decreasing boundary conditions are made to be satisfied for the closed channels. It can be shown by an extension of the standard analysis[4] that the radiative scattering amplitude in which the momentum of the incident electron, $\vec{k}_0$, is in the direction $\hat{k}_0$ and the final momentum, $\vec{k}_N$, after exchange of N photons, is in the direction $\hat{k}_N$, is related to the open channel S-matrix elements by

$$f_{0\to N}(\hat{k}_0,\hat{k}_N) = f_{\text{Coulomb}}(\Theta)\delta_{N,0} \tag{9}$$

$$+ \sum_{l_0 m_0 l m} \frac{2\pi i}{\sqrt{(k_0 k_N)}}\, i^{l_0-l} e^{i(\sigma_{l_0}+\sigma_l)} Y_{l_0 m_0}^{*}(\hat{k}_0)\, Y_{lm}(\hat{k}_N)\left[\delta_{N,0}\,\delta_{ll_0}\,\delta_{mm_0} - S_{0\,l_0 m_0}^{Nlm}\right]$$

$\sigma_l$ are the $l^{\text{th}}$ partial wave Coulomb phase shifts and $f_{\text{Coulomb}}(\Theta)$ is the ordinary Rutherford scattering amplitude where $\Theta$ is the angle between the incident and the scattered directions, for N=0. Hence from (9) the elastic cross-section modified by the field is

$$\frac{d\sigma^{(0)}}{d\Omega}(\Theta) = \frac{d\sigma_{\text{Coulomb}}(\Theta)}{d\Omega} + |f_{rad}^{(0)}(\Theta)|^2 + 2\text{Re}[f_{\text{Coulomb}}^{*}(\Theta)f_{rad}^{0}(\Theta)] \tag{10}$$

where $f_{rad}^{(0)}(\Theta)$ is the term with N=0 in second line of (9). Similarly for the inelastic processes of stimulated absorption (N<0) and emission (N>0) the differential cross-sections are

$$\frac{d\sigma^{(N)}}{d\Omega}(\Theta) = \frac{k_N}{k_0}|f_{rad}^{(N)}(\hat{k}_0,\hat{k}_N)|^2 \tag{11}$$

where $f_{rad}^{(N)}(\hat{k}_0,\hat{k}_N)$ is given by the second line of (9) for $N\neq 0$. Hence, from (11) the angle-integrated total cross-sections of N photon absorption or emission are given by

$$\sigma^{(N)}(\hat{k}_0) = \frac{4\pi^2}{k_0{}^2} \sum_{lm} \left| \sum_{l_0 m_0} i^{l_0} e^{i\sigma_{l_0}} S_{0\,l_0 m_0}^{\,Nlm}\, Y^{*}_{l_0 m_0}(\hat{k}_0) \right|^2 \tag{12}$$

## Results and Discussion

Fig. 1 shows the radial dependence of the channel-coupling potentials as a function of increasing radial distance r for $\lambda$=0,1,2 and 3. The ordinate is shown in units of $\frac{Z}{\alpha_0}$ and the abscissa in $\rho = \frac{r}{\alpha_0}$ where $\alpha_0 = \frac{A_0}{c}\omega = \frac{F_0}{\omega^2}$ is the classical radius of vibration of the electron in the field of peak strength $F_0$. It is to be noted that for $\lambda$=0, at small r the laser field lifts the usual divergence of the Coulomb potential at r=0 and gives a constant value $-\frac{Z}{\alpha_0}$ which decreases in magnitude with increasing field strength, $F_0$, and/or decreasing frequency. Higher multipole-potentials are seen to decrease both in maximum strengths as well as in ranges; this is essential for the successful implementation of the standard close-coupling method in practice. In fig. 2 we show the angular dependence of the coupling-potential for $\lambda$ = 0,1,2 and 3.

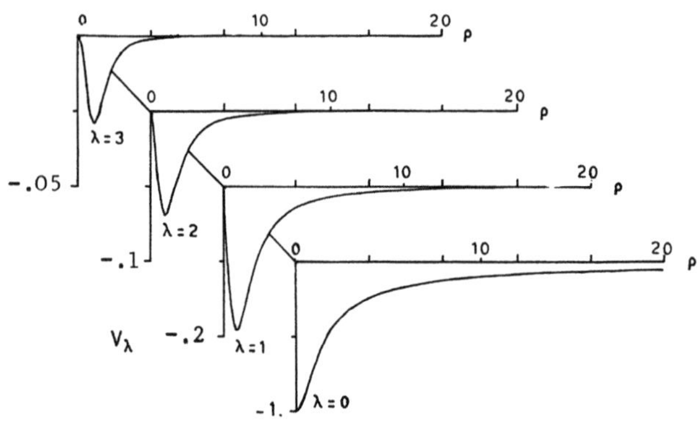


FIGURE 1. Radial dependence of the channel-coupling potentials for the multipoles $\lambda$ =0,1,2 and 3. The ordinate is in units of $\frac{Z}{\alpha_0}$ in (au) and the abscissa is in $\rho = \frac{r}{\alpha_0}$ (au); $\alpha_0 = \frac{F_0}{\omega^2}$(au); $\phi = 0$ and $\theta = \frac{\pi}{2}$ ; p = 0

Both elastic and inelastic radiative e + $H^+$ scattering cross-sections are considered below. We first present the result of the field-modified elastic scattering in Fig. 3 as a function of the incident electron energy. For the sake of comparison we have shown the ratio of the field-modified scattering cross-section to that in the absence of the field - the latter being given by the Rutherford scattering formula. In this case the incident electron momentum is in the direction $\Omega_0 = (\Theta_0{=}90°, \Phi_0{=}0°)$ where the z-axis is chosen to be the field propagation direction and the scattered direction is chosen to be $\Omega$ = (90°,90°). The peak field strength is $F_0$ = .005au and the frequency of the laser is $\hbar\omega$ = 0.46 Ryd = 6.256 eV. The most prominent features of this result are (a) the existence of a series of very clear resonance structure and (b) away from the resonances the field-modified elastic cross-sections are rather closely given by unmodified Rutherford cross-section. The resonance structures are indications for the fact that the scattering electrons become captured by the proton

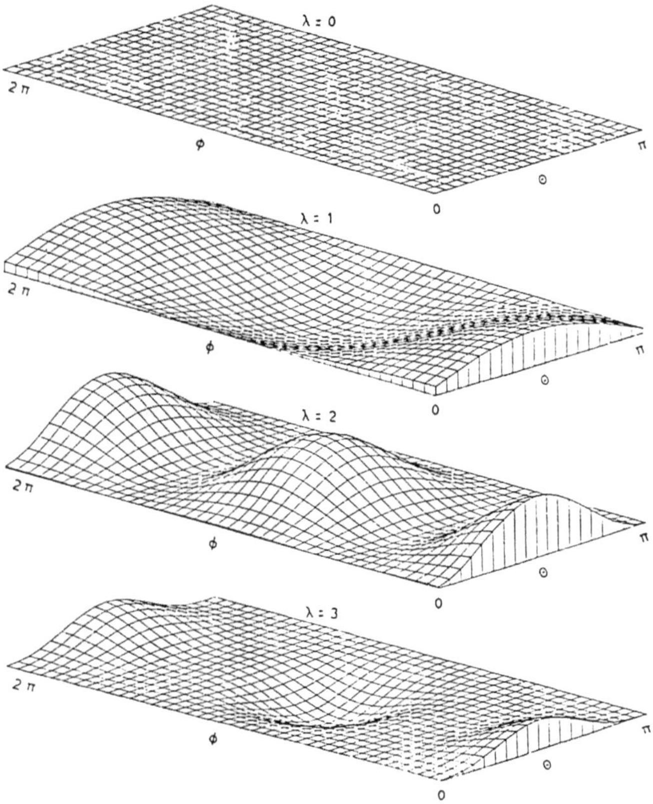


FIGURE 2. The angular dependence of the coupling potentials for the multipoles $\lambda = 0,1,2$ and 3 at a fixed radial distance $\rho = \frac{r}{a_0} = 1$ (au)

in the bound states of the H-atom and are then reemitted in the same incident energy channel. This is a resonant induced phenomenon. Thus at a given laser frequency which matches the energy difference between the incident electron energy (positive) and a Rydberg level energy (negative) the laser field can force the electron to emit a photon and cause it to be captured temporarily in the Rydberg-state till the subsequent absorption of a photon permits the electron to escape from this state into the continuum again. The delay introduced by this capture-escape episode shows up as a resonance in the scattering signal. This resonance of course can be tuned not only by varying the frequency at a fixed incident electron energy but also by varying the incident energy at a given laser frequency.

Fig. 3 shows the first such "capture-escape resonances" corresponding to n=2,3 and 4 principal quantum numbers of the intermediate neutral H-atom due to the "reaction" wich can be schematized as

$$\hbar\omega + e + H^+ \rightarrow H(n) + 2\hbar\omega \rightarrow \hbar\omega + e + H^+ \qquad (13)$$

We note the analogy of this process with the capture-escape resonance processes via the formation of negative ions, discussed by us previously[5,6,7,8].

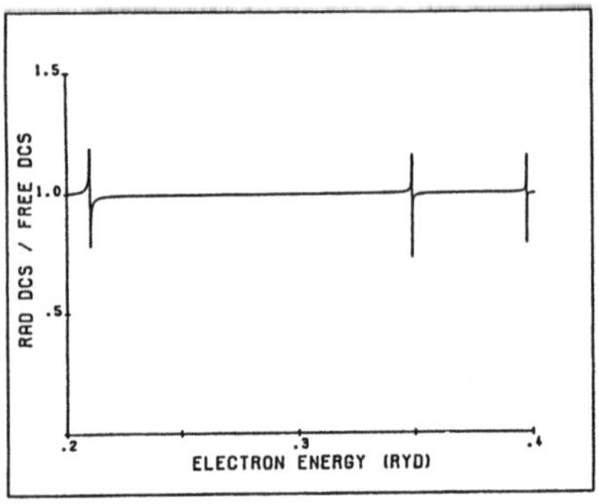


FIGURE 3. The ratio of the field-modified elastic $e + H^+$ scattering cross-section in a circularly polarized field of the field strength $F_0$=.005au and $\hbar\omega$ = .46 Ryd = 6.256 eV, with respect to the ordinary Rutherford cross-sections, shown as a function of incident electron energy. The incident momentum is in the direction $\Omega_0 = (\Theta=90°, \Phi=0°)$ and the final momentum is in the direction $\Omega = (\Theta=90°, \Phi=90°)$. The field propagation direction is along the z-axis. Note the occurence of the capture-escape Rydberg resonances with respect to the n=2,3 and 4 states of neutral H.

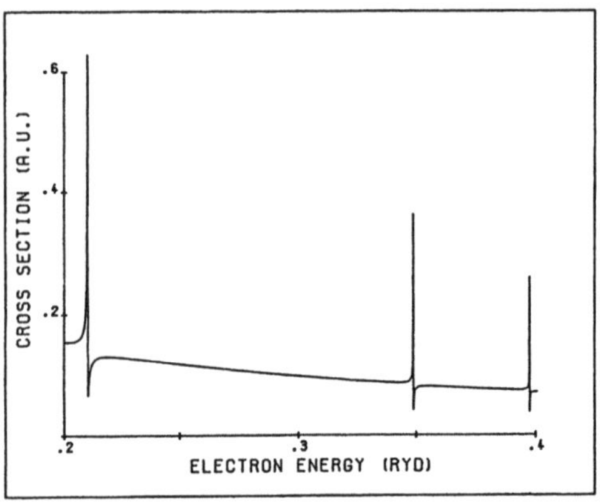


FIGURE 4. The one-photon angle-integrated total absorption cross-section as a function of the incident electron energy in the presence of a circularly polarized field. Field strength $F_0$ = .005 au, $\hbar\omega$ = .46 Ryd = 6.256 eV. Incident electron direction $\Omega_0 = (90°, 0°)$. The field propagation direction is along the z-axis.

In fig. 4 we present the radiative inelastic cross-section for the absorption of a photon. This process can be thought of as a field-induced-inverse-Bremsstrahlung process in a strong field at low electron energies. In the absence of the field no such channel is possible and hence the background to this signal, in the one-photon absorption channel, is intrinsically zero. Here too we obtain a very prominent occurrence of the sequences of "capture-escape-Rydberg resonances".

Fig. 5 shows the n=2 resonance in magnification. It has a width of ~5 meV. We note that with increasing field strength this and the other resonances for higher n tend to broaden. In particular the higher Rydberg resonances which are weak at a given intensity begin to appear significantly with increasing field strength. We also remark that the width of a resonance can be thought of as a measure of the ionization

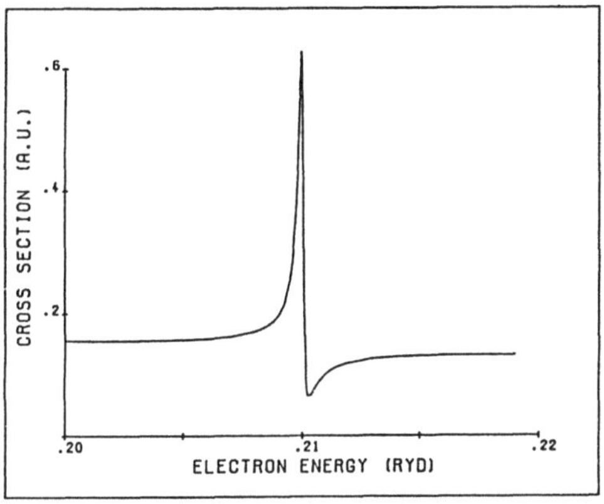


FIGURE 5. Magnification of the one-photon angle-integrated total absorption cross section for $e + H^+$ scattering in a circularly polarized field in the region of the capture-escape Rydberg resonance with n=2 state of neutral H. Incident electron momentum along $\Omega_0 = (90°, 0°)$. The laser propagation direction along z-axis. Field strength $F_0 = .005$ au, $\hbar\omega = .46$ Ryd = 6.256 eV.

rate of that particular Rydberg state of the neutral Hydrogen atom in the presence of the field.

In conclusion, we have shown that the powerful technique of the close-coupling method can be applied very effectively to obtain fully quantitative information of radiative $e + H^+$ scattering in the presence of a strong laser field. The results reveal the existence of the "capture-escape-Rydberg resonances" in the elastic scattering and in the photon absorption cross-sections.

Acknowledgements

This work has been supported by Deutsche Forschungsgemeinschaft under the project no. SFB-216-M2.

References


1. W. Henneberger: Phys. Rev. Lett. 21, 838 (1968)
2. F.H.M. Faisal: J. Phys. B6, L89 (1973)
3. P.G. Burke and M.J. Seaton: in Methods in Computational Physics, Vol. 10, 1 (1971)
4. N.F. Mott and H.S. Massey: in Theory of Atomic Collisions, 3rd.ed., (Oxford, 1965), p 380
5. F.H.M. Faisal: in Fundamentals of Laser Interactions, ed. by F. Ehlotzky, Lecture Notes in Physics, (Springer, Heidelberg, 1985), p 18
6. L. Dimou and F.H.M. Faisal: in Collisions and Half-Collisions with Lasers, eds. N.K. Rahman and C. Guidotti, (Harwood, New York, 1984), p 121
7. F.H.M. Faisal: Comments on At. Mol. Phys. 15, 119 (1984)
8. F.H.M. Faisal: J. de Phys. 46, Colloque C1, Suppl. no 1, eds. J.L. Picque', G. Spiess, F.J. Wuilleumier, 1985 p c1-223

# Electron-Proton and Electron-Hydrogen Atom Collisions in the Presence of a Radiation Field

*A. Maquet*

Laboratoire de Chimie Physique, Université Pierre et Marie Curie, 11, Rue Pierre et Marie Curie, F-75231 Paris Cedex 05, France

## 1. Introduction.

We report the results of recent perturbative calculations related to the collisions of charged particles, in the presence of a radiation field. More precisely, we will discuss and compare the main features of the collision cross-sections of electrons with either a H atom or a bare nucleus, in the presence of an external field. These processes correspond to well-defined test cases illustrating the effect of a radiation field on electron-atom collisions and potential scattering (infinite range potential). Our results concerning this latter case can be of some help towards a better understanding of the dynamics of laser-assisted $e^-$-ion collisions.

The processes considered here obey the following equations, which depict respectively:

i) one-photon free-free transitions (FFT) occurring during the course of a $e^-$-H atom collision:

$$e^-(E_i,\mathbf{k_i}) + H(1s) \pm \gamma(\omega,\epsilon) \longrightarrow e^-(E_f,\mathbf{k_f}) + H(1s), \qquad (1)$$

here $E_i$ ($E_f$) and $\mathbf{k_i}$ ($\mathbf{k_f}$) are the energies and wave vectors of the incoming (outgoing) projectile electron, $\pm\gamma(\omega,\epsilon)$ represents the absorption (+) or the emission (-) of one photon with frequency $\omega$ and polarisation $\epsilon$; $E_i$ and $E_f$ are connected via the energy conservation relation: $E_f = E_i \pm \hbar\omega$. We report here the results of a recent perturbative calculation, valid for collisions involving fast electrons, which allows us to discuss the occurrence of resonances and of deep minima in the dispersion curves and angular distributions for the cross sections [1].

ii) excitation of a H atom occurring during the course of a $e^-$- H atom collision, in the presence of a radiation field:

$$e^-(E_i,\mathbf{k_i}) + H(1s) \pm \gamma(\omega,\epsilon) \longrightarrow e^-(E_f,\mathbf{k_f}) + H^*(n,\ell), \qquad (2)$$

with the same notations than before. Here the energy conservation relation reads: $E_f + \epsilon_n = E_i \pm \hbar\omega + \epsilon_1$ , where $\epsilon_n$ denotes the energy of the hydrogenic state $(n,\ell)$.

Given their potential interest in laser-assisted collision studies, such processes have given rise to numerous theoretical works performed either in a perturbative [2-4], or a non-perturbative context [5,6]. We will briefly report the results of another perturbative calculation performed recently and will discuss the main features of the corresponding cross-sections[7].

iii) <u>two-photon transitions in a Coulomb potential:</u>

$$e^-(E_i,\mathbf{k_i}) + H^+ \pm \gamma(\omega_1,\boldsymbol{\epsilon_1}) \pm \gamma(\omega_2,\boldsymbol{\epsilon_2}) \longrightarrow e^-(E_f,\mathbf{k_f}) + H^+, \qquad (3)$$

this general equation represents the exchange of two distinct photons between the electron-proton system and the radiation field. Belonging to this class of processes is the recently reported two-photon bremsstrahlung,[8], which corresponds to an interaction of the $e^-$-$H^+$ system with the vacuum (empty modes of the field). The case of stimulated processes (absorption or stimulated emission) is easily obtained by a suitable specialization of Eq.(3),[9,10].

The perturbative calculations of the cross-sections related to the processes referenced in Eqs. (1)-(3), have several common features as well as distinct ones. Thus, radiative transitions occurring during the course of $e^-$-H atom collisions and resulting in the emission or absorption of one photon, can be considered as a first-order transition into the continuous spectrum of the $H^-$ ion. Given the difficulties inherent to the description of the continuous spectrum of a two-electron system, this picture leads to very difficult numerical calculations[11]. However, if the projectile is fast enough, a simpler approach may be contemplated, since it is then permissible to represent it as a plane wave (Born approximation). The corresponding transition amplitude then becomes second-order, two interaction hamiltonians being responsible for the transition. Indeed, one has:

$$H_e = e^2/r_{01} - e^2/r_0 \ ; \quad r_{01} = |\mathbf{r_0} - \mathbf{r_1}|. \qquad (4a)$$

$$H_r = (e/mc)\mathbf{p}.\mathbf{A}; \qquad (4b)$$

where $H_e$ denotes the interaction between the projectile and the atom and $H_r$ is associated to the electron-field interaction. The index 0 labels the projectile electron, the index 1 labels the atomic electron, **p** is the momentum operator and **A** is the vector potential of the field. The relevant transition amplitudes, associated to the Eqs. (1) and(2), are of the general form:

$$T^{(2)} = \sum_n \frac{\langle f|H_e|n\rangle\langle n|H_r|i\rangle}{(E_i - E_n)} + \text{sym.}(H_e \longleftrightarrow H_r), \qquad (5)$$

On the contrary, in the case of two-photon FFT in a Coulomb potential, Eq. (3), the relevant transition hamiltonian is $H_r$, and one has:

$$T'^{(2)} = \sum_n \frac{\langle f|H_r|n\rangle\langle n|H_r|i\rangle}{(E_i - E_n)} \quad . \qquad (6)$$

In both cases the contribution of the field can be easily factorized and the amplitudes are ultimately expressed in terms of second-order atomic matrix elements containing the Coulomb Green's function: $G_C(\Omega) = (\Omega - H_C)^{-1}$, where $H_C$ is the Coulomb hamiltonian. The computations we present here rely on the use of different representations of $G_C$ which have proven to be most useful in various multiphoton calculations in hydrogen.

## 2. $e^-$-H atom radiative collisions.

Within the framework of the approximation retained here (Born approximation, exchange neglected) the second order transition amplitude, Eq. (5) actually involves four terms. These may be associated to the diagrams displayed below. The ones numbered I and II will be referred to as "direct" ones since they account for the coupling of the field with the projectile electron. On the contrary, the diagrams III and IV depict the interaction of the field with the H-atom and will be accordingly termed as "atomic".

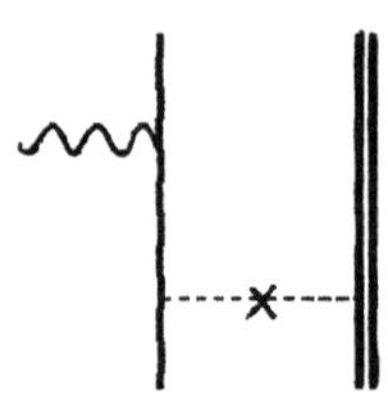

I

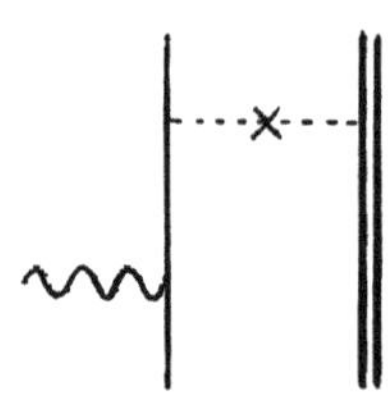

II

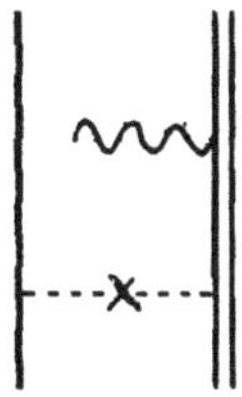

**III**

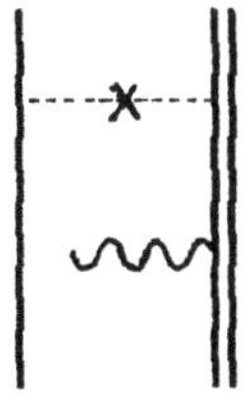

**IV**

The summation and integration over the projectile (plane-wave) states, as well as the factorization of the field-dependent terms are easily performed and one is left with atomic matrix elements of the following form (atomic units will be used throughout):

$$M_I + M_{II} = \langle n|(e^{i\mathbf{\Delta}.\mathbf{r}} - 1)|1s\rangle, \qquad (7)$$

$$M_{III} = \langle n|e^{i\mathbf{\Delta}.\mathbf{r}} G_C(E_{1s} \pm \omega)\mathbf{r}.\boldsymbol{\epsilon}|1s\rangle, \qquad (8a)$$

$$M_{IV} = \langle n|\mathbf{r}.\boldsymbol{\epsilon} G_C(E_{1s} \pm \omega) e^{i\mathbf{\Delta}.\mathbf{r}}|1s\rangle, \qquad (8b)$$

where $\mathbf{\Delta} = \mathbf{k_i} - \mathbf{k_f}$ and $|n\rangle$ represents the final atomic state. The computation of the generalized form factors in Eq.(7) does not present any particular difficulty [12], the crux of the calculus lying into the evaluation of the "atomic" 2$^{nd}$ order amplitude Eqs. (8). Depending on the process actually considered, we have used different representations of the Coulomb Green's function.

<u>A) One-photon free-free transitions:</u>

The relevant atomic matrix elements are obtained from Eqs. (8) by a simple specialization $|n\rangle \longrightarrow |1s\rangle$. Then, by using Hostler's integral representation of $G_C(\Omega)$,[13] it is a relatively simple task to derive a closed form expression of the amplitude which is then expressed in terms of Appell's hypergeometric functions of two variables $F_1$, see Ref.[1]. Note that the numerical computation of these analytical expressions can be easily performed on a microcomputer (we actually did use a 512k Macintosh).

<u>B) Excitation of the H-atom:</u>

In the general case, in which the final atomic state is not the ground state, analytical calculations become extremely intricate. We thus relied on other representations of $G_C$, using independently both a sturmian [14-16], and a $L^2$ basis [17] expansion. This gave us the opportunity to cross-check our numerical results [7].

The use of these powerful techniques allowed us to improve previous calculations on the same subject and to discuss in greater details the main features of these processes. One of the main conclusion which may be drawn from our results is that the contribution of the "atomic" terms (diagrams III and IV) plays an essential part in the overall behavior of the cross-sections. This role is particularly important in the following instances:

i) The atomic contribution is dominant at small scattering angle, or equivalently, for small momentum transfer $\Delta$. As a matter of fact, it may be easily shown that the electronic term, see Eq. (7), identically vanishes in the limit $\Delta \to 0$.

ii) The atomic contribution becomes also dominant if the laser frequency nearly matches an atomic resonance, i.e., if: $\omega \sim \omega_n = 1/2 - 1/2n^2$. For such frequencies, one of the atomic matrix element becomes very large and dominates the other terms in the total transition amplitude.

iii) In the soft-photon limit, on the contrary, the "direct" term becomes by far dominant. This is clearly displayed in the case of FFTs, when considering the respective dependences of the amplitudes with respect to the frequency. Indeed, one finds in this limit:

$$\left(\frac{d\sigma}{d\Omega}\right) \approx 8\pi\alpha \frac{k_f (\mathbf{\Delta}.\boldsymbol{\epsilon})^4}{k_i \omega^4 \Delta^4} [16(\Delta^2+4)^{-2}-1]^2. \qquad (9)$$

This result, which displays the divergence of the FFT cross-sections in the limit $\omega \to 0$, is a form of the Low theorem, related to the so- called infrared divergence of the bremsstrahlung cross-section [18].

iv) The angular dependence, as well as the dispersion curves for the cross-sections display two kinds of deep minima:

a) One type of minima has a purely kinematical origin. For transitions between s-states, one easily shows that the overall transition amplitude contains the factor $(\mathbf{\Delta}.\boldsymbol{\epsilon})$. So, depending on the scattering geometry, the cross-section can identically vanish if the condition $\mathbf{\Delta}.\boldsymbol{\epsilon} = 0$ is fulfilled.

b) The other type of minima arises from destructive interferences occurring between the "direct" and "atomic" amplitudes which can have opposite signs. The position of these minima cannot be easily predicted,since it depends on the relative magnitudes and signs of the amplitudes, which in turn, depend in a complicated way on the projectile energies (in the initial and final states), the laser frequency and the geometry of the experiment.

v) Stimulated emission can be strongly favored with respect to absorption ( and vice-versa ), depending on the laser frequency and on the scattering geometry. For instance, if the overall factor $\mathbf{\Delta}.\boldsymbol{\epsilon} = 0$, in the absorption case, the same condition cannot be met for the emission at the same scattering angle. In this particular geometry, stimulated emission is thus strongly favored.

vi) The Bethe-Born approximation, which amounts to approximate $e^{i\mathbf{\Delta}.\mathbf{r}} \approx 1 - i\mathbf{\Delta}.\mathbf{r}$, is not adequate to describe such processes. As a matter of fact, it leads to neglect the "direct" term which is often dominant and cannot account properly for the above-mentioned destructive interferences.

One of the main conclusion which emerges from this study is thus that the dynamic polarizability of the atomic target should be properly taken into account in an accurate treatment of these collision processes in the presence of a radiation field [19]. It should be noted however, that its influence can be less marked when considering total cross-sections.

## 3. Two-photon FFT in a Coulomb potential.

Two-photon transitions in a Coulomb potential are described with the help of second-order amplitudes of the general form displayed in Eq. (6). After factorizing out the field dependence, one is left with atomic matrix elements of the general form:

$$M_{f/i} = \langle \mathbf{k_f}^{(+)} | \boldsymbol{\epsilon}_2.\mathbf{p}\, G_C(\Omega)\, \boldsymbol{\epsilon}_1.\mathbf{p}\, | \mathbf{k_i}^{(-)} \rangle, \qquad (10)$$

where $|\mathbf{k}^{(\pm)}\rangle$ are Coulomb wave functions with asymptotic momentum $\mathbf{k}$ and the correct outgoing or ingoing wave properties and $G_C(\Omega)$ is the Coulomb Green's function. In sharp contrast with the case of bound-bound or bound- free transitions the computation of such second-order free-free matrix elements presents very severe difficulties which may be ascribed to the oscillatory nature as well as to the angular momentum infinite degeneracy of the continuum wave functions. Such difficulties can nevertheless be partially overcome in the special test case of the Coulomb potential, thanks in particular to its O(4) symmetry properties which make feasible a closed form analytical calculation of second-order amplitudes. These properties are conveniently exploited in momentum-space calculations by using Schwinger's integral representation of $G_C(\Omega)$ [20]. After some lengthy algebra the two-photon transition amplitude can be expressed in terms of single variable integrals which have been computed numerically with great accuracy [9,10]. This allowed us to obtain the cross-sections for several (stimulated or spontaneous) two-photon processes. In particular, we have been able to study the dependence of the differential cross-sections with respect to the various parameters governing the collision dynamics ($\mathbf{k_i}$, $\mathbf{k_f}$) as well as the characteristics of the laser light ($\omega,\epsilon$). The main results we have obtained may be summarized as follows:

i) in the low-frequency limit, the cross-sections for (stimulated) emission or absorption become equiprobable and diverge as $\omega \to 0$. They are then proportional to the Rutherford scattering cross-section (Low theorem).

ii) in the Born approximation, the above results remain qualitatively valid.

iii) our perturbative results go over smoothly into the low intensity limits of the non perturbative theories, both in the low-frequency [21] and high-frequency [22] regimes.

iv) one of the main parameters governing the dynamics of the process is, beyond the details of the calculation, the dot product $\mathbf{\Delta}.\boldsymbol{\epsilon}$. The differential cross-sections often exhibit a minimum in scattering geometries such that $\mathbf{\Delta}.\boldsymbol{\epsilon} = 0$.

v) at optical frequencies and for relatively low incoming electron energies no approximate calculation [see i) and ii)] can provide reliable results and the full calculation is necessary.

## 4. Conclusion.

We have reported accurate, perturbative calculations of various radiative (FFT) processes occurring during the course of collisions of fast electrons with H atoms or with a bare nucleus. The Coulomb interaction has been taken into account exactly throughout the calculation, by using adequate representations of the Coulomb Green's function. Our results should be useful as reference marks when discussing more sophisticated models

**Acknowledgements.** The work reported here result from fruitful collaboration with J. Broad (Bielefeld), A. Dubois (Paris), M. Gavrila (FOM, Amsterdam), S. Jetzke (Bielefeld) and V. Véniard (Paris and FOM, Amsterdam).

## References.

[1]. A. Dubois, A. Maquet and S. Jetzke, to appear in Phys. Rev. A (1986).

[2]. V. M. Buimistrov, Phys. Lett. **30A**, 136 (1969), see also: V. M. Buimistrov and L.I. Trakhtenberg, Sov. Phys. JETP, **42**, 54 (1976).

[3]. N. K. Rahman and F. H. M. Faisal, J. Phys. B **9**, L275 (1976); Phys. Lett. **57A**, 426 (1976).

[4]. S. Jetzke, F. H. M. Faisal, R. Hippler and H. O. Lutz, Zeit. Phys. A, **315**, 271 (1984).

[5]. B. A. Zon, Sov. Phys. JETP, **46**, 65 (1977).

[6]. F. W. Byron and C. J. Joachain, J. Phys. B, **17**, L295 (1984); C. J. Joachain, in "Fundamentals of Laser Interactions", edited by F. Ehlotzky, (Springer, Berlin, 1985)p. 37.

[7]. S. Jetzke, J. Broad and A. Maquet, to be published.

[8]. J. C. Altman and C. A. Quarles, Phys. Rev A **31**, 2744 (1985).

[9]. V. Véniard, A. Maquet and M. Gavrila, in "Fundamental Processes in Atomic Collision Physics", edited by H. Kleinpoppen, J. S. Briggs and H. O. Lutz, (Plenum, New York,1985), p. 641; A. Maquet and V. Véniard, in "Fundamentals of Laser Interactions", edited by F. Ehlotzky,(Springer, Berlin,1985), p. 275; see also: V. Véniard, Thesis (Université Pierre et Marie Curie, Paris, 1986), unpublished.

[10]. M. Gavrila, A. Maquet and V. Véniard, Phys. Rev. A **32**, 2534 (1985).

[11]. M. J. Conneely and S. Geltman, J. Phys. B **14**, 4847 (1981).

[12]. S. Jetzke and J. T. Broad, J. Phys. B **19**, L199 (1986).

[13]. L. Hostler, J. Math. Phys. **5**, 591 (1964).

[14]. L. Hostler, J. Math. Phys. **11**, 2966 (1970).

[15]. A. Maquet, Phys. Rev. A **15**, 1088 (1977).

[16]. A. Lami and N. K. Rahman, J. Phys. B **16**, L201 (1983).

[17]. J. T. Broad, Phys. Rev. A **31**, 1494 (1985).

[18]. F. E. Low, Phys. Rev. **110**, 974 (1958).

[19]. M. Ya. Amusia, A. S. Baltenkov and A. A. Paiziev, Sov. Phys. JETP Lett. **24**, 332 (1976); G. Wendin and K. Nuroh, Phys. Rev. Lett. **39**, 48 (1977); for a recent review, see for instance: M. Ya. Amusia, Comments At. Mol. Phys. **11**, 123 (1982).
[20]. J. Schwinger, J. Math. Phys. **5**, 1606 (1964).
[21]. N. M. Kroll and K. M. Watson, Phys. Rev. A **8**, 804 (1973).
[22]. M. Gavrila and J. Z. Kaminski, Phys. Rev. Lett. **51**, 613 (1984).

# Electron Scattering in Intense, High-Frequency Laser Fields

*M. Gavrila*[1], *J.Z. Kaminski*[2], *and M.J. Offerhaus*[1]

[1]FOM-Institute for Atomic and Molecular Physics, Kruislaan 407, NL-1098 SJ Amsterdam, The Netherlands
[2]Institute of Theoretical Physics, Warsaw University, PL-00-681 Warsaw, Poland

## I. Introduction

In view of the continued progress in the development of intense, high-frequency lasers (see Refs. [1], [2]) a theory was recently presented for electron-atom collisions in such fields [3], [4]. The theory was first applied to one-electron systems, i.e. to the collision of an electron with an atomic potential. We shall review in the following results obtained in exploratory calculations done for the Coulomb [5] and Yukawa [6] potentials. The Coulomb case, apart from its intrinsic interest, is relevant for scattering from ions, as occurring in laser-driven fusion plasmas. The Yukawa case is relevant to scattering from neutral atoms, if finer details (such as the deviation of the latter from the atomic selfconsistent potential, the polarization potential, etc.) are ignored.

The radiation field was assumed to be a purely monochromatic plane wave (frequency $\omega$), linearly polarized (real polarization vector $\vec{\varepsilon}$), in the dipole approximation. In the *high-intensity, high-frequency regime* we are considering, it was shown [3],[4] that the semiclassical time-dependent Schrödinger equation reduces in *first approximation* to the time-independent equation:

$$[\tfrac{1}{2}\vec{p}^{\,2} + V_0(\vec{\alpha}_0,\vec{r})]\,\psi = E\psi \quad , \tag{1}$$

where

$$\vec{\alpha}_0 = \alpha_0\vec{\varepsilon} \;, \qquad \alpha_0 = I^{\frac{1}{2}}\,\omega^{-2} \;, \tag{2}$$

and I is the intensity. (We are using atomic units; the a.u. of intensity is $3.51 \cdot 10^{16}$ W/cm$^2$.) Eq.(1) contains the static potential $V_0(\vec{\alpha}_0,\vec{r})$, which we denote as the "dressed" potential. It is given by

$$V_0(\vec{\alpha}_0,\vec{r}) = \frac{1}{\pi}\int_{-1}^{+1} V(\vec{r}+\vec{\alpha}_0 u)\,\frac{du}{(1-u^2)^{\frac{1}{2}}} \;. \tag{3}$$

Here $V(\vec{r})$ is the original potential acting on the electron. The dressed potential depends on the laser parameters $\omega$ and I only through the parameter $\alpha_0$. The values of $\alpha_0$ of present interest extend from 0 to about 100.

Eq.(1) determines the scattering to lowest order in our regime. Note that the scattering can only be elastic, showing that the system cannot undergo multiphoton free-free transitions. These become possible only in the second approximation of our high-intensity, high-frequency regime. However, these will not be discussed here and we shall restrict ourselves to the elastic case.

If the original potential V(r) is spherically symmetric, the dressed potential $V_o(\vec{\alpha}_o,\vec{r})$ has axial symmetry around $\vec{\alpha}_o$ (or $\vec{e}$). This considerably complicates the scattering problem since the azimuthal quantum number $\ell$ is no longer conserved. Here we shall describe a simplified approach which consists of modelling $V_o(\vec{\alpha}_o;\vec{r})$ by its spherical average

$$\bar{V}_o(\alpha_o,r) = \frac{1}{4\pi} \int V_o(\vec{\alpha}_o,\vec{r})d\Omega_{\vec{\alpha}_o} \quad . \tag{4}$$

The scattering problem is thus reduced to a phase-shift calculation. The use of $\bar{V}_o$ instead of $V_o$, while washing out some interesting features of the exact calculation, nevertheless keeps many of the salient features of the problem, especially at low energies where $\alpha_o p \lesssim 1$ (with p the electron momentum). This has been substantiated in the Coulomb case by comparison with the recently completed exact calculation for $V_o$ [7].

The exact dressed potential, Eq.(3), can be looked upon as created by a linear distribution of "charges" extending from $-\alpha_o$ to $+\alpha_o$ along $\vec{e}$, with density

$$\sigma(\xi) = \frac{1}{\pi\alpha_o} \left[ 1 - \left(\frac{\xi}{\alpha_o}\right)^2 \right]^{-\frac{1}{2}} \quad , \tag{5}$$

the unit of "charge" generating the potential V(r). On the other hand, as is easily seen, $\bar{V}_o(\alpha_o,r)$ is the potential energy due to a spherically symmetric distribution of "charges" extending up to $r = \alpha_o$ , of radial density:

$$\tau(r) = 2\sigma(r) \quad , \tag{6}$$

with the function $\sigma$ defined by Eq.(5). Our scattering problem resembles thus that of an electron probing an extended charge nucleus [8] of density $\tau(r)$ and radius $\alpha_o$.

## II. Coulomb potential

We shall first consider the case of the Coulomb potential $V(r) = -Z/r$. By applying Eqs.(3) and (4) one finds

$$\bar{V}_o(\alpha_o, r) = \frac{-Z}{\pi\alpha_o\rho}\left[2\arcsin\rho - \rho\ln\frac{1-(1-\rho^2)^{\frac{1}{2}}}{1+(1-\rho^2)^{\frac{1}{2}}}\right], \quad \rho \leq 1$$

$$= -\frac{Z}{\alpha_o\rho}, \quad \rho \geq 1, \tag{7}$$

where $\rho = r/\alpha_o$. For $r \geq \alpha_o$, $\bar{V}_o$ coincides with the original Coulomb potential. At the origin $\bar{V}_o$ has a *logarithmic singularity*. This is much weaker than that of the Coulomb potential.

Because of the long-range behavior of the potential, Eq.(7), the scattering amplitude is given by the theory for *modified Coulomb scattering*:

$$f(\theta) = f_c(\theta) + f'(\theta), \tag{8}$$

$$f_c(\theta) = \frac{(-\gamma)}{2p\sin^2\frac{\theta}{2}}\exp(-i\gamma\ln\sin^2\frac{\theta}{2} + 2i\sigma_o), \tag{9}$$

$$f'(\theta) = \frac{1}{2ip}\sum_\ell (2\ell+1)e^{2i\sigma_\ell}(e^{2i\delta_\ell} - 1)P_\ell(\cos\theta). \tag{10}$$

Here $\theta$ is the scattering angle, $\sigma_\ell$ is the Coulomb phase, $f_c$ is the Coulomb amplitude, $\delta_\ell$ is the extra phase due to $\bar{V}_o - V$ and $f'$ is the corresponding amplitude. The modified cross section is

$$\frac{d\sigma}{d\Omega} = \frac{d\sigma_c}{d\Omega} + 2\,\mathrm{Re}\, f_c^* f' + |f'(\theta)|^2, \tag{11}$$

where the first term at the right is the Rutherford cross-section and the second represents the interference of the Coulomb and short-range amplitudes. Eqs.(8)-(11) are similar to those used in the analysis of the classical proton-proton collision experiments at low (nuclear) energies with the short-range nuclear force taken into account, see Ref. [8].

In Fig. 1 we present the angular dependence of the ratio $R = (d\sigma/d\Omega)/(d\sigma_c/d\Omega)$ of the modified cross-section, Eq.(11), to the Rutherford cross-section, at $Z = 1$, for a number of electron energies E and values of $\alpha_o$.

In all cases $R \to 1$ as $\theta \to 0$. This is due to the fact that $f_c(\theta)$ and $d\sigma_c/d\Omega$ become infinite as $\theta \to 0$, whereas $f'(\theta)$ stays finite. Moreover, for $\theta \to 0$, R has infinitely many oscillations of increasing frequency and decreasing amplitude. These are due to the presence of the Coulomb phase factor, see Eq.(9), in the interference term of eq.(11) (but not in $d\sigma_c/d\Omega$). The amplitude of these *Coulomb-interference oscillations* decreases for $\theta \to 0$ because the relative importance of the interference

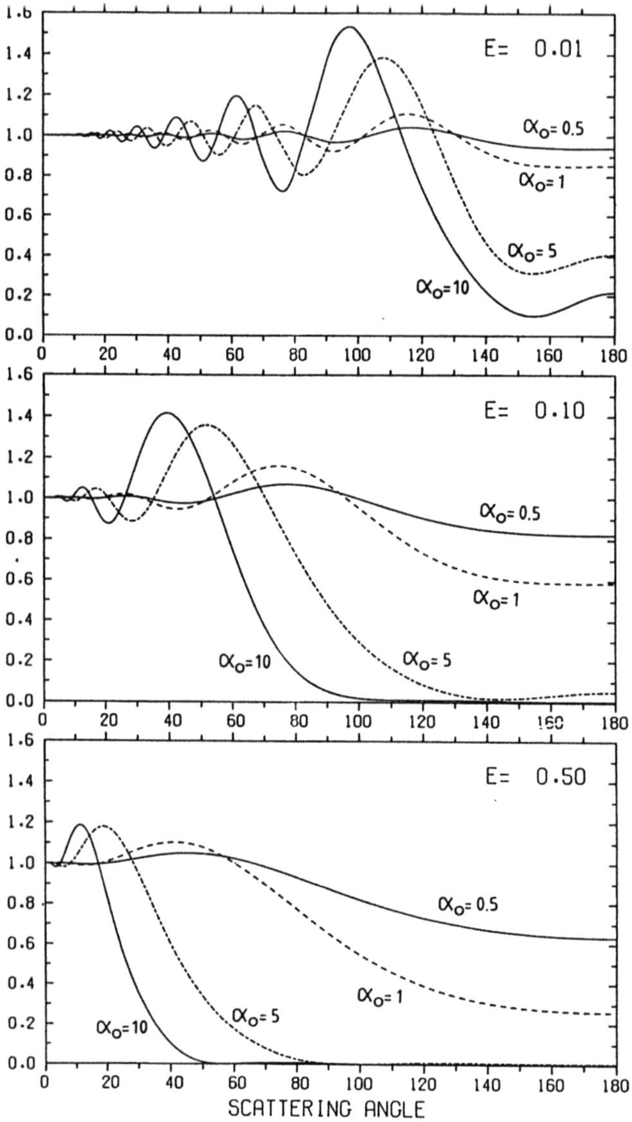


*Fig. 1. Ratio R of the elastic scattering cross-section for the averaged dressed Coulomb potential, Eq.(7), to the Rutherford cross-section. Nuclear charge Z = 1. Values of the electron energy E (in Ry) and of $\alpha_o$, as indicated.*

term in Eq.(11) diminishes compared to $d\sigma_C/d\Omega$. Note that, as seen in Fig. 1, the Coulomb-interference oscillations occur in an accessible range of experimental parameters.

For large scattering angles Fig. 1 shows that the modified cross-section is considerably reduced with respect to the original value, because of destructive interference in Eq.(11). The scattering ampli-

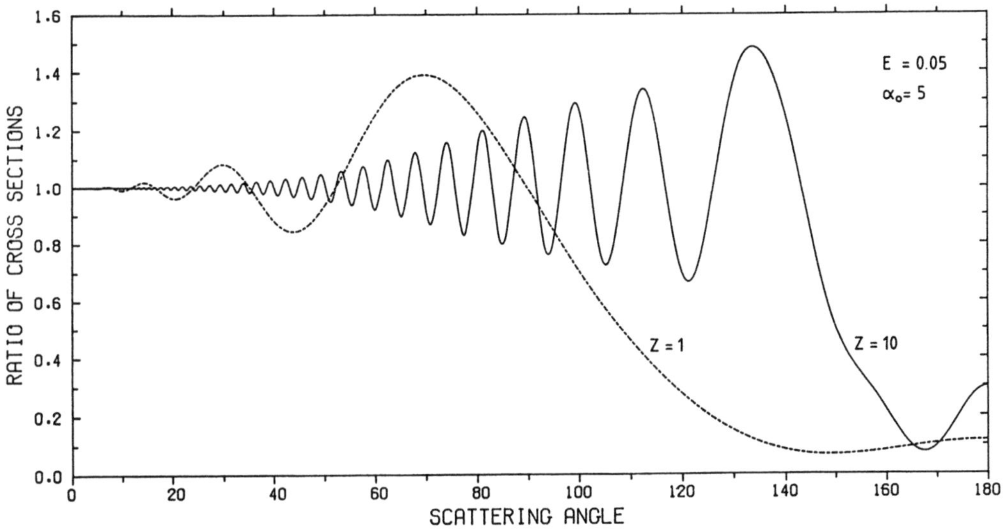


*Fig. 2. Ratio of cross-sections R for Z = 1 and Z = 10, at E = 0.05 Ry and $\alpha_0 = 5$.*

tude for nuclear charge Z is related to that for nuclear charge 1 by the scaling law:

$$f(Z,\alpha_0,E,\theta) = Z^{-1}\, f(1,\alpha_0 Z,E/Z^2,\theta) \quad . \tag{12}$$

In Fig. 2 we display the effect of the increase of the nuclear charge Z at given E and $\alpha_0$.

## III. Yukawa potential

We now turn to the Yukawa potential $V(r) = -Z\, e^{-\mu r}/r$. By inserting it in Eqs.(3) and (4) we find

$$\bar{V}_0(\alpha_0,r) = -\frac{Z}{\pi\mu\alpha_0}\,\frac{1}{r}\int_0^1 \frac{1}{u(1-u^2)^{\frac{1}{2}}} \times \left(e^{-|r-\alpha_0 u|\mu} - e^{-(r+\alpha_0 u)\mu}\right)du. \tag{13}$$

Consequently, $\bar{V}_0(\alpha_0,r)$ has different analytic forms according as $r < \alpha_0$ or $r > \alpha_0$:

$$\bar{V}_0(\alpha_0,r) = -\frac{2Z}{\pi\mu\alpha_0}\,\frac{1}{r} \tag{14}$$

$$\times\left\{e^{-\mu r}\int_0^{r/\alpha_0} \operatorname{sh}\alpha_0\mu u\,\frac{du}{u(1-u^2)^{\frac{1}{2}}} + \operatorname{sh}\mu r\int_{r/\alpha_0}^1 e^{-\alpha_0\mu u}\,\frac{du}{u(1-u^2)^{\frac{1}{2}}}\right\}\ , \qquad r \leqq \alpha_0\ ;$$

$$= -g(\alpha_0\mu)\,\frac{Z}{r}\,e^{-\mu r}\ , \qquad r \geqq \alpha_0\ .$$

We have denoted

$$g(x) = \frac{2}{\pi x} \int_0^1 \text{sh}\, xu \frac{du}{u(1-u^2)^{\frac{1}{2}}} \quad . \tag{15}$$

Note that for $\alpha_o \to 0$, Eq.(14) goes over into the original Yukawa potential. For $\mu \to 0$, on the other hand, it agrees with the corresponding expression for the Coulomb case, Eq.(7).

The radial dependence of the averaged dressed potential $\bar{V}_o(\alpha_o,r)$ is represented for $\mu = 2$ and various values of $\alpha_o$ in Fig. 3. Note that near the origin, for all values of $\alpha_o$, $\bar{V}_o(\alpha_o,r)$ is weaker than $V(r)$, as in the case of the Coulomb potential.

Fig. 3 shows that at an r-value somewhat below $\alpha_o$, the curve of $\bar{V}_o(\alpha_o,r)$ intersects that of $V(r)$; from here on $\bar{V}_o$ is more negative than $V$, and for $r > \alpha_o$ it becomes a Yukawa potential of charge $Zg(\alpha_o\mu)$.

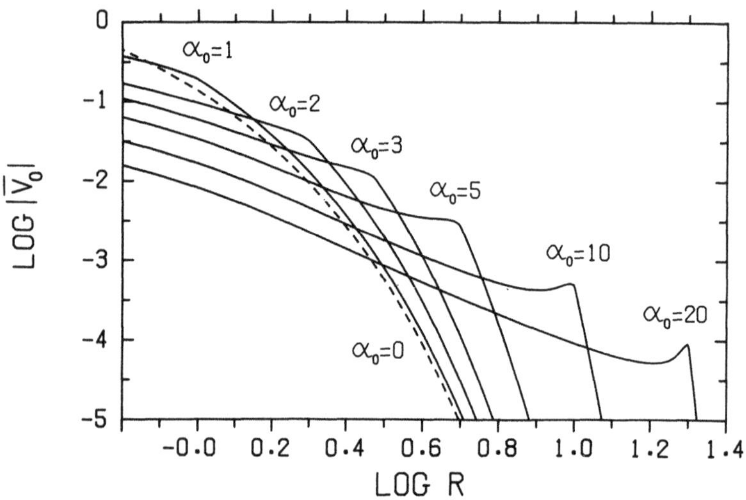


*Fig. 3. Radial dependence of the averaged dressed Yukawa potential, Eq.(14), for $\mu = 2$ and various values of $\alpha_0$.*

The factor $g(x)$ is always larger than 1; for example at $x = 20$ (e.g. $\mu = 2$ and $\alpha_o = 10$), $g(x) = 2.2 \times 10^3$. At these distances however, $\bar{V}_o$ is very weak anyway.

At high $\alpha_o$, $\bar{V}_o(\alpha_o,r)$ eventually becomes small with respect to the kinetic energy in all relevant parts of space, and we expect the Born approximation to become valid. For this we need the Fourier transform of $\bar{V}_o(\alpha_o,r)$. This can be obtained from the Fourier transform of $V_o(\vec{\alpha}_o,\vec{r})$ of Eq.(3), which is given by $\mathcal{V}_o(\vec{\alpha}_o,\vec{k}) = \mathcal{V}(k) J_o(\vec{\alpha}_o\vec{k})$. Here $\mathcal{V}(k)$ is the transform of the original Yukawa potential, and $J_o(x)$ is the Bessel function of zeroth order. Taking into account Eq.(4), the transform of $V_o(\alpha_o,r)$ is

$$\mathcal{V}_o(\alpha_o,k) = \mathcal{V}(k)Q(\alpha_o k) \quad , \qquad \text{where} \tag{16}$$

$$Q(u) = \frac{1}{u}\int_0^u J_o(x)\,dx \quad . \tag{17}$$

Since $\mathcal{V}(k) = -(Z/2\pi^2)(k^2+\mu^2)^{-1}$, the Born approximation for the cross-section becomes

$$\frac{d\sigma^B}{d\Omega} = \frac{4Z^2}{(\Delta^2+\mu^2)^2}\,Q^2(\alpha_o\Delta) \quad , \tag{18}$$

with $\Delta = 2p\sin\theta/2$, the momentum transfer and $\theta$ is the scattering angle. For $\alpha_o = 0$ this reduces to the Yukawa case, since $Q(0) = 1$. Note also that at $\theta = 0$, Eq.(18) yields $d\sigma^B/d\Omega = 4Z^2\mu^{-4}$ for any E and $\alpha_o$. We shall discuss below to what extent Eq.(18) is approached in the numerical computation at increasing E and/or $\alpha_o$.

Our numerical results for the differential cross-section at $Z = 1$ are shown in Figs. 4 and 5. We have chosen for illustration two representative values of $\mu$: $\mu = 2$ and $\mu = 0.5$. The transition between these characteristic types of cross-section behavior lies at about $\mu = 1.3$. The range $\mu > 1.3$ also corresponds to the values $\mu = 0.98\,Z^{1/3}$ for which, as Z varies, the Yukawa potential yields fairly good K and L shell binding energies [9].

In Figs. 4 and 5 each row refers to one energy value E, given in rydbergs, and in a sequence of five frames we show the dependence of the cross-section on the scattering angle $\theta$ for growing values of $\alpha_o$ (full curves). For those $\alpha_o$ for which the Born approximation comes near the exact cross-section, we have represented it too (dashed curves). At large $\alpha_o$, the agreement between the exact calculation and the Born approximation is in some cases so good that only one distinct curve can be traced.

The *dependence on* E of the cross-section, at given $\mu$ and $\alpha_o$, shows the known pattern. At sufficiently low E only the $\ell = 0$ phase shift contributes significantly and the scattering is nearly isotropic (e.g. Fig. 4, top left). As the energy increases more phase shifts contribute and anisotropy sets in (e.g. Fig. 4, bottom left). At still larger values of E the cross-section clearly approaches the Born approximation (especially for the larger values of $\alpha_o$, as seen in Figs. 4 and 5). Note that the Born approximation Eq.(18) contains the function $Q^2(u)$, which has an oscillatory behavior similar to $J_o^2(x)$, and makes the cross-section vanish at certain angles. This behavior was emphasized earlier [10], and can be barely seen on the scale of Figs. 4 and 5 (see the cases $E = 0.1$ and $0.5$ of Fig. 5).

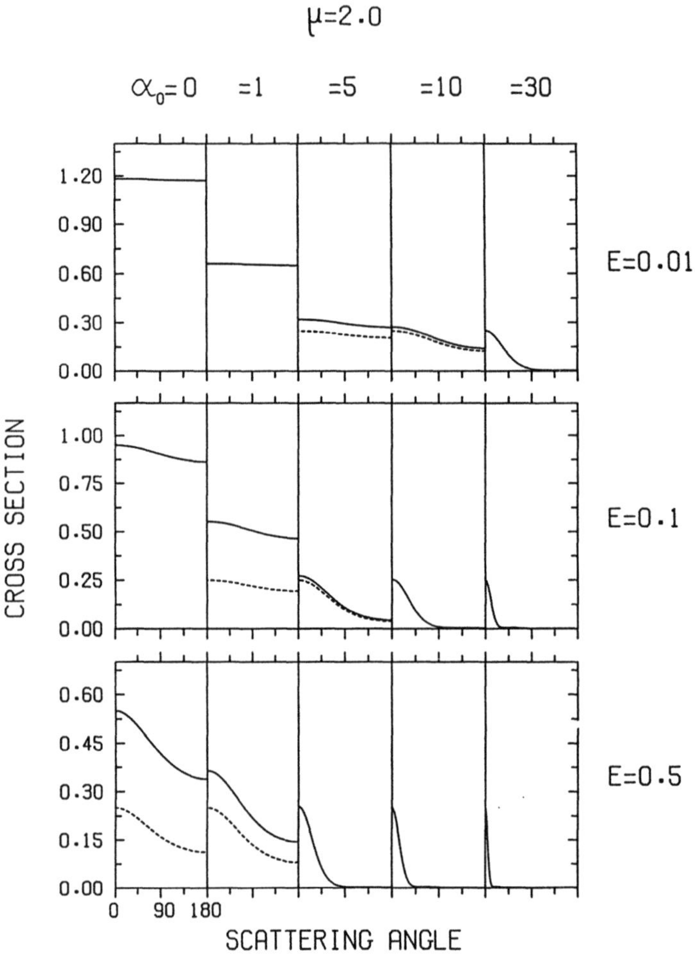


*Fig. 4. Modified elastic scattering cross-section for the dressed Yukawa potential with $\mu = 2$, see Eq.(14), as function of the scattering angle $\theta$, for various energies E (in rydbergs) and values of $\alpha_0$. $\theta$ varies in each frame from 0° to 180°.*

The *dependence on* $\alpha_0$ at fixed $\mu$ and E reflects the deformation of the dressed potential $\bar{V}_0(\alpha_0,r)$. For increasing $\alpha_0$, different patterns arise according as $\mu \gtrsim 1.3$ (Fig. 4) and $\mu \lesssim 1.3$ (Fig. 5). In both figures we look at the frames for $E = 0.01$ and increasing $\alpha_0$ (top row). In the frames on the left the $\ell = 0$ wave dominates, and the scattering is nearly isotropic. However, the cross-section may decrease with $\alpha_0$, as in Fig. 4, or increase, as in Fig. 5. (An explanation for this is given in [6].)

The Born approximation is reached for any E if $\alpha_0$ is increased to the point that $\bar{V}_0(\alpha_0,r)$ is sufficiently depressed with respect to E. This is apparent in all rows in Figs. 4 and 5. (The statement is not

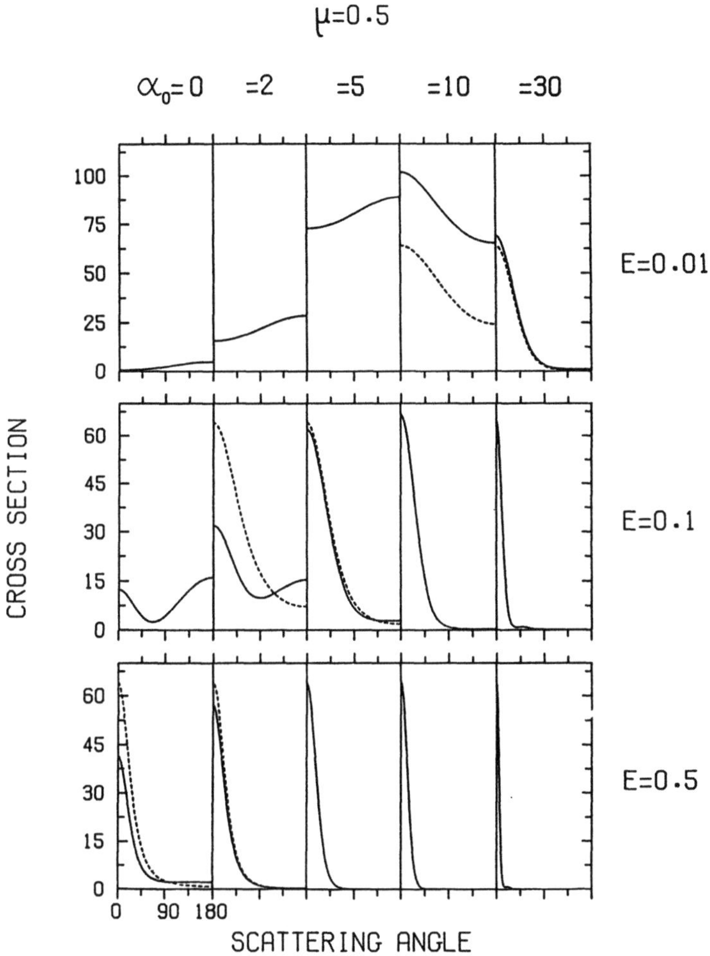


*Fig. 5. Same as Fig. 4, but with* $\mu = 0.5$ .

always true for a Coulomb potential ; this will be discussed elsewhere.) We also note that the cross-sections for forward scattering tend in the Born approximation to $4\mu^{-4}$ which is independent of E and $\alpha_o$, as discussed above in connection with Eq.(18).

The effect of changing the charge of the potential from 1 (as chosen in Figs. 4,5) to Z is given by the scaling law for the scattering amplitude:

$$f(Z,\mu,\alpha_o,E,\theta) = Z^{-1} f(1,\mu/Z,\alpha_o Z,E/Z^2,\theta) \quad . \tag{19}$$

This generalizes Eq.(12).

## IV. Conclusions

In the high-frequency, high-intensity regime the electron cannot absorb or emit photons during the collision, and, with a potential to model the atom, only elastic scattering can happen. The exploratory calculations we have presented reveal, under circumstances, large departures from laser-free scattering, and certain salient features expected to be present also in a more accurate handling of the problem. Such are the appearance of "Coulomb interference oscillations" and, for the Yukawa case, the validity of the Born approximation at sufficiently large values of $\alpha_0$, whatever the energy.

## References


1. M. Gavrila, in: Atomic Physics, Vol.9, eds. R. van Dyck and E.N. Fortson (World Scientific, Singapore, 1985) p.523.
2. C.K. Rhodes, Science 229, 1345 (1985).
3. M. Gavrila and J.Z. Kaminski, Phys.Rev.Lett 52, 613 (1984), and to be published.
4. M. Gavrila, in: Fundamentals of laser interactions, ed. F. Ehlotzky (Lecture notes in physics, vol.229, Springer, 1985), p.3.
5. M.J. Offerhaus, J.Z. Kaminski and M. Gavrila, Phys.Lett. 112 A, 151 (1985).
6. M. Gavrila, M.J. Offerhaus and J.Z. Kaminski, Phys.Lett., in press.
7. J. van de Ree, J.Z. Kaminski and M. Gavrila, to be published.
8. L.R. Elton, Nuclear Sizes (Oxford University Press, 1961).
9. C.S. Lai, Phys.Rev. A 23, 455 (1981).
10. F.H. Faisal, Phys.Lett. 50 A, 193 (1974).

# Applications of Generalized Coherent States

*J. Bergou*[1], *S. Varró*[1], *and F. Ehlotzky*[2]

[1]Central Research Institute for Physics, Hungarian Academy of Sciences, P.O. Box 49, H-1525 Budapest, Hungary
[2]Institute for Theoretical Physics, University of Innsbruck, A-6020 Innsbruck, Austria

## 1. Introduction

Generalized coherent states have the advantage of forming on the one hand a complete and orthonormal set of states, which can be conveniently applied in perturbation calculations, and, on the other hand, these states preserve the quasi-classical features of a quantum mechanical process. We shall demonstrate this by considering two typical examples from laser and plasma physics.

## 2. Scattering in a Quantized Radiation Field

As a first example, we consider potential scattering of electrons in a quantized radiation field /1/. Following BERGOU and VARRO /2/ we introduce generalized coherent states for electrons of mass M and charge -e in a quantized, linearly polarized, monochromatic radiation field of frequency $\omega$ in the dipole approximation. We start from the Schrödinger equation

$$i\hbar\partial_t|\Psi\rangle = H|\Psi\rangle, \quad H = H_e + H_f + H_I$$
$$H_I = -(e/Mc)\vec{A}\cdot\vec{p}, \quad \vec{A} = \alpha\vec{\varepsilon}(a^+ + a), \quad \alpha = c\left(\frac{2\pi\hbar}{\omega V}\right)^{1/2}, \tag{1}$$

and we look for stationary solutions of the form

$$|\Psi\rangle = |\phi\rangle e^{-iEt/\hbar}, \quad |\phi_{\vec{p},n}\rangle = |\vec{p}\rangle D_{\sigma_p}|n\rangle, \tag{2}$$

where $|\vec{p}\rangle$ are momentum eigenstates of the electron and $|n\rangle$ are number states of the field oscillator. The unitary displacement operator $D_\sigma$, accomplishing the particle-field coupling and thereby creating generalized coherent states, is defined by

$$D_{\sigma_p} = \exp[\sigma_p(a^+ + a)], \quad \sigma_p = \frac{e\alpha}{mc\hbar\omega}\vec{p}\cdot\vec{\varepsilon}. \tag{2a}$$

The energy eigenvalues corresponding to (2) are given by

$$E_{\vec{p},n} = \frac{\vec{p}^2}{2m} + \hbar\omega(n + 1/2). \tag{2b}$$

In (1) we have neglected the $A^2$-part of the interaction, which only yields an overall Stark shift. The states (2) fulfil the orthonormality and completeness relations

$$\langle\phi_{\vec{p}',n'}|\phi_{\vec{p},n}\rangle = \delta_{n',n}\,\delta(\vec{p}' - \vec{p}), \quad \sum_n\int d^3p|\phi_{\vec{p},n}\rangle\langle\phi_{\vec{p},n}| = 1. \tag{3}$$

Now we use these states to treat potential scattering of electrons in a quantized radiation field. The transition matrix element $T_{fi}$ of this process can be decomposed into an infinite sum of incoherent contributions $T_n$, depending on the number n of quanta $\hbar\omega$ emitted or absorbed during the scattering event. Thus we obtain in first and second order Born approximation with respect to the scattering potential,

$$T_n^{(1)} = -2\pi i U(\vec{Q})\langle n_f|D_{\sigma_i-\sigma_f}|n_i\rangle\delta(E_f^{(n)} - E_i), \tag{4}$$

where the energies of the initial and final states are given by

$$E_i = \frac{\vec{p}_i^{\,2}}{2m} + n_i\hbar\omega\ , \quad E_f^{(n)} = \frac{\vec{p}_f^{\,2}}{2m} + n_f\hbar\omega\ , \quad n_f = n_i - n. \tag{4a}$$

In (4), the multiplication rule of the D-operators has been used, and $U(\vec{Q})$ is the Fourier transform of the scattering potential as a function of the momentum transfer $\vec{Q} = \vec{p}_f - \vec{p}_i$. In the second order Born approximation we get

$$T_n^{(2)} = -2\pi i \int d^3p' \frac{U(\vec{p}_f - \vec{p}\,')U(\vec{p}\,' - \vec{p}_i)}{E' - E_i^{(0)} - k_{eff}\hbar\omega + i\eta} \langle n_f|D_{\sigma_i - \sigma_f}|n_i\rangle \delta(E_f^{(n)} - E_i), \tag{5}$$

where a detailed analysis shows that the effective number of radiation quanta present in the intermediate states can be approximated by

$$k_{eff} \cong n(\vec{p}\,' - \vec{p}_i)\cdot\vec{\varepsilon}\,[\vec{Q}\cdot\vec{\varepsilon}\,(1 + \gamma n/2n_i)]^{-1}, \ \frac{n}{n_i} < 1, \ |\gamma| \cong 1. \tag{6}$$

This procedure can be carried through to all orders of the Born-series by introducing renormalized particle momenta and energies,

$$\tilde{\vec{p}}_n = \vec{p} - n\hbar\omega m\vec{\varepsilon}\,[\vec{Q}\cdot\vec{\varepsilon}\,(1 + \gamma n/2n_i)]^{-1}, \quad \tilde{E}_n = \frac{\tilde{\vec{p}}_n^{\,2}}{2m}. \tag{6a}$$

Retaining linear terms in $n/n_i$, the total renormalized transition amplitude $T_n$ for a particular nonlinear process of the order $n$ ($n > 0$ stimulated absorption, $n < 0$ stimulated emission) can thus be written

$$T_n = -2\pi i[t_n(\tilde{E}_i,\vec{Q}) + n\hbar\omega\frac{\gamma n}{2n_i}\frac{\partial t_n}{\partial \tilde{E}_i}]\,\delta(\tilde{E}_f - \tilde{E}_i), \tag{7}$$

where

$$t_n(\tilde{E}_i,\vec{Q}) = t_{el}(\tilde{E}_i,\vec{Q})\langle n_f|D_{\sigma_i - \sigma_f}|n_i\rangle, \tag{7a}$$

with $t_{el}(\tilde{E}_i,\vec{Q})$ being the renormalized elastic amplitude at the energy $\tilde{E}_i$. For $n_i \to \infty$ we recover from (7) and (7a) the Kroll-Watson relation /3/, in which case $\langle n_f|D_{\sigma_i-\sigma_f}|n_i\rangle \to J_n(2\sigma_{fi}\sqrt{n_i})$, and $J_n$ is an ordinary Bessel function of the order n. Hence, for an infinite number $n_i$ of quanta in the radiation field mode, the classical field representation is recovered. Somewhat more complicated results have been obtained by ROSENBERG /4/.

## 3. Scattering in a Magnetic Field and in a Microwave

In our second example, we consider potential scattering of electrons in a constant homogeneous magnetic field oriented along the z-axis, and in a monochromatic circularly polarized microwave propagating in the z-direction and treated in the dipole approximation. The starting point of our investigation is basically SCHRÖDINGER's original work of 1926 /5/, in which he generates a coherent state from a displaced harmonic oscillator ground state $|o\rangle$. This idea can be generalized to excited states $|n\rangle$. The corresponding wave packets then follow classical trajectories. We take up this concept, to generate a class of generalized coherent states for electrons in a magnetic field and in a microwave, by applying appropriate displacement operators to the Landau states $|l,m\rangle$ in an (x,y) symmetric gauge. These generalized coherent states form an orthonormal and complete set and may be conveniently applied as "in" and "out" states in perturbation theory, in particular, in the treatment of electron scattering.

First, we solve the Schrödinger equation for an electron in the following external field configuration

$$\vec{A}^{ext} \equiv (-\tfrac{1}{2}By + \tfrac{cF}{\omega}\cos\omega t,\ \tfrac{1}{2}Bx + \tfrac{cF}{\omega}\sin\omega t,\ o), \tag{8}$$

where F and B are the electric and magnetic field strengths, respectively, and $\omega$ is the frequency of the microwave. The corresponding solution of the Schrödinger equation may be written in the form

$$|\Psi\rangle_t = |p_z\rangle_t D(\sigma)|\Phi\rangle_t e^{-i\Delta Et/\hbar}, \tag{9}$$

where $|p_z\rangle_t$ is a free particle state, describing the electron motion in the z-direction. $D(\sigma)$ is a displacement operator, similar to (2a), which couples the particle to the microwave

$$D(\sigma) = e^{\sigma A^+ - \sigma^* A}, \quad \sigma = i\lambda_0\sqrt{\gamma}e^{i\omega t} . \tag{10}$$

The operators $A^+$ and A describe collective excitations and deexcitations of the electron motion in the (x,y)-plane. These operators are related to the corresponding operators $(a^+,a)$ and $(b^+,b)$ of the motion in the x- and y-directions by

$$A^+ = \frac{1}{\sqrt{2}}(a^+ + ib^+), \quad A = \frac{1}{\sqrt{2}}(a - ib). \tag{10a}$$

Moreover, we have introduced in (10) the parameters

$$\lambda_0 = \frac{eF}{M\omega c}\frac{\omega}{\Delta\omega}\bar{\lambda}, \quad \bar{\lambda} = \frac{c}{\omega}, \quad \Delta\omega = \omega - \omega_c \neq 0 ,$$
$$\gamma = \frac{M\omega_c}{2\hbar} = \lambda_B^{-2}, \quad \omega_c = \frac{eB}{Mc} , \tag{10b}$$

where $\lambda_0$ is the classical amplitude of the electron oscillation in the microwave, with $\Delta\omega$ being the detuning of the microwave frequency $\omega_0$ from the cyclotron frequency $\omega_c$. $\lambda_B$ is a characteristic length in our magnetic field problem. Finally, we have in (9) the state $|\Phi\rangle_t$, which is an exact solution of the Schrödinger equation for a particle in a magnetic field alone, and $\Delta E$ is an overall energy shift, which is of little concern to us here. For $|\Phi\rangle_t$ we could take the complete set of stationary Landau states

$$|\Phi\rangle_t = |l,m\rangle_t = |l,m\rangle e^{-i\omega_{lm}t}, \tag{11}$$

where l and m are the radial and azimuthal quantum numbers, respectively, and $\omega_{lm}$ are the corresponding eigenfrequencies of the states $|l,m\rangle$. In plane polar coordinates $(\rho,\phi)$, we obtain /6/

$$\langle x,y|l,m\rangle = \Phi_{l,m}(\rho,\phi) = \left(\frac{\gamma}{\pi}\right)^{1/2}\left[\frac{l!}{(l+|m|)!}\right]^{1/2} e^{im\phi}$$
$$\times e^{-\xi^2/2}\,\xi^{|m|/2}\,L_l^{|m|}(\xi), \quad \xi = \gamma\rho^2 , \tag{11a}$$

where $L_l^{|m|}(\xi)$ are associated Laguerre polynomials, and

$$\omega_{l,m} = \omega_c[l + \tfrac{1}{2}(m + |m| + 1)], \quad l = 0,1,2,\ldots \quad m = 0,\pm1,\pm2,\ldots \tag{11b}$$

Instead of using the number states (11) and (11a,b), we next introduce, as a first step, generalized coherent states for electrons in a magnetic field alone. Special cases of these states have been considered by MALKIN and MAN'KO /7/ and by VARRÓ /8/. These states are defined by

$$|\Phi\rangle_t = |\alpha,\beta;l,m\rangle_t \equiv e^{-iH_B t/\hbar} D_a(\alpha_1) D_b(\beta_1)|l,m\rangle , \tag{12}$$

where $\alpha \equiv \frac{1}{2}(\alpha_1 - i\beta_1)$ and $\beta \equiv \frac{1}{2}(\alpha_1 + i\beta_1)$ are arbitrary complex parameters and the displacement operators $D_a$ and $D_b$ are given by

$$D_a(\alpha_1) = e^{\alpha_1 a^+ - \alpha_1^* a}, \quad D(\beta_1) = e^{\beta_1 b^+ - \beta_1^* b}, \tag{12a}$$

while $H_B$ is the Hamiltonian of an electron, embedded in a magnetic field

$$H_B = \frac{\hbar\omega_c}{2}(a^+a + b^+b + 1) + \frac{i\hbar\omega_c}{2}(ab^+ - a^+b). \tag{12b}$$

In a second step, we couple a particle described by (12) to the microwave by applying the corresponding displacement operator $D(\sigma)$ given by (10) and (10a,b). This yields the generalized coherent states, which we are looking for,

$$D(\sigma)|\Phi>_t = D(\sigma)|\alpha,\beta;l,m>_t . \qquad (13)$$

Now we define the classical complex position of an electron in its transverse motion in the external fields by $u(t) = x_c(t) - iy_c(t)$ and obtain as solution of the Lorentz equation of motion

$$u(t) = i\lambda_o e^{-i\omega t} + i\frac{v_o}{\omega_c} e^{-i\omega_c t} + u_o . \qquad (14)$$

Then we can show that with each arbitrary pair of classical initial values $(u_o, v_o)$ a pair of complex parameters $(\alpha,\beta)$ can be uniquely associated as follows

$$\alpha = i\frac{v_o}{\omega_c}\left(\frac{\gamma}{2}\right)^{1/2} , \quad \beta = u_o^*\left(\frac{\gamma}{2}\right)^{1/2} . \qquad (14a)$$

Hence we shall denote our generalized coherent states by

$$D|\alpha,\beta;l,m>_t \equiv |u(t);l,m> , \qquad (15)$$

and the time development of the probability distributions of these states follow the corresponding classical trajectories, namely,

$$|<\vec{x}_\perp|u(t);l,m>|^2 = |\Phi_{l,m}(\vec{x}_\perp - \vec{x}_c(t))|^2 , \qquad (15a)$$

where the Landau states $\Phi_{l,m}$ are given by (11a). The states (15) moreover fulfil the orthonormality and completeness relations

$$<u(t);l,m|u(t);l',m'> = \delta_{l',l}\,\delta_{m',m} , \quad \sum_{l,m} |u(t);l,m><u(t);l,m| = 1 . \qquad (15b)$$

As an example of the application of the states (9), (13) and (15), we consider electron scattering by a screened Coulomb potential $V(r) = (A/r)\exp(-k_D r)$, where $k_D$ is the Debye wave number. The interaction with the potential will be treated in the first order Born approximation. Furthermore, we only consider transitions $l = m = 0 \rightarrow l' = m' = 0$, while the classical electron trajectory changes from $u(t)$ to $u'(t)$ during the scattering process. Corresponding to the use of generalized coherent states, the following quasiclassical transition matrix elements and cross-sections can be derived, which describe various external field-induced nonlinear processes,

$$T_{fi} = -2\pi i \sum_{n,s=-\infty}^{+\infty} \delta\left(\frac{p_z'^2 - p_z^2}{2M} + n\hbar\omega + s\hbar\omega_c\right) M_{fi}^{(n,s)} , \qquad (16)$$

and

$$\sigma^{(n,s)} = \left(\frac{\pi}{\gamma}\right) \frac{(2\pi)^2}{v_z' v_z} |M_{fi}^{(n,s)}|^2 , \qquad (17)$$

where $v_z = p_z/m$ and $v_z' = p_z'/m$ are the particle velocities in the z-direction before and after the scattering. The square of the moduli of the matrix elements $M_{fi}^{(n,s)}$ is determined by

$$|M_{fi}^{(n,s)}|^2 = \left(\frac{A}{\pi\hbar}\right)\left|\frac{u_o'}{u_o}\right|^2 \left|\frac{v_o'}{v_o}\right|^2 e^{-\gamma|u_o'-u_o|}\, e^{-(\gamma/\omega_c^2)|v_o' - v_o|}\, |I_{n,s}|^2$$
$$I_{n,s} = \int_0^\infty x\,dx \frac{e^{-x^2}}{\alpha^2 + x^2} J_n(ax) J_{n+s}(bx) J_s(cx) , \qquad (17a)$$

with

$$\alpha^2 = \frac{k_D^2 + q^2}{4\gamma} , \; a = 2\lambda_o\gamma^{1/2} , \; b = 2|u_o' u_o|^{1/2}\gamma^{1/2} , \; c = \frac{2}{\omega_c}|v_o' v_o|^{1/2}\gamma^{1/2} . \qquad (17b)$$

Here we should note that $q = (p_z' - p_z)/\hbar$ accounts for the longitudinal momentum transfer, and $\lambda_B = \gamma^{-1/2}$ determines the widths of the probability distributions of the coherent states. The $J_n$ are again ordinary Bessel functions. The integral $I_{n,s}$ in (17a) yields explicit analytic expressions for the limiting cases of strong magnetic fields and/or weak screening ($\alpha^2 << 1$) and for weak magnetic fields and/or strong screening ($\alpha^2 >> 1$). The corresponding results are natural generalizations of formulae derived earlier by VARRO and EHLOTZKY /9/ and in the absence of the microwave ($a = 0$)

by VENTURA /10/. In the work of these authors the quasi-classical limit of highly excited Landau states was considered. The details of the investigations of section 3 will be published elsewhere /11/.

## 4. Conclusion

As demonstrated in the foregoing sections, generalized coherent states are very convenient for the description of quantum mechanical scattering processes in external fields, reflecting at the same time their quasi-classical features. Other applications could possibly be found in the treatment of the Kepler-Coulomb problem /12/.

## Acknowledgments

This work was supported by the Österreichische Forschungsgemeinschaft under Projekt No. 06/0400 and 06/0479 and by the Exchange Agreement between the Austrian and the Hungarian Academies of Sciences.

## References

1. J. Bergou and F. Ehlotzky: Phys. Rev. A 33, 3054 (1986)
2. J. Bergou and S. Varró: J. Phys. B: At. Mol. Phys. 15, L179 (1982)
3. N.M. Kroll and K.M. Watson: Phys. Rev. A 8, 804 (1973)
4. L. Rosenberg: Phys. Rev. A 21, 157 (1980)
5. E. Schrödinger: Naturwiss. 14, 664 (1926)
6. L.D. Landau and E.M. Lifshitz: In Quantum Mechanics (Pergamon, London 1958)
7. I.A. Malkin and V.I. Man'ko: Sov. Phys. JETP 28, 527 (1969)
8. S. Varró: J. Phys. A: Math. Gen. 17, 1631 (1984)
9. S. Varró and F. Ehlotzky: J. Phys. B: At. Mol. Phys. 18, 3395 (1985)
10. J. Ventura: Phys. Rev. A 8, 3021 (1973)
11. S. Varró and F. Ehlotzky: submitted for publication to Phys. Rev. A
12. C.C. Gerry: Phys. Rev. A 33, 6 (1986)

# High Intensity Effects in Multiphoton Free-Free Transitions

S. Bivona[1], R. Daniele[2], G. Ferrante[2], and F. Trombetta[2]

[1]Dipartimento di Energetica e Applicazioni di Fisica, viale delle Scienze, F-90128 Palermo, Italy

[2]Istituto di Fisica, via Archirafi 36, I-90123 Palermo, Italy

## I. Introduction

In this report we present a number of theoretical results concerning the multiphoton free-free transitions (MFFT), in which the intensity of the radiation field is found to play a key role. The emphasis is on high values of the field intensity because, as it is now largely known, in the theory of MFFT the field may be included exactly. Thus the theoretical treatment of MFFT opens the possibility of investigating in rigorous way high-intensity effects, and the related information may prove useful for the more general context of atomic physics driven by intense, nonresonant lasers. This remarkable peculiarity of the MFFT theory extends to include also the treatment of the statistical properties of the radiation, and has already provided new and unexpected information (see, for instance, refs. 1 and 2).

The results reported below exhibit as a rule two characteristic features: i) at high intensities they depart strongly from those predicted by lowest order perturbation theory; and ii) some useful approximations and models, frequently used to treat this kind of problem , are found to break down.

In the following Section II we discuss in particular some MFFT results obtained recently with the ideal model of a coherent field. The Section III is instead devoted to results concerning stochastic radiation fields. Due to the limited space available, as a rule, derivations and technical details are reduced or left out in the original references[3-6].

## II. Free-Free Transitions in the Presence of Strong Coherent Radiation Fields

In this Section we review some high intensity effects occurring in MFFT assisted by a coherent field. The field is described by its classical electric field $\underline{E}(t) = \underline{E}_\circ \sin \omega t$, taken homogeneous and in dipole approximation. In the first Born approximation (FBA), the differential cross-section (DCS) for transition from the initial momentum $\hbar \underline{k}_i$ to $\hbar \underline{k}_f$ is[7]

$$\frac{d\sigma}{d\Omega} = \sum_n \left[ k_f(n)/k_i \right] J_n^2 \left[ \lambda(n) \right] \frac{d\sigma(n)}{d\Omega} = \sum_n \left[ \frac{d\sigma}{d\Omega} \right]_n \tag{2.1}$$

where (atomic units are used throughout)

$$k_f(n) = k_i \left[ 1 + 2(n\omega/k_i^2) \right]^{1/2} , \quad \lambda(n) = \underline{E}_\circ \cdot \left[ \underline{k}_f(n) - \underline{k}_i \right] / \omega^2 \tag{2.2}$$

$J_n$ is the Bessel function of integer order and $d\sigma(n)/d\Omega$ is the field-free FBA, computed at the final particle momentum $\hbar \underline{k}_f(n)$. As $\underline{\alpha}_\circ = \underline{E}_\circ/\omega^2$ and $\underline{v}_\circ = \underline{E}_\circ/\omega$ are respectively the classical excursion and velocity of a particle embedded in an electromagnetic wave, the field-particle coupling $\lambda(n)$ can be written as

$$\lambda(n) = (\underline{v}_\circ \cdot \Delta \underline{p})/\omega = \Delta\varepsilon/\omega \tag{2.3}$$

$\Delta \underline{p}$ being the momentum transfer in the collision and $\Delta\varepsilon$ the related particle energy variation. The multiphoton regime is then described by $\lambda(n) >> 1$, while significant departure from predictions based on lowest order perturbation theory starts with $\lambda \gtrsim 1$.

For free-free transitions an approximation is possible in which not only the field but also the collision event is treated in entirely classical terms. In this case, the generally assumed picture is that of a collision taking place instantaneously in the time scale of the field, the latter providing the scattered particle only with a shifted momentum. In other words, the collision is assumed to occur as in absence of the field, but with a shifted particle momentum containing the field phase, and eventually an average over that phase is performed.

This classical approximation has the advantage that calculations based on it are considerably more straightforward than the quantum mechanical ones, without loss of accuracy in many cases. However, we have recently found that for parallel geometry ($\underline{v}_i || \underline{E}_\circ$) when the oscillatory velocity $v_o$ is almost equal to the incident one $v_i$, the classical treatment becomes unreliable; for usual values of the relevant parameters and within the FBA, the condition $v_o \simeq v_i$ implies strong fields.

When the field strength is such that $v_o \simeq v_i$, for times of the order of the field period the particle has a resulting velocity which may be very small, and accordingly the collision time is largely increased. When it occurs, the assumptions on which the approximation rests are clearly violated and the classical instantaneous collision treatment is expected to break down.

This breakdown is shown by calculating the absorption coefficient within the quantum mechanical treatment[3]

$$\alpha_q = (N_e N \omega v_i / I) \sum_{n=1}^{\infty} n\,(\sigma_n^+ - \sigma_n^-) \tag{2.4}$$

and the classical one

$$\alpha_{cl} = (N_e / 2\pi I) \int_0^{2\pi} d\vartheta \; d\varepsilon(\vartheta)/dt \tag{2.5}$$

where the field-dependent energy rate is given by

$$d\varepsilon(\vartheta)/dt = \left\{ NRL \sin\vartheta / 4 v_i |L^3| \right\} \left\{ \log\left(1 + Z^2/b^2\right) - Z^2/(Z^2+b^2) \right\}$$

and

$$L = 1 + R\sin\vartheta \;, \quad Z = 2v_i L \;, \quad b = 1/r_0 \;, \quad R = v_o/v_i$$

$N_e$ and N are respectively the number densities of the incoming particle beam and the scattering centers; n is the number of exchanged photons, having total cross-sections (TCS) $\sigma_n^+$ for absorption and $\sigma_n^-$ for emission and $r_o$ range of the potential.

In Fig.1 we show our results for $\alpha_q$ and $\alpha_{cl}$; they are everywhere almost equal, except when $v_o \simeq v_i$ ($R \simeq 1$), in which the two curves depart considerably.

Another strictly related high-intensity result concerns the so-called sum rule for the MFFT. This is derived within a low-frequency limit, defined by

$$\varepsilon_i = k_i^2/2 \gg n\omega \tag{2.6}$$

for the relevant multiphoton exchanges, so that by (2.2)

$$K_f(n) \simeq K_i \;, \quad \lambda(n) \simeq \lambda(0) \tag{2.7}$$

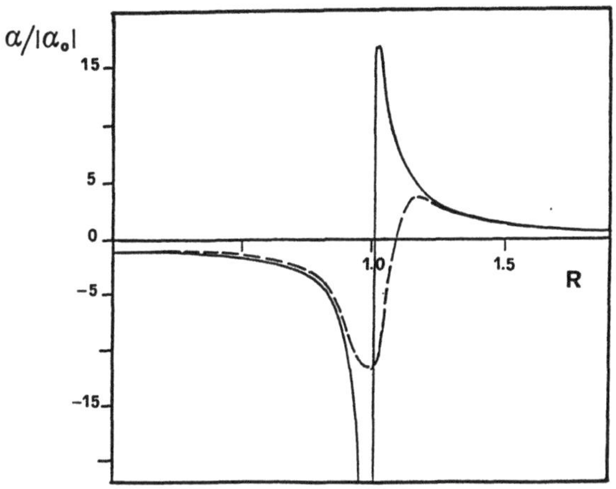


*Fig. 1 - Absorption coefficient (in units of $|\alpha_0| = 10^{-7} cm^{-1}$) vs $R = v_0/v_i$. Quantum mechanical calculation - dashed curve; classical one - full curve. Field polarization parallel to the incoming particle momentum. Field photon energy 0.117 eV. Initial particle energy 100 eV. Scattering potential - screened coulomb potential, with screening radius equal to 50 $a_0$ ($a_0$ Bohr radius). (After ref.3).*

and by (2.1)

$$\frac{d\sigma}{d\Omega} \simeq \frac{d\sigma(0)}{d\Omega} \sum_n J_n^2[\lambda(0)] = \frac{d\sigma(0)}{d\Omega} \tag{2.8}$$

where a well-known closure relation of the Bessel functions has been used and a smooth energy-dependence of the field-free DCS assumed. Eq. (2.8) states that the angular distribution of the scattered particles is not affected by the field, which just discretizes the final particle energies. By (2.8), a sum rule also follows for the TCS

$$\sigma = \sum_n \int d\Omega \left(\frac{d\sigma}{d\Omega}\right)_n \simeq \int d\Omega \left(\frac{d\sigma(0)}{d\Omega}\right) = \sigma(0) \tag{2.9}$$

As pointed out in ref.4, this derivation implicitly relies on a low intensity assumption; in fact, a correct expansion of $\lambda(n)$ when the low frequency condition (2.6) is fulfilled gives (in the parallel geometry, $\underline{E}_0 \| \underline{k}_i$)

$$\lambda(n) = -2\,(E_0 k_i/\omega^2)\sin^2(\vartheta/2) + n\,(v_0/v_i)\cos\vartheta + \ldots = \lambda(0) + n(v_0/v_i)\cos\vartheta \tag{2.10}$$

$\Theta$ being the scattering angle. Actually, for small scattering angles (which provide the leading contribution for long-range potentials), the usually retained term $\lambda(0)$ yielding the sum rule (2.8), is smaller than the next term of the expansion. Further, if $v_\circ \gtrsim v_i$, $\lambda(0)$ may be neglected and

$$\frac{d\sigma}{d\Omega} \simeq \frac{d\sigma(0)}{d\Omega} \sum_m J_m^2 (m v_\circ / v_i) \qquad (\vartheta \ll 1 \, , \; v_\circ \gtrsim v_i) \tag{2.11}$$

According to the well-known property of a Bessel function of having its maximum when its index approximately equals the argument, eq.(2.11) implies an enhancement of each multiphoton process when $v_\circ \simeq v_i$ and a breakdown of the sum rule, which is just given by the n=0 term alone. This violation easily extends to the TCS as well, particularly for energetic particles and moderately long-range potentials (for very long ranges the n=0 term is the leading one and the sum rule is recovered).

In Fig.2 we show the TCS vs the field intensity for various number of exchanged photons, computed by the exact formulas, which confirm the estimates given above. The breakdown of the sum rule (2.9) appears to pose an interesting problem in the MFFT, because an essentially classical parameter (the velocity $v_\circ$) is found to play an important role in a quantum-mechanical treatment of the collision process. On the other hand, the breakdown is apparently due to a lengthening of the collision time and to a corresponding strengthening of the particle-potential interaction, intuitively expected when $v_\circ \simeq v_i$. Thus, it indirectly calls for a new analysis of the criteria of validity of the first Born approximation in strong fields. An evaluation of the pertinent higher order of the Born series in the high intensity domain could prove very useful.

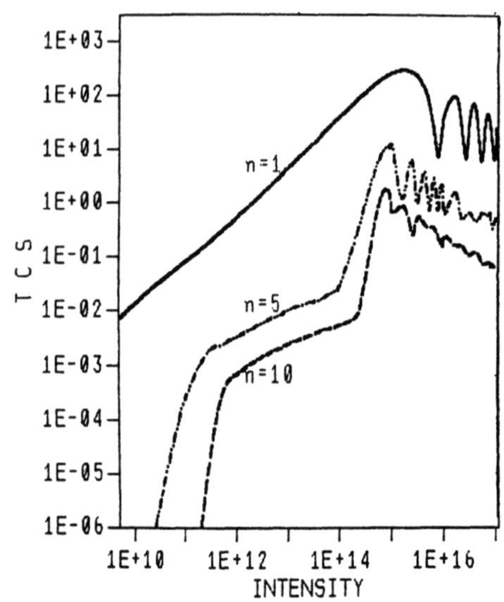


*Fig. 2 - Total cross-sections (in $\pi a_\circ^2$ units) vs the field intensity (in $W/cm^2$) for 1, 5, 10 photon absorbed. Field photon energy 1.17 eV. Other parameters as in Fig. 1.*

## III. Free-Free Transitions in the Presence of Strong Stochastic Fields

The experiments involving multiphoton processes are generally performed with multimode lasers, because of the required high intensities. Thus, a realistical theoretical treatment must include this information as well, and it is now well known that fluctuations of the field parameters may change the results in a very significant way. Again, MFFT transitions provide a process in which many laser models can be included in the theory beyond perturbative schemes, and the high intensity regimes can be analyzed also with respect to field fluctuations.

Here, we report a few results obtained for two laser models: (a) the chaotic (CH) model, in which both the phase and amplitude of the field fluctuate as a consequence of the beating of the many operating modes (this being a suitable model for describing pulsed multimode lasers with a large number of modes) and (b) the phase diffusion (PD) model, in which the amplitude is well stabilised but the phase diffuses as a consequence of quantum noise (continuous-wave lasers). Many references of interest and the pertinent derivations for the above field models can be found in refs. 2 and 5.

To analyze high-intensity effects in collisions in the presence of stochastic radiation fields, two quantities are usually considered: (a) the coherence (or statistical) factor and the scattering linewidth. For n photons exchanged, with reference to differential cross-sections, the coherence factor is defined as

$$\chi_n^x = (d\sigma/d\Omega)_n^x \,/\, (d\sigma/d\Omega)_n^{PC} \tag{3.1}$$

where x denotes the field model and PC the ideal (purely coherent) one. At the lowest order of perturbation theory (in the field), the n-photon process exhibits an $I^n$-dependence on the field intensity I, so that an average over the field fluctuations gives[8]

$$\langle I^n \rangle_{CH} = n! \langle I \rangle^n \,, \quad \langle I^n \rangle_{PD} = \langle I^n \rangle_{PC} = \langle I \rangle^n \tag{3.2}$$

Thus, in the low-intensity regime

$$\chi_n^{CH} = n! \quad , \quad \chi_n^{PD} = 1 \tag{3.3}$$

It must be remarked, however, that accounting for the field fluctuations by means of an average over the probability distribution only holds for vanishing field spectrum bandwidth (i.e., for field correlation times much larger than any other time entering the process). Otherwise an ab initio calculation of the transition probabilities in the presence of the fluctuating field is required. In that case the results (3.3) may be not true even for weak fields.

The DCS for the MFFT in the presence of a chaotic field of vanishing bandwidth is given by

$$(d\sigma/d\Omega)^{CH} = \sum_m \left[ k_f(m)/k_i \right] e^{-\lambda_0^2/2} I_m(\lambda_0^2/2) \left[ d\sigma(m)/d\Omega \right] \quad (3.4)$$

where $\lambda_0 = \lambda(\bar{E}_0/E_0)$, $\bar{E}_0$ being the square root of the quadratic field amplitude and $\lambda$ is given in (2.2); $I_n$ is the Bessel function of imaginary argument. The coherence factor in this case is then given as

$$\chi_m^{CH} = e^{-\lambda_0^2/2} I_m(\lambda_0^2/2) / J_m^2(\lambda) \quad (3.5)$$

which exhibits the n! behaviour only for weak fields. As shown in Fig.3, $\chi_m^{CH}$ has a deep minimum (less than unity) when $\lambda = \lambda_0 \simeq n$, because in such a region the purely coherent field is more effective than the chaotic one. For larger intensities, it becomes an oscillating function. Thus, the usual interpretation according to which the chaotic field enhances the multiphoton processes because of the spikes in its intensity is valid only for weak fields, when, however, such processes have very low

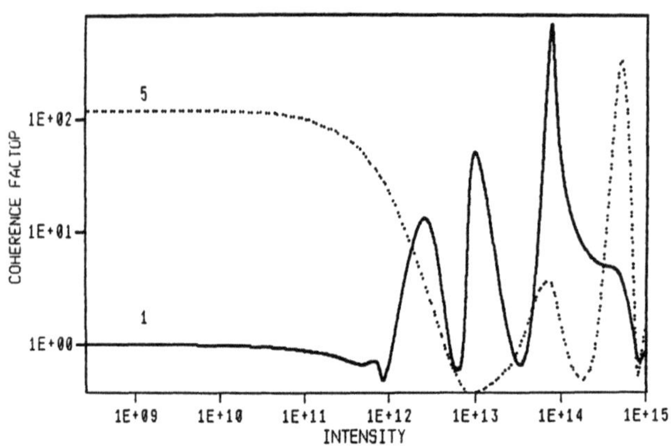


*Fig. 3 - Coherence factors for the chaotic field (see eq.(3.6) of the main text) vs the field intensity (in W/cm²) for 1 and 5 photons absorbed. The scattering angle is 45°. Other parameters as in Fig. 2.*

probability. Coherence factors for finite bandwidths give even more information and are presently under consideration.

We now consider briefly the scattering linewidths, defined as the half-width at half-maximum (HWHM) of the scattering lines. The well-known results of perturbation theory are given by[8]

$$\Gamma_{CH}^{(n)} = n\,\Delta\omega \qquad \Gamma_{PD}^{(n)} = n^2\Delta\omega \tag{3.6}$$

$\Gamma_x^{(n)}$ being the HWHM for the process with n photons exchanged, with the field described by the model x and having an HWHM in its frequency spectrum equal to $\Delta\omega$. Our results are instead reported in Fig.4, clearly showing the intensity-dependent broadening of the scattering lines, both for the chaotic field (dashed lines) and the phase diffusion one (continuous lines) for various number of exchanged photons. $\Gamma_{CH}^{(n)}$ increases faster than $\Gamma_{PD}^{(n)}$, due to the stronger fluctuations ($\delta I = \langle I^2\rangle - \langle I\rangle^2 = \langle I\rangle$). Further, for strong fields, the n-dependence becomes very weak, for both models.

The above-discussed laser models are both markovian, in the sense that the fluctuating quantities are assumed to have a short memory. In turn, this means that the random "forces" driving them to fluctuate are $\delta$-correlated in time. Of course, this is essentially an assumption on

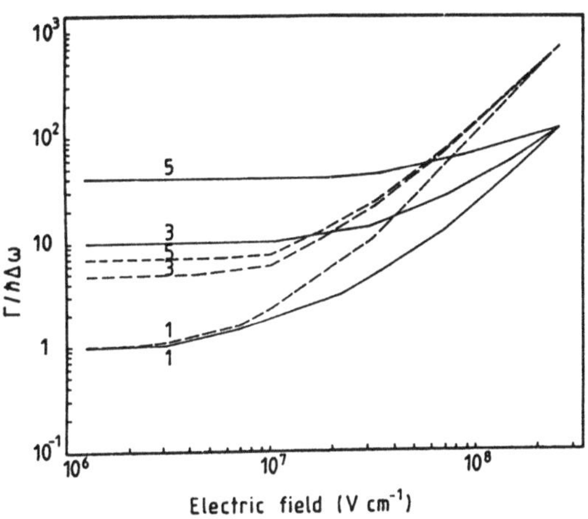


*Fig. 4 - Scattering linewidths (normalized to the field spectrum one) vs the electric field amplitude (in V/cm) for the chaotic field (dashed lines) and the phase diffusion (continuous lines) field, for various numbers of exchanged photons (numbers over the lines). The field spectrum bandwidth is such that $\hbar\Delta\omega = 10^{-4}$ eV. Other parameters as in Fig. 3. (After ref. 5).*

the times of the process. It generally implies a lorentzian spectrum for the field, which is known to be not very realistic. We shall now show that for strong fields the markovian assumption (and the ensuing lorentzian spectrum) may become invalid.

Let us assume the field phase is fluctuating according to the Langevin equation

$$\ddot{\varphi} = -\beta\dot{\varphi} + \mathcal{F}(t) \tag{3.7}$$

where $\mathcal{F}(t)$ is a gaussian random force with zero average and $\delta$-correlated

$$\langle \mathcal{F}(t)\,\mathcal{F}(t')\rangle = 2b\beta^2\,\delta(t-t') \tag{3.8}$$

$b^{-1}$ and $\beta^{-1}$ may be taken as the correlation times of the phase and its derivative, respectively. If $\beta \gg b$ (i.e., if one neglects the $\dot{\varphi}$ -correlation time), eq.s (3.7)-(3.8) give

$$\dot{\varphi} = F_\varphi(t) \quad , \langle F_\varphi(t)\,F_\varphi(t')\rangle = 2\,\Delta\omega\,\delta(t-t') \tag{3.9}$$

where $\Delta\omega = \lim b\ (\beta \to \infty)$ represents the field spectrum bandwidth. Eq.(3.9) describes the well-known Wiener-Levy (markovian) process. Instead, for finite $\beta$, by (3.7), $\dot{\varphi}$ is an Ornstein-Uhlenbeck process and $\varphi$ is no more markovian.

The transition probability per unit time in the presence of a PD field whose phase fluctuates according to (3.7)-(3.8) is found as[6]

$$W = 2\,\mathrm{Re}\sum_{\ell,k} J_\ell(\lambda)\,M_{\ell k}(-i\omega_{fi})\,J_k(\lambda)\,C(\underline{Q}) \tag{3.10}$$

where the matrix M is given by the continued-matrix fraction

$$M(s) = \left[ s - i\omega B - A\,\frac{b\beta}{s+\beta - i\omega B - A\,\dfrac{2b\beta}{s+2\beta - i\omega B - A\ldots}\,A}\,A\right]^{-1}$$

and

(3.11)

$$B_{mk} = -m\,\delta_{mk} \quad , \quad A_{mk} = i\omega\left[(\lambda/2)(\delta_{m,k-1}+\delta_{m,k+1}) - m\,\delta_{mk}\right]$$

(3.12)

$$C(\underline{Q}) = \left|\int d\underline{r}\, e^{i\underline{Q}\cdot\underline{r}}\,V(\underline{r})\right|^2 \quad , \quad \underline{Q} = \underline{k}_f - \underline{k}_i \quad , \quad \omega_{fi} = (k_f^2 - k_i^2)/2$$

Truncating M(s) after the first step including a finite $\beta$, one gets a quasi-markovian limit for the present problem. It corresponds to account to the first order for memory effects in the field phase. In this problem, the criterion of validity of the quasi-markovian limit reads, by (3.11)

$$\frac{b\beta}{(\beta+\omega)^2}\left(m^2+\lambda^2/2\right) \ll 1 \tag{3.13}$$

which is essentially a condition on the field intensity, entering $\lambda$. If $\beta \to \infty$ (markovian limit), the inequality (3.13) always holds, but for finite $\beta$ the memory of the process is actually short only for weak fields.

It may be shown that the condition (3.13) can be put in the form $\beta \gg \Gamma^{(n)}_{PD}(\lambda)$, this latter being related to the scattering linewidth. This condition suggests that the field intensity, broadening the scattering lines, reduces the lifetime of the particle energy states to such an extent to become comparable with the correlation time of the phase derivative, usually neglected to arrive to the Markov approximation. Thus, for a strong field, the markovian approximation may become of dubious use.

In conclusion, the reported results concerning multiphoton free-free transitions suggest the need of critically reconsidering assumptions, models and results which have been originally derived within perturbative treatments and/or for weak fields situations. Familiar approximations and methods need to be reassessed ab initio if they are to be used in strong field contexts, or should be replaced by the proper generalizations and extensions.

Acknowledgements

This work has been partially supported by the Italian Ministry of Education, the National Group of Structure of Matter and the Sicilian Committee for Nuclear and Structure of Matter Researches.

References

1. J.H. Eberly and J. Krasinski, in: Advances in Multiphoton Processes and Spectroscopy, Ed. S.H. Lin, World Scientific, Singapore (1984), vol. 1, pag. 1.
2. F. Trombetta, R. Daniele, F. Morales and G. Ferrante, in: Quantum Optics, Ed.s J. Fiutak and J. Mizerski, World Scientific, Singapore, (1986), pag. 197.

3. S. Bivona, R. Zangara and G. Ferrante, Phys. Lett. 110A, 375 (1985).
4. R. Daniele, G. Ferrante, F. Morales and F. Trombetta, J. Phys. B: At. Mol. Phys. 19, L133 (1986).
5. F. Trombetta, G. Ferrante, K. Wodkiewicz and P. Zoller, J. Phys. B: At. Mol. Phys. 18, 2915 (1985).
6. F. Trombetta, G. Ferrante and P. Zoller, "High Intensity Effects in Multiphoton Processes Induced by Non Markovian Fluctuating Fields", Optics Commun. (in press).
7. N.M. Kroll and K.M. Watson, Phys. Rev. A8, 804 (1973).
8. G.S. Agarwal, Phys. Rev. A1, 1445 (1970).

# New Results for Perturbative Calculations of Free-Free Scattering

*E.J. Robinson and C. Chang**

Physics Department, New York University, New York, NY 10003, USA

The past decade and one half has seen considerable effort, especially on the part of theorists, devoted to the problem of free-free radiative scattering of electrons [1-18]. Unlike earlier treatments, which were primarily perturbative, work in this period has primarily emphasized nonperturbative approximations. In the present contribution, we return to the limit where perturbation theory is valid, and consider three disparate aspects of the problem.

In what is probably the simplest approximation, the effect of the atom on the scattering of the electron is simulated by a fixed potential V(r). To take the interaction with the radiation field into account in this single-electron model, one simply calculates matrix elements of the appropriate transition operator (the dipole approximation is made here and throughout the paper) between initial and final unbound states of a particle subject to the assumed potential. However, the target also interacts with the field. As a first approximation, one may treat this effect by postulating the existence of an induced dipole oscillating at the field frequency and interacting with the free electron [19-24]. That is, the effective transition operator is taken to be,

$$\vec{d}=\vec{r}-\alpha(\omega)\vec{r}f(r)/r^3, \tag{1}$$

where $\alpha$ is the atomic polarizability at frequency $\omega$, f is a cut-off function (arbitrarily taken to be of the form $1-\exp\{-(r/r_0)^3\}$, where $r_0$ is a characteristic target radius) which corrects for penetration of the projectile into the core. Atomic units are used here and in all subsequent parts of the paper. Accordingly, we calculate matrix elements of $\vec{d}$ instead of those of $\vec{r}$ alone.

To perform the calculations, we must choose a particular V(r). Our selection is one which has been used previously and which typically makes predictions that are in good agreement with experiment and/or more detailed (many-body) theories [25-29]. This potential is

$$V=V_{HS}-\{\exp(-r/r_p)-1\}/r/-\alpha\{1-\exp(-r/r_p)\}/(r^2+r_0^2)^2, \tag{2}$$

where $V_{HS}$ is the Hartree-Fock-Slater potential for an electron interacting with an ionic core [30]. The second term removes the Coulomb tail of VHS, and the third term, proportional to the static polarizability of the atom, is a representation of the adiabatic $r^{-4}$ polarization potential. We note that polarizabilities enter the present model in two rather different ways.

*Present Address: Physics Department, Choson University, Kwanju, Chunnum, Korea

Numerical calculations of single photon absorption cross-sections were performed for the specific case of 1s2s $^3S_1$ helium, where $\alpha$ is large (so that we anticipate an appreciable effect).

The relative matrix elements of $\vec{r}$ and $\vec{r}\ f(r)\ /r^3$ scale roughly as $\omega^2$ for very low frequencies, so that we expect the correction to be minor for light from $CO_2$ laser. At shorter wavelengths, allowing for field-induced target distortion (TD) makes for a dramatic change in the absorption. For example, for the $Nd^{+++}$ laser ($\lambda$=1.06μ), the cross-section is increased by a factor of order $10^3$. Moreover, for the calculations which we performed, total cross-sections averaged over incident projectile direction, one must retain twice as many angular momentum states to achieve convergence of the partial wave expansion with TD as without it. Since high L corresponds to large impact parameter, we find that $\vec{r}/r^3$ is effectively of longer range than $\vec{r}$, while the functional forms of the two operators suggest the reverse. This apparent paradox can be resolved by recalling that for high angular momentum, the radial wave functions are essentially the same as those for free particles, for which matrix elements of $\vec{r}$ must vanish identically for initial and final states with different energies. This result, which forbids a completely free particle from absorbing or emitting a photon, does not apply to the TD contribution.

Table I lists total single-photon cross-sections at intensity one MW/CM$^2$, and $Nd^{+++}$ frequency, for a range of electron energies, with and without TD. We present results averaged over all incident directions. In addition to the enormous change in the order of magnitude of the absorption (which results primarily from a large $\alpha(\omega)$), TD creates structure in what was otherwise a featureless spectrum. It should be obvious from the entries in the table that these oscillations are too large to be simply the result of interference between the two terms in $\vec{d}$. Instead, they are manifestations of variations with energy in the matrix elements of $f(r)\vec{r}/r^3$ alone.

Table I. Total cross-sections averaged over incident projectile directions, for electrons radiatively scattered by 1s2s $^3S_1$ He in 1 MW/CM$^2$ intensity of 1.06μ wave length light. Results compare single-photon total cross-sections without TD ($Q_0$) to cross-sections Q which include TD.

| Incident Electron Energy [Ryd] | $Q_0[a_o^2]$ | $Q[a_o^2]$ |
|---|---|---|
| 0.1 | $9.0\times10^{-5}$ | $1.0\times10^{-1}$ |
| 0.2 | $6.1\times10^{-5}$ | $5.0\times10^{-2}$ |
| 0.3 | $5.1\times10^{-5}$ | $5.0\times10^{-2}$ |
| 0.4 | $4.7\times10^{-5}$ | $7.3\times10^{-3}$ |
| 0.5 | $4.2\times10^{-5}$ | $1.0\times10^{-2}$ |
| 0.6 | $4.0\times10^{-5}$ | $3.9\times10^{-3}$ |
| 0.7 | $3.5\times10^{-5}$ | $2.9\times10^{-2}$ |
| 0.8 | $3.4\times10^{-5}$ | $4.3\times10^{-2}$ |
| 0.9 | $3.3\times10^{-5}$ | $5.0\times10^{-2}$ |
| 1.0 | $3.2\times10^{-5}$ | $5.5\times10^{-2}$ |

The second matter which we shall discuss is whether free-free processes tend to occur with higher probability for linearly or circularly polarized light, in the case of multiphoton transitions. In the early 1970's this question was explored in detail for multiphoton ionization, the bound-free counterpart of the process under discussion here [33-35].

The bound-free and free-free cases have important differences. For photoionization, the material system is impolarized and the square of the electronic angular momentum is well defined. In the free-free problem, there is a preferred direction in the initial state (the direction of motion of the incident electron) while there is a superposition of angular momentum components. The preferred direction, but not the combination of L-states, is removed by our procedure of calculating total cross-sections averaged over electron directions.

The question of which kind of light polarization is more conducive to free-free absorption can be unambiguously answered in two limits - at low frequencies (LF) or where the first Born approximation (BA) is valid for the basic scattering problem [1-5,9,11].

We refer specifically to the perturbative versions of these approximations- that is, only terms through Nth-order in the field strength are retained in a calculation of an n- photon cross-section. Moreover, for the LF case, we mean the "extreme" soft-photon approximation, in which the energy variation of radiation-free T- matrix elements is neglected. In both BA and LF limits, one finds $R = Q_L/Q_C$, the ratio of the indicated cross-sections for linearly and circularly polarized light, is given by

$$Q_L/Q_C = (2N-1)!!/N! \,, \tag{3}$$

for an N- photon process.

To see whether this result applies more generally, at least approximately, we calculated the cross-section for two-photon absorption by an electron in the model potential $V = -\lambda\delta(r-a)$, for which the solution may be obtained in closed form. We used the differential equation method to evaluate the infinite summations characteristic of second-order perturbation theory. That is, the amplitude for the process is (we use the velocity gauge where the solution is simpler) proportional to

$$M_2 = \frac{1}{c^2} \sum_k \frac{\langle f|P_\varepsilon|k\rangle\langle k|P_\varepsilon|0\rangle}{E_0 - E_k + \omega + i\beta} \tag{4}$$

where $|f\rangle$, $|k\rangle$, $|0\rangle$, are final, intermediate and initial state wave functions, $P_\varepsilon$ the component of momentum along the polarization vector of the light, and c the speed of light. The parameter $\beta$ is a positive infinitesimal which exhibits the outgoing nature of the relevant Green's function. The summation on $|k\rangle$, which includes an integration over the continuum, may be reduced to a single term,

$$M_2 > = \langle f|P_\varepsilon|1\rangle, \tag{5}$$

where $|1\rangle$ satisfies $\left(\frac{P^2}{2} + V(r) - E_0 - \omega - i\beta\right)|1\rangle = -P_\varepsilon|0\rangle. \quad (6)$

If V(r) is the delta shell potential, this equation and the integral in (5) may be evaluated analytically in the partial wave representation.

Table II. Polarization Dependence of two-photon free-free absorbtion cross-sections for an electron interacting with the model potential $V=-\lambda\delta(r-a)$ with $\lambda = 1$ atomic unit and $a = 0.5\ a_{Bohr}$. Table entries include $Q_c$, $Q_L$, the cross-sections for circularly polarized and linearly polarized light, $Q_L^{(B)}$, the linearly polarized cross-section calculated in the first Born approximation and the ratio $R=Q_L/Q_c$. All Q's are for a fixed intensity and represent total cross-sections averaged over incident diredtions. Cross-sections are in arbitrary units.

| Electron Energy [Atomic Units] | Photon Energy [Atomic Units] | $Q_L^{(B)}$ | $Q_L$ | $Q_C$ | R |
|---|---|---|---|---|---|
| $5.0\times10^{-2}$ | $5.0\times10^{-2}$ | $5.3\times10^{11}$ | $1.46\times10^{13}$ | $1.0\times10^{13}$ | 1.40 |
| $1.0\times10^{-2}$ | $5.0\times10^{-2}$ | $9.9\times10^{11}$ | $1.52\times10^{13}$ | $1.04\times10^{13}$ | 1.47 |
| $5.0\times10^{-2}$ | $1.0\times10^{-1}$ | $1.4\times10^{9}$ | $3.66\times10^{10}$ | $2.80\times10^{10}$ | 1.31 |
| $1.0\times10^{-1}$ | $1.0\times10^{-1}$ | $2.0\times10^{9}$ | $2.76\times10^{10}$ | $1.96\times10^{10}$ | 1.41 |
| $1.5\times10^{-1}$ | $1.0\times10^{-1}$ | $2.7\times10^{9}$ | $2.71\times10^{10}$ | $1.87\times10^{10}$ | 1.45 |
| $2.0\times10^{-1}$ | $1.0\times10^{-1}$ | $3.6\times10^{9}$ | $2.82\times10^{10}$ | $1.92\times10^{10}$ | 1.47 |
| $5.0\times10^{-2}$ | $1.5\times10^{-1}$ | $4.9\times10^{7}$ | $1.28\times10^{9}$ | $1.03\times10^{9}$ | 1.25 |
| $1.0\times10^{-1}$ | $1.5\times10^{-1}$ | $5.9\times10^{7}$ | $7.92\times10^{8}$ | $5.90\times10^{8}$ | 1.34 |
| $1.5\times10^{-1}$ | $1.5\times10^{-1}$ | $7.4\times10^{7}$ | $6.96\times10^{8}$ | $4.96\times10^{8}$ | 1.40 |
| $2.0\times10^{-1}$ | $1.5\times10^{-1}$ | $9.1\times10^{7}$ | $6.74\times10^{8}$ | $4.68\times10^{8}$ | 1.44 |
| $2.5\times10^{-1}$ | $1.5\times10^{-1}$ | $1.1\times10^{8}$ | $6.79\times10^{8}$ | $4.64\times10^{8}$ | 1.46 |
| $3.0\times10^{-1}$ | $1.5\times10^{-1}$ | $1.3\times10^{8}$ | $6.95\times10^{8}$ | $4.71\times10^{8}$ | 1.48 |
| $5.0\times10^{-2}$ | $3.0\times10^{-1}$ | $1.8\times10^{5}$ | $5.04\times10^{6}$ | $4.33\times10^{8}$ | 1.16 |
| $1.0\times10^{-1}$ | $3.0\times10^{-1}$ | $1.8\times10^{5}$ | $2.39\times10^{6}$ | $1.93\times10^{8}$ | 1.24 |
| $1.5\times10^{-1}$ | $3.0\times10^{-1}$ | $1.9\times10^{5}$ | $1.72\times10^{6}$ | $1.32\times10^{6}$ | 1.30 |
| $2.0\times10^{-1}$ | $3.0\times10^{-1}$ | $2.1\times10^{5}$ | $1.44\times10^{6}$ | $1.07\times10^{6}$ | 1.35 |
| $2.5\times10^{-1}$ | $3.0\times10^{-1}$ | $2.3\times10^{5}$ | $1.30\times10^{6}$ | $9.34\times10^{6}$ | 1.39 |
| $3.0\times10^{-1}$ | $3.0\times10^{-1}$ | $2.5\times10^{5}$ | $1.22\times10^{6}$ | $8.62\times10^{6}$ | 1.41 |
| $3.5\times10^{-1}$ | $3.0\times10^{-1}$ | $2.7\times10^{5}$ | $1.17\times10^{6}$ | $8.16\times10^{6}$ | 1.44 |
| $4.0\times10^{-1}$ | $3.0\times10^{-1}$ | $3.0\times10^{5}$ | $1.15\times10^{6}$ | $7.87\times10^{6}$ | 1.46 |

The entries in Table II are $Q_L$, $Q_C$ and their ratio for a range of electron and photon energies. We also give the corresponding values of $Q_L^{(B)}$, the cross-section for linearly polarized light determined in the BA. We find that the results tend to be close to the limit of (3), $3!!/2! = 1.5$, except where the photon energy exceeds that of the electron. This result is apparently inherited from the LF approximation, since the BA is evidently completely invalid for the electron energies presented in the Table. It appears that $Q_L/Q_C \rightarrow 1$ as the ratio of incident electron energy to photon energy $\rightarrow 0$. A unit ratio $Q_L/Q_c \rightarrow 1$ seems to be a lower limit. We are unable to state whether this is a universal tendency or an artifice of the model.

The final point we wish to address involves a method for solving (6) numerically. For the bound-free case, there is some folklore to the effect that the inhomogeneous differential equation method is of limited value when there is a nearly resonant intermediate state or when $|1\rangle$ decays very slowly with r or not at all. Since both these features are present here, free-free transitions would appear to be a worst case limit for these difficulties. We shall demonstrate that one may surmount anticipated numerical problems by working in a mixed momentum-acceleration gauge representation.

Indeed, numerical problems occur even in the evaluation of single-photon free-free matrix elements, in the momentum gauge, for example. To evaluate these amplitudes, one numerically integrates from the origin to some large radius $r_0$, and completes the integration to infinity analytically, using the known asymptotic forms of radial wave functions.

The analytic extension is rendered convergent by appending a factor $\exp[-\lambda r]$ to the integrand and passing to the limit $\lambda \to 0$. (This trick is equivalent to recognizing that the final state is actually a wave packet-a mixture of momentum components.) Problems arise because the numerical and analytic parts of the integration may nearly annul one another. The small remainder is then extremely sensitive to even minor numerical errors. This effect, which is inherited from the free particle case where the cancellation is exact, becomes especially serious for high angular momentum.

One may circumvent this difficulty in single photon problems by working in the acceleration gauge. This obviates the need for an analytic extension to infinity, since the integral converges for finite upper limit [12]. One might therefore suspect that this representation would suffice for the two-photon case as well. Unfortunately, use of the acceleration gauge replaces one numerical problem by another.

The acceleration gauge interaction Hamiltonian is given by

$$H_a' = c^{-2}\,\nabla V\cdot\vec{Z}+c^{-4}\,(\vec{Z}\cdot\nabla)^2\,V/2+c^{-2}A^2/2 \tag{7}$$

where $\vec{A}$ and $\vec{Z}$ are the vector and Hertz potentials of the field and V the potential driving the electron in the absence of radiation. While transition matrix elements of $A^2$ vanish in dipole approximation, those of the term quadratic in $\vec{Z}$ don't, and there is often substantial cancellation between this contribution and the second-order $\nabla V\cdot\vec{Z}$ term. Moreover, for realistic model potentials, which approach $r^{-1}$ as $r\to 0$, the $(\vec{Z}\cdot\nabla)^2$ V term will contain a $\delta(\vec{r})$ contribution, which means that one requires accurate knowledge of the wave function at the origin, notoriously difficult to obtain in numerical calculations.

The superior numerical accuracy that the acceleration gauge yields in single-photon calculations may be achieved in the two-photon case by using a mixed approach, together with the differential equation method for evaluating infinite sums. Although only the two-photon case will be discussed explicitly, it will be obvious that the ideas may be extended to higher order.

Since $E_f$ and $E_0$ lie on the energy shell, we may rewrite (4) as

$$M_2=c^{-2}((E_f-E_0)/2\,\omega)\sum\frac{\langle f|P_\varepsilon|k\rangle\langle k|P_\varepsilon|\,0\rangle}{(E_0-E_k+\omega+i\beta)}, \tag{8}$$

or since $(E_f-E_0)=(E_f-E_k)+(E_k-E_0)$,

$$M_2=(c^{-2}/2\omega)\ \Sigma\ \frac{<f|[H_0,P_\varepsilon]|k><k|P_\varepsilon|0>+<f|P_\varepsilon|k><k|[H_0,P_\varepsilon]|0>}{E_0-E_k+\omega+i\beta}\ . \tag{9}$$

We note that the commutator $[H_0, P_\varepsilon]$ is proportional to the component of $\nabla V$ parallel to the polarization direction, i.e., it is the linear term in $H'_a$. Defining the functions $|1>$ and $<1'|$ according to

$$|1>=\Sigma\ |k><k|P_\varepsilon|0>/(E_0-E_k+\omega+i\beta), \tag{10}$$

$$<1'|=\Sigma<f|P_\varepsilon|k><k|/(E_0-E_k+\omega+i\beta), \tag{11}$$

which satisfy

$$(H_0-E_0-\omega-i\beta)|1>=-P_\varepsilon|0>\ , \tag{12}$$

$$<1'|(H_0-E_0-\omega-i\beta)=-<f|P_\varepsilon, \tag{13}$$

we have

$$M_2=(c^{-2}/2\omega)\ \{<f|[H_0,P_\varepsilon]|1>+<1'|[H_0,P_\varepsilon]|0>\}\ . \tag{14}$$

We experienced no numerical difficulties in implementing this procedure. We solved (12) and (13) in the partial wave representation, and performed the integrations in (14) by standard methods. As in single-photon acceleration gauge calculations of matrix elements, the radial integrals converge without an analytic extension to infinity.

As a way of convincing ourselves (and others) that the method is indeed correct, we calculated two-photon absorption cross-sections, averaged over incident electron directions, for a range of photon and electron energies, for targets of atomic He, C, Ne, and Ar in their ground states, and compared results to the "extreme" soft photon approximation. In most cases, the two agreed within a few percent. Cross-sections for two systems, He and C at the frequency of the ruby laser, which represent, respectively, typical levels of agreement and the worst case, are presented in Table III.

That the agreement is poorest at low electron energies is due to the breakdown of the soft-photon approximation. The percent discrepancy is reduced monotonically as the electron energy rises, reflecting the greater reliability of the LF approximation. We note in passing that the agreement improves despite an increase in the number of partial waves required for convergence, as well as possible reduced accuracy for the numerical methods (associated with more rapidly oscillating radial wave functions.)

This work was supported in part by the U.S. Air Force Systems Command (Aeronautical Systems Division) and the New York University Research Challenge Fund.

Table III. Two-photon free-free total cross-sections, averaged over incident projectile directions, for electrons interacting with He and C targets and ruby laser light of intensity 1 MW/cm$^2$. Results represent a comparison of numerical calculations in the mixed gauge (Q) and the soft-photon approximation ($Q_s$).

| Incident Electron Energy [Ryd] | He $Q[a_0^2]$ | He $Q_s[a_0^2]$ | C $Q[a_0^2]$ | C $Q_s[a_0^2]$ |
|---|---|---|---|---|
| 0.1 | $5.75x10^{-14}$ | $5.41x10^{-14}$ | $1.16x10^{-13}$ | $8.94x10^{-14}$ |
| 0.2 | $9.92x10^{-14}$ | $9.58x10^{-14}$ | $2.02x10^{-13}$ | $1.70x10^{-13}$ |
| 0.3 | $1.53x10^{-13}$ | $1.49x10^{-13}$ | $3.25x10^{-13}$ | $2.88x10^{-13}$ |
| 0.4 | $2.15x10^{-13}$ | $2.11x10^{-13}$ | $4.80x10^{-13}$ | $4.41x10^{-13}$ |
| 0.5 | $2.84x10^{-13}$ | $2.80x10^{-13}$ | $6.66x10^{-13}$ | $6.25x10^{-13}$ |
| 0.6 | $3.59x10^{-13}$ | $3.55x10^{-13}$ | $8.78x10^{-13}$ | $8.34x10^{-13}$ |
| 0.7 | $4.37x10^{-13}$ | $4.33x10^{-13}$ | $1.11x10^{-12}$ | $1.07x10^{-12}$ |
| 0.8 | $5.18x10^{-13}$ | $5.14x10^{-13}$ | $1.37x10^{-12}$ | $1.32x10^{-12}$ |
| 0.9 | $6.01x10^{-13}$ | $5.96x10^{-13}$ | $1.63x10^{-12}$ | $1.59x10^{-12}$ |
| 1.0 | $6.83x10^{-13}$ | $6.79x10^{-13}$ | $1.91x10^{-12}$ | $1.86x10^{-12}$ |

## References


1. N. Kroll, K. Watson: Phys. Rev. A 8, 804 (1973).
2. H. Krüger, C. Jung: Phys Rev. A 17, 1706 (1978)
3. L. Rosenberg: Adv. At. Mol. Phys. 18, 1 (1982)
4. E. Kelsey, L. Rosenberg:Phys. Rev A 19, 756 (1978)
5. L. Rosenberg: Phys. Rev.A 21, 157 (1980)
6. J.L. Gersten, M.H. Mittleman: Phys. Rev. A 13, 123 (1976)
7. L. Rosenberg: Phys. Rev. A 21, 1939 (1980)
8. M. Gavrila, M.J. VanDerweil:Comments At. Mol. Phys. 8 1 (1978)
9. R. Shakeshaft, E. J. Robinson:Phys.Rev. A 25, 1977 (1982)
10. A. Weingartshofer, J.K. Holmes, G. Caudle, E.M. Clarke, H. Krüger: Phys. Rev. Lett. 39 269 (1977)
11. H. Krüger, M. Schultz: J. Phys. B 9, 1899 (1976)
12. S. Geltman: J. Quant. Spect. Rad Trans. 13, 601 (1973)
13. F.W. Byron, C.J. Joachain: J. Phys. B 17, L295 (1984)
14. P. Zoller: J. Phys. B 13, L249 (1980)

MIX
Papier aus verantwortungsvollen Quellen
Paper from responsible sources
FSC® C105338

If you have any concerns about our products,
you can contact us on
**ProductSafety@springernature.com**

In case Publisher is established outside the EU,
the EU authorized representative is:
**Springer Nature Customer Service Center GmbH**
**Europaplatz 3, 69115 Heidelberg, Germany**

Printed by Libri Plureos GmbH
in Hamburg, Germany